SOFT SHORE PROTECTION

Coastal Systems and Continental Margins

VOLUME 7

Soft Shore Protection

An Environmental Innovation in Coastal Engineering

Edited by

Constantine Goudas

Studium of Mechanics,
Department of Mathematics, University of Patras, Greece

George Katsiaris

Studium of Mechanics,
Department of Mathematics, University of Patras, Greece

Vincent May

School of Conservation Sciences,
Bournemouth University, United Kingdom

and

Theophanis Karambas

Department of Civil Engineering,
Aristotle University of Thessaloniki, Greece

KLUWER ACADEMIC PUBLISHERS
DORDRECHT / BOSTON / LONDON

A C.I.P. Catalogue record for this book is available from the Library of Congress.

ISBN 1-4020-1153-9

Published by Kluwer Academic Publishers,
P.O. Box 17, 3300 AA Dordrecht, The Netherlands.

Sold and distributed in North, Central and South America
by Kluwer Academic Publishers,
101 Philip Drive, Norwell, MA 02061, U.S.A.

In all other countries, sold and distributed
by Kluwer Academic Publishers,
P.O. Box 322, 3300 AH Dordrecht, The Netherlands.

Printed on acid-free paper

Printed in the Netherlands.

TABLE OF CONTENTS

INTRODUCTION

Global warming, melting polar caps, rising sea levels and intensifying wave-current action, factors responsible for the alarming phenomena of coastal erosion on the one hand and adverse environmental impacts and the high cost of 'hard' protection schemes, on the other, have created interest in the detailed examination of the potential and range of applicability of the emerging and promising category of 'soft' shore protection methods. 'Soft' methods such as beach nourishment, submerged breakwaters, artificial reefs, gravity drain systems, floating breakwaters, plantations of hydrophylous shrubs or even dry branches, applied mostly during the past 20 years, are recognised as possessing technical, environmental and financial advantageous properties deserving more attention and further developmental experimentation than has occured hitherto.

On the other hand, 'hard' shore protection methods such as seawalls, groins and detached breakwaters, artefacts borrowed from port design and construction technology, no matter how well designed and well implemented they may be, can hardly avoid intensification of the consequential erosive, often devastating, effects on the down-drift shores. Moreover, they often do not constitute environmentally and financially attractive solutions for long stretches of eroding shoreline. Engineers and scientists practising design and implementation of shore defence schemes have been aware for many years of the public demand for improved shore protection technologies. They are encouraging efforts that promise enrichment of those environmentally sound and financially attractive methods that can be safely applied.

The objectives of the First International Conference on Soft Shore Protection held in October 2000 in Patras, Greece, the proceedings of which are reported in this volume, were to offer the opportunity to scientists and engineers to present their ideas and work and at the same time, to alert the international community to the need to intensify and/or sponsor research and pilot *in situ* experiments on promising methods of 'soft' protection.

The concept of 'soft' intervention, or 'resilience' to marine wave and current action requires clarification, both for specialists and the general public. Marine action, a natural and unchanging phenomenon, develops the capability when exceeding certain thresholds to transport the sediment of the seabed, by either lifting grains (suspended matter) or displacing the bedload in small steps. The sediment transport mechanism in the surf zone operates through either the 'longshore' water current motion, driven by wave action and other causes, or the 'rip' currents produced by the water setup during wave activity. Suspended matter distributed within the entire moving water body is not discussed here due to the complexity of this phenomenon.

Any arrangement able to reduce the velocity or modify the direction of the motion of longshore or rip currents, and thus reduce or eliminate their sediment transport capability, is a common shore protection methodology. Breakwaters, groins, jetties, and piers have this property and hence are correctly classified as shore protection arrangements. The fact that they are 'borrowed' from port construction technology is not a disadvantage *per se*, since cost effectiveness and environmental acceptability constitute the criteria of suitability for application. These criteria support the shift 'away from hard protection structures (e.g., seawalls, groins) toward soft protection measures (e.g. nourishment)', as the UNEP Intergovernmental Panel on Climate Change observed recently (see *Summary for Policymakers, Climate Changes 2001: Impacts, Adaptation, and Vulnerability*, approved by IPCC Working Group II in Geneva, 13–16 February 2001, web site address http://www.usgcrp.gov/ipcc/wg2spm.pdf). Furthermore, on 27 September 2000, the Commission to the Council and the European Parliament published its report on "Integrated Coastal Zone Management: A Strategy for Europe", (http://europa.eu.int/comm/environment/iczm/comm2000.htm) a strategy which aims to ensure that the coastal zone is planned and managed by partnership between the government and civil society. Alternative methods of shore protection are expressly encouraged in this approach.

This imperative for the future of coastal regions and their sustainable development finds the world unprepared, the authorities apprehensive, and the engineers and contractors unwilling. Very little research and experiment time is dedicated to or attention given to the dissemination of knowledge about soft protection measures. Notwithstanding the expected reaction by 'hard' advocates, movement towards developing and establishing wise soft-protection practices is imminent. This conference, announced a year before the UN IPCC report and held four months before it, is an expression of interest in this imperative.

Some obvious theoretical considerations concerning the logic behind soft protection measures are perhaps useful for the general reader. Not all the water body participating in longshore or rip current motion is responsible for bedload transport. Only a very thin water layer, not exceeding 5 cm in thickness, moving immediately above the seabed and transferring to the seabed constituents part of its kinetic energy, is the agent of sediment transport and the related erosion. Hence neutralization of the erosive action of this water layer is of prime concern for shore protection arrangements. Complete interception of wave-current energy is unnecessary as well as harmful. Indeed, intercepting and absorbing wave and current water energy that does not erode a shore implies elimination of the benevolent function of natural pollutant recycling performed by the sea. It has been shown in fact, that interception and absorption of all sea water energy is an anti-environmental act and, as such, should be prohibited altogether.

A broader definition of 'soft' protection arrangements is, therefore, the interception and cut-off, diversion, elimination or simply suitable modification of water current motion within the thin erosive water layer acting upon bedload. Depending upon the requirements of the arrangement (such as stability, integrity, functionality, and so on), the layer of water whose energy is allowed to be intercepted can be several times the size of the energy transfer layer, indicating the need for minimal intervention in the water motion.

As already mentioned, drainage systems using pumps or gravity to eliminate setup during wave storms aim at eliminating rip currents at their point of generation. Longshore currents are not intercepted at all and such arrangements are soft. Floating breakwaters, while eliminating wave action and hence intensification of longshore currents and generation of water setup and rip currents, are also soft. Detached low-height above-seabed breakwaters and low-height groins are also soft arrangements. In many areas, important soft interventions use beach nourishment, a practice deserving general attention because of its generic character. The use of nourishment has been criticised, but may be most effective for some sites with harder structures. There has long been a principle that *effective* coast protection 'hard' structures are those which are protected by stable 'soft' beaches. Finally, there are some areas where no intervention may be most effective, a solution which challenges planners and local communities to change the way they use coastal lands.

The collection of papers presented in the conference appears in this volume in the same order of presentation and without other classification. The only exception is the handout material relating to the Short Course on Beach Nourishment which is presented at the end of this volume. The readers will benefit from this material, which was prepared by the expert instructors, Professors Bob Dean and Richard Davis.

Identification of soft protection methods and evaluation of their quality (effectiveness, environmental impacts, cost, and so on) has been left to the authors and to the good judgment of the reader. This is the first collection of technical information relating exclusively to soft shore protection and hopefully covering all available experience.

'Hard' protection advocates will find understanding listeners to their criticism. Soft shore protection is only starting to develop and certainly needs the input of their experience in the future to soft shore protection methods as well. These advocates, now under attack by governments and courts in many countries, may find a good ally in soft protection and also wisdom in the saying 'Man masters nature not by force but by understanding' (Jacob Bronowski, 1908–74).

LIST OF CONTRIBUTORS, ADDRESSES & E-MAILS

P. L. Aminti, Dipartimento di Ingegneria Civile, Università degli Studi di Firenze, Via di S. Marta, 3 - 50137 Florence, Italy. aminti@dicea.unifi.it

H. D. Armono, Center for Coastal Dynamics BPDP BPPT, Ministry for Research and Technology, 11th Floor Bldg 2 BPP Teknologi, Jl. M.H. Thamrin no. 8, Jakarta 10340, Indonesia. cveob@nus.edu.sg

R. Besson, Regione Lazio, Dipartamento Opere Publiche, Via Capitan Bavastro 110, 00154 Rome, Italy. raimondo.besson@llpp.regione.lazio.it

C. Cammelli, Dipartimento di Scienze della Terra, Università degli Studi di Firenze, Via Iacopo Nardi, 2 - 50132 Florence, Italy. ccammelli@yahoo.com

F. L. Chiocci, University of Rome 'La Sapienza', Dipartamento di Scienze della Terra, P.le A. Moro 5, Box 11, 00185 Rome, Italy. francesco.chiocci@ uniroma1.it

S. Christopoulos, HYDROMARE, Ethn. Antistaseos 3A, Kalamaria, Thessaloniki, 551 34, Greece. christopoulos@civil.auth.gr

L. E. Cipriani, Regione Toscana - Dipartimento delle Politiche Territoriali e Ambientali, Via di Novoli, 26 - 50127 Florence, Italy. l.cipriani@regione.toscana.it

T. T. Coates, HR Wallingford, Howbery Park, Wallingford. Oxon, UK, OX108BA. info@hrwallingford.co.uk

J. S. Damgaard, HR Wallingford, Howbery Park, Wallingford. Oxon, UK, OX108BA. pipelines@hrwallingford.co.uk

R. A. Davis, Jr., Coastal Research Laboratory, Department of Geology, University of South Florida, Tampa, FL 22620, USA. rdavis@chuma.cas.usf.edu

R. Dean, Department of Civil and Coastal Engineering, University of Florida, Gainesville, FL 32605, USA. dean@coastal.ufl.edu

P. Delimani, Department of Civil Engineering, Democritus University of Thrace, 67100 Xanthi, Greece. xeidakis@civil.duth.gr

Y. Goda, Emeritus Professor, ECOH Corporation, Ueno-Takeuchi Bld., 2-6-4 Kita-Uneno, Taito-ku, Tokyo 110-0014, Japan. goda@ecoh.co.jp

C. L. Goudas, Department of Mathematics, University of Patras, 26500 Patras, Greece. goudasconstantine@hotmail.com and stud-mec@otenet.gr

I. Hasegawa, Yokohama National University, Ecoh Corporation, 2-6-4 Kita-Uneno, Taito-ku, Tokyo 110-0014, Japan. katano@ecoh.co.jp

Dick Holmberg, Holmberg Technologies, Inc. 1775 Chadwick Rd. Englewood, FL 34223, USA. DHolmbrg@ aol.com

P. Holmes, Department of Civil & Environmental Engineering 331, South Kensington Campus, Imperial College, London, UK, SW7 2AZ. p.holmes@ic.ac.uk

C. D. Istiyanto, Center for Coastal Dynamics BPDP BPPT, Ministry for Research and Technology, 11th Floor Bldg 2 BPP Teknologi, Jl. M.H. Thamrin no. 8, Jakarta 10340, Indonesia. banten@bppt.go.id

R. E. Jorissen, Ministry of Transport, Public Works & Water Management, National Institute for Coastal and Marine Management/RIKZ, PO Box 20907, 2500 EX The Hague, the Netherlands. jorissen@rikz.rws.minvenw.nl

N. Kafoussias, Dept. of Mathematics, University of Patras, 26500 Patras, Greece. nikaf@ math.upatras.gr

G. Karahalios, Dept. of Physics, University of Patras, 26500 Patras, Greece. karahal@wcl.ee.upatras.gr

Th. V. Karambas, Aristotle University of Thessaloniki, Dept. of Civil Engineering, Division of Hydraulics and Environmental Engineering, Thessaloniki, 54006, Greece. karambas@civil.auth.gr

A. Katano, Ecoh Corporation, 2-6-4 Kita-Uneno, Taito-ku, Tokyo 110-0014, Japan. katano@ecoh.co.jp

K. Katoh, Port and Harbour Research Institute, Hydraulic Division, Ministry of Transport, 3-1-1 Nagase, Yokosuka, 239, Japan. katano@ecoh.co.jp

G. A. Katsiaris, Dept. of Physics, University of Patras, 26500 Patras, Greece. gkatsiaris37@hotmail.com

M. Kok, Twente University, Department of Civil Engineering, PO Box 217, 7500 AE Enschede, the Netherlands. M.Kok@sms.utwente.nl

Ch. Koutitas, Aristotle University of Thessaloniki, Dept. of Civil Engineering, Division of Hydraulics and Environmental Engineering, Thessaloniki, 540 06, Greece. koutitas@civil.auth.gr

G. Labeas, Dept. of Mathematics, University of Patras, 26500 Patras, Greece. labeas@xtreme.com

G. B. La Monica, University of Rome 'La Sapienza', Dipartamento di Scienze della Terra, P.le A. Moro 5, Box 11, 00185 Rome, Italy. giovannibattista.lamonica@ uniroma1.it

B. López de San Román-Blanco, HR Wallingford Howbery Park, Wallingford. Oxon, OX108BA, UK. blb@hrwallingford.co.uk

P. Lupino, Regione Lazio, Dipartamento Opere Publiche, Via Capitan Bavastro 110, 00154 Rome, Italy. p.lupino@flashnet.it

I. Mariño-Tapia, Institute of Marine Studies, Plymouth Environmental Research Centre, University of Plymouth, Drake Circus, Plymouth, UK PL4 8AA. imarino-tapia@plymouth.ac.uk

C. Massalas, Dept. of Mathematics, University of Ioannina, 45-110 Ioannina, Greece. cmasalas@cc.uoi.gr

V. May, School of Conservation Sciences, Bournemouth University, Fern Barrow, Poole, UK, BH12 5BB. vmay@bournemouth.ac.uk

K. Nakamura, Department of Ocean Civil Engineering, Kagoshima University, 1-21-40 Kirimoto, Kagoshima, 890-0065 Japan. sato@oce.kagoshima -u.ac.jp

Y. Nir, coastal and marine geologist consultant, Rehovot 76100, Israel. jacobni@netvision.net.il

R. Nishi, Department of Ocean Civil Engineering, Kagoshima University, 1-21-40 Kirimoto, Kagoshima, 890-0065 Japan. sato@oce.kagoshima-u.ac.jp

P. O'Brien, Coastal Engineering Solutions Pty Ltd, Brisbane, Qld, Australia. cesvic@ cesm.com

B. Ontowirjo, Center for Coastal Dynamics BPDP BPPT, Ministry for Research and Technology, 11th Floor Bldg 2 BPP Teknologi, Jl. M.H. Thamrin no. 8, Jakarta 10340, Indonesia. cveob@nus.edu.sg

M. Perpignan, harbour, marine and coastal engineering consultant, PO Box 17, 3152 Tel Aviv, Israel. jacobni@netvision.net.il

G. Pnevmatikos, Dept. of Physics, University of Patras, Patras, Greece. pnevma@ physics.upatras.gr

E. Pranzini, Dipartimento di Scienze della Terra, Università degli Studi di Firenze. Via Iacopo Nardi, 2 - 50132 Florence, Italy. epranzini@unifi.it

S. Rahman, Bangladesh Water Development Board, Hs. 122, Rd. 1, Block F, Banani, Dhaka - 1213, Bangladesh. cerp@dotbd.com

H. P. Riedel, Coastal Engineering Solutions, 561 Burwood Rd ., Hawthorn, Vic., 3122 Australia. cesvic@cesm.com

P. E. Russell, Institute of Marine Studies, University of Plymouth, Drake Circus, Plymouth, UK. P.Russell@plymouth.ac.uk

M. Saari, Jaakko Pöyry Development Ltd., PO Box 4, Jaakonkatu 3, FIN-01621 Vantaa, Finland. markus.saari@poyry.fi

T. Sasaki, Department of Ocean Civil Engineering, Kagoshima University, 1-21-40 Kirimoto, Kagoshima, 890-0065 Japan. sato@oce.kagoshima -u.ac.jp

M. Sato, Department of Ocean Civil Engineering, Kagoshima University, 1-21-40 Kirimoto, Kagoshima, 890-0065 Japan. sato@oce.kagoshima -u.ac.jp

R. Smith, Coastal Engineering Solutions, 561 Burwood Rd, Hawthorn, Vic., 3122 Australia. cesvic@cesm.com

E. Tzirtzilakis, Dept. of Mathematics, University of Patras, 26500 Patras, Greece. nikaf@math.upatras.gr

B. G. van Vuren, Delft University of Technology, Section of Hydraulic Engineering, PO Box 5048, 2600 GA Delft, the Netherlands. B.G.vanVuren@ct.tudelft.nl

G. S. Xeidakis, Department of Civil Engineering, Democritus University of Thrace, 67100 Xanthi, Greece. xeidakis@civil.duth.gr

M. Xenos, Dept. of Mathematics, University of Patras, 26500 Patras, Greece. mikexenos@ hotmail.com

S. Yanagishima, Port and Harbour Research Institute, Hydraulic Division, Ministry of Transport, 3-1-1 Nagase, Yokosuka, 239, Japan. katano@ecoh.co.jp

SOFTER SOLUTIONS TO COASTAL EROSION: MAKING THE TRANSITION FROM RESISTANCE TO RESILIENCE

VINCENT MAY

Bournemouth University, School of Conservation Sciences, Talbot Campus, Fern Barrow, Poole, Dorset, United Kingdom, BH12 5BB

Abstract

The transition from hard resistant coast protection to soft protection requires transitions in four inter-related principles. First, the design, implementation and evaluation of the softer methods need to be appropriate to the severity of the erosion risk. Second, the costs and benefits need to be assessed in ways that take account of changes in levels of risk, frequencies of damaging events and intangible costs and benefits. Third, the coastal communities must be confident that the softer methods will provide protection. Fourth, the systems of softer protection need to ensure the resilience of vulnerable human communities and ecosystems. This paper reports the results of a series of softer protection schemes and discusses lessons learnt in using softer rather than hard solutions. The uncomfortable conclusion for traditional approaches is that people are part of coastal ecosystems and must be part of the solution. Sustainable coastal protection depends upon behavioural changes in all those involved with the coast and its management. Softer solutions must be subject to the following tests:

1. Is the softer protection measure able to protect against extreme events? Does softer mean safer?
2. Who loses and/or gains?
3. What is the timescale over which losses or gains are to be assessed?
4. How does the human community itself assess the risk? What is it prepared to accept and at what cost? Is it able or prepared to move?

The paper uses these tests to assess the effectiveness of the transition to softer solutions from a number of small rural sites within which important human and natural resources are at risk if erosion continues. The evaluation of transitions from hard protection approaches to softer methods has been carried out by observing the decision-making process and public involvement in action. Qualitative research methods are used since the populations involved are typically small in number. Six test sites in southern England were used. They include rapidly eroding sandy and shingle beaches and soft cliffs, with different levels of backshore use and property values.

There is no consistent pattern of behaviour in the six test sites. A very important role is played by the site manager and project team depending upon the local politics and the development of trust in the 'experts'. The extremes range from substantial public resistance requiring formal public enquiries and conflict between landowners, coast protection agencies and users of the site to compliance with the approach adopted by the site owner to protection against erosion. The latter approach requires time and patience in the explanation and discussion of alternatives, but more substantial changes

C. L. Goudas et al. (eds.), Soft Shore Protection, 1-16.

in approaches to the problems are also required.

Adoption of softer protection depends in part on the process of decision-making and public participation. The engineering design needs to be embedded within the socio-economic and politico-cultural processes. The sensitivity of these processes and systems need to be assessed. In particular, it is essential to identify the thresholds at which demands for greater resistance overcome the softer approaches. In the past, resilience was one, but not the only, common way of dealing with coastal risk. As a perspective on managing coastal erosion, it needs to be re-learnt.

1. Introduction

Coast protection ranges from 'hard' engineered solutions such as sea walls, groynes and harbour walls to 'soft' protection solutions such as beach replenishment and salt marsh management. A consistent tenet of coastal engineering holds that the best defence for any coast or engineered structure is a beach which is wide and high enough to absorb wave energy. Unfortunately, extreme events can damage, overtop or destroy any well-designed and normally effective coast protection. With rising sea levels, existing protection structures and beaches are increasingly likely to be overtopped. Both the costs of protection and the value of property at risk are rising. Environmental arguments about the impacts of hard structures on coastal systems have gained in importance. Some coast protection authorities and engineers have moved gradually from 'hard' engineered solutions towards 'softer' alternatives, which include beach nourishment and even managed retreat, for example, allowing coastal areas to be regained by the sea from which they were reclaimed in the past. The problem for the introduction of softer solutions to coast protection arises fundamentally not from the engineering or technical issues, but from the perception, attitudes and involvement of human communities in the management of the coastal ecosystem of which they are part. Softer solutions raise serious concerns for many coastal communities about the extent to which they provide protection against the loss of or damage to them and their property and livelihood. This paper considers how communities have responded to proposals for softer solutions to coast protection and outlines ways in which the transition from resistance to resilience may be restricted or brought about effectively.

2. Understanding transitions towards softer protection

The study on which this paper is based examined small rural sites within which important human and natural resources are at risk if erosion continues. They represent a range of approaches to coast protection where hard engineering solutions are inappropriate either because of cost or because of the overall character of the site. Typically, there are strong ecological arguments for minimal intervention in the coastal system. The sites are also important for recreation in non-urban settings. The evaluation of transitions from hard protection approaches to softer methods has been carried out by observing the decision-making process and public involvement in action. Qualitative research methods were used since the populations involved are typically small in number. The approach tracked the decisions made by the individuals and groups involved and/or at risk at each site. It identified their objectives and considered ways in which these were met or were not achieved. Although not all participants in the process attained their objects or agreed with some aspects of the final decisions, they had access to both formal and informal processes of decision development and decision-making. The research process was designed to use interviews, analysis of formal records of councils and other organisations, the formal planning processes and the consultation process, and media reporting. Through this combination of methods, it has been possible to address a number of questions, the answers to which allow tests of transition from resistance to resilience to be made. Key questions were:

- Who were involved, what were their roles and their objectives?
- What did they achieve, what was critical for success and how long did it take?
- What objectives were not attained and why?
- What action was taken, by whom and in what time sequence?
- What effects did the action have on stakeholders and habitats?
- Where external effects changed conditions during the project, how adaptable were the stakeholders, advisers and decision-makers?

Dronkers and de Vries (2000) have commented that systematic analysis of coastal development history from an integrated natural-social scientific approach has hardly been undertaken for any coastal system in the world. They pose similar but not identical questions to those used for this study. The approach towards understanding of the patterns of adoption of softer protection has required embedding science within social science. As a result, the study had adopted an historical approach to coastal project evaluation such as that advocated by Dronkers and de Vries (2000) and considered how the lessons learnt from this could be applied to the transition to softer shore protection. From the questions outlined above, four tests of the transition from resistance to resilience were identified and applied to each site.

- Test 1. Is the softer protection able to protect against extreme events? Does softer mean safer?
- Test 2. Who loses and/or gains?
- Test 3. What is the timescale over which losses or gains are to be assessed?
- Test 4. How does the human community itself assess the risk? What is it prepared to accept and at what cost? Is it able or prepared to move?

3. Test sites

Five sites were originally included in the study, but a sixth was added in its later stages. The sites lie along the southern coast of England and are all within 15 minutes drive of large urban centres (details of the key features of each site are listed in Table 1). With the exception of Hengistbury Head, each site is the responsibility of a single local authority and typically the number of landowners is between one and twenty. Where there are several landowners, they are usually house owners. The areas protected by the sites include much larger populations (in the case of Pagham up to 10 000), although typically the population directly at risk within the next 25 years is less than 50.

Table 1
Site characteristics (present)

Location	Hengistbury Head	Pagham	Birling Gap	Brownsea	Ringstead	Studland
Site characteristics	Cliffs, sand and shingle beach, patchy dunes	Shingle beach plain and ridge	Cliffs and platform in chalk, with shingle/sand pocket beach	Clay and sand cliffs capped by gravels	Clay cliffs	Dunes
Maximum height (m)	35	5	30	20	15	5
Retreat rate – 100 year timescale (m a^{-1})	0.4	0.1	0.9	0.1 Locally accretion	0.3	0.8 Accretion > 1.0
Retreat rate – 10 year timescale (m a^{-1})	0.7	1.0, but locally accretion > 1.0	0.9	0.1 locally more rapid. Accretion reduced	0.3	0.8 Accretion > 1.0
Maximum retreat in single event (m)	2	10	4	2	4	10
Present defences	Rock rubble groynes, beach replenishment	Wooden groynes, beach reconstruction after storms	None	Timber palisades at selected sites	Shingle replenishment and rock rubble groyne	Gabions and timber palisades in selected location
First date of any protection	1935	1950s	None	1980s	1960s	1980s
Conservation status	SSSI, AAI, GCR	Part GCR	SSSI, AONB, GCR, HC, AAI	SSSI	SSSI, GCR, SAC, AONB, HC, WHS	NNR, SSSI, GCR, AONB, HC
Economic status	with >1 million visitors annually	Coast protection against flooding of village with total population >10,000	Cottages and hotel Recreational beach	National Trust property	10 residential and holiday properties + caravan park	National Trust property with > 1 million visitors annually

Key:

AAI	Area of Archaeological Importance	AONB	Area of Outstanding Natural Beauty
GCR	Geological Conservation Review site	HC	Heritage Coast
NNR	National Nature Reserve	SAC	Special Area of Conservation (EU Habitats Directive)
SSSI	Site of Special Scientific Interest	WHS	World Heritage Site (part)

Hengistbury Head, to the east of Bournemouth, divides into two parts, low cliffs which are subject to a south-west Atlantic swell and a sand spit at the mouth of Christchurch harbour which is sheltered from south-west waves but is occasionally eroded by short period waves from the south-east. It is currently protected by combinations of hard and soft protection and provides a very good example of a cooperative approach which recognises the recreational importance of the site as well as its protective role. Rock groynes, buried rock ramparts, gravel blankets, beach reconstruction and beach replenishment have been used in combination to reduce the risks of erosional loss. There is a longstanding management plan for the site which has been revised progressively over the last 20 years. Of equal importance, however, is the perceived need to minimise erosion, which would allow a breach in the cliffs and potentially flooding of Christchurch Harbour and the surrounding town.

Pagham lies to the north-east of a low-lying headland (Selsey Bill) which during the 1950s was notorious as the most rapidly retreating shoreline in Britain (Duvivier 1961). That shoreline was hardened with seawalls and groynes and the policy of resistance has worked. It is a shingle beach, partly protected by wooden groynes, and provides the main protection for the community from flooding by the sea. However, it requires consistent maintenance and the effect of reducing the retreat has been very largely to cut off the supply of shingle to the adjacent beach to the northeast at Pagham. This is a familiar story. The narrow shingle bank develops into a series of recurves (Robinson 1955). The shingle beach and its narrow beach plain at Pagham have provided protection for the land and property behind them. Soft protection was in place. However, with the reduced sediment budget and the onset of a series of southeasterly storms during the mid-1990s, the ridge has been overtopped and breached. On every high spring tide which coincides with high wave energies, it is now necessary to bulldoze the beach into a higher than natural ridge in order to ensure that the properties which were protected remain so. It could be argued that property development should have been prevented in the risk area, but at the time, this was not perceived as a serious risk. After all, the erosion of the worst site in the coast of southern England had been controlled.

No scenario had been envisaged in which the sediment budget and increased storminess together with rising sea levels would join to produce the increased risk that now prevails and will worsen. Since Duvivier's 1961 paper, the sea level in this area has risen by about 5 mm a^{-1} (i.e. 0.21 m). It is not surprising, therefore, that this community has little faith in the effectiveness of the softer protection. It is positively antagonistic towards the designation of part of the shingle feature as a Site of Special Scientific Interest when this means that protection works may be prevented by the scientific interest of the site. If the original coastal planning had recognised the intimate link between the erosion of the headland and the sustainability of the natural protection, there would have been fewer houses and thus fewer people and properties at risk. The long-term scenario is that hard protection will be put in place because the recurring costs of temporary reconstruction of the protective shingle barrier are rising, the risks are rising and the population will not move. The most likely force for abandonment of such areas will come from the refusal of insurers to provide insurance against losses from flooding. Under current English law, insurers have to provide insurance and they have agreed to continue to do this for the next few years. They expect more investment in flood protection will have been made, but in the meantime, it is likely that premiums for insurance will rise.

Birling Gap, on the Sussex coast west of Eastbourne, is formed by vertical chalk cliffs with a narrow shingle beach at their foot. It has retreated at about 0.9 m a^{-1} for at least the last 150 years. The small community, which lives on the cliff-top, feels very strongly that the erosion should be controlled. Proposals to provide some hardened protection covered by a restored and replenished beach were opposed by the Coast Protection Authority, the Planning Authority, the National Trust (who own much of the site) and English Nature. The site was the subject of a Public Enquiry in summer 2000 established by the (then) Department for the Environment, Transport and the Regions (DETR), a practice used to resolve a planning/development conflict. The arguments are primarily focussed on interpretation of the planning legislation and its application. The community displayed a serious lack of trust in the public authorities. As a result, the community itself made proposals for resistance to erosion. The conclusion of the Public Enquiry was that these proposals should not be allowed. If resistant structures were put in place, they would add potentially to the problems rather than resolve them. The conclusions were based on considering the site within the wider coastal system and accepting that sustainability of the overall system was more important than the needs of the more localised site.

Brownsea Island is owned by the National Trust, and has suffered erosion of its cliffs in recent years. The potential loss here is not of buildings or homes but of land which contains important historical and ecological features. Low technology solutions have been used which are in keeping with the natural character of the site, but in some parts of the site, while these have partly reduced the cliff erosion problem, they have accelerated the beach erosion. These semi-soft protection measures represent a response which sees coastal hardening, however simple, as providing protection. Allowing erosion to continue whilst some local protection was used would probably produce sustainable long-term solutions. The issue at this site is mainly understanding the complex and adaptive nature of the processes in the site and the historical development of the present coastline.

At Ringstead on the Dorset coast, the erosion hazard affects a small community which includes a caravan holiday park largely protected for many years by a shingle beach. The shingle beach was not fed by a significant level of cliff erosion, being mainly flint derived from earlier phases of erosion of the chalk. In the past, significant erosion of the cliffs and extensive landslides have taken place. However, the settlement enjoyed comparative shoreline stability until the mid-1990s. Because there are two areas of intertidal platforms, the shoreline forms three separate beaches: a western beach which is seriously depleted from time to time, a central beach in a slight embayment where the village is mainly located and an eastern beach which is mainly below cliffs of Kimmeridge clay and a very large landslide of Portlandian limestone and chalk. Under conditions of southerly gales (approximately 1 in 200 events), wave energy penetrated between the two platforms into the central beach and removed much of the existing shingle. The cliff retreated and the underlying, mainly clay, platform was lowered by about one metre.

Although the local authority did not propose to carry out coastal works, the community itself put pressure on the authority to carry out works under the Coast Protection Act 1949 and a scheme was drawn up. This involved three elements: beach replenishment, cliff drainage and a groyne at the eastern end of the central beach which would help retain shingle under southerly or south-westerly wave conditions. Because of the biological importance of the intertidal platforms, it was agreed that there should be no importing of shingle from the sea and therefore terrestrial gravels were used rather than marine gravels. The gravels were of comparable size. The main difference, however, was that these terrestrial gravels are mainly platy in shape and quite distinctive from the rounded or sub-rounded shingle of the main beach. The cost-benefit analysis carried out in preparation for the scheme showed a balance of benefit over cost which arose mainly from the amenity and tourist value of the beach and the coastal land. Public meetings demonstrated that the community would not accept no action, but was prepared to accept a scheme that did not significantly change the overall characteristics of this site aesthetically. Following the implementation of the scheme, changes in the beach were monitored and within six weeks of the completion of the beach replenishment, large quantities of the material had moved from east to west into the western beach where they formed 40 per cent of the surface material. Within one year, the new terrestrial material was buried to a depth of at least 0.5 metres. The central beach that had been replenished has retained sufficient shingle over most of its length to provide the protection to the foot of the cliff that was anticipated. However, on the eastern side of the groyne, there has been significant lowering of the beach and retreat of the lower part of the cliff in places in excess of two metres.

Since this scheme was constructed, there has been a change in the approach to funding. Shoreline Management Plans (SMPs) have identified management policies for the whole of the English coast. These identify the preferred action from a short list of "Do Nothing, Managed Retreat, Hold the Line or Advance the Line".

Studland is the largest area of natural dunes on the coast of southern England and is a very important recreational site. Owned by the National Trust, much of the site is either a National Nature Reserve or a Site of Special Scientific Interest. During the 1990s, it suffered a series of erosional events which cut the dunes back by up to 10 metres within 24 hours. Following investigations into the erosional processes, the National Trust undertook a series of actions which were designed to reduce the immediate risks of damage from similar events, to provide a medium-term resilience to such events and which would allow progressive re-siting of the most valuable assets of the site. The process involved discussions and agreement with both statutory organisations and with individuals who have commercial and property interests in the site. The National Trust decided that it would combine this strategy with an open approach to the media and general public. It would explain why changes were required and it utilised the local university's coastal zone management students in both the technical and informational aspects of the project. Although the National Trust as landowner has taken responsibility for the project and has provided the funding, it has also involved a wide range of stakeholders in both the management of its erosion, but also in the wider management plan for the site (May 1999).

4. Applying the transition tests

The four tests of transition show that there is no consistent pattern of behaviour in the test sites (Table 2, and 3). A very important role is played by the site manager and project team depending upon the local politics and the development of trust in the 'experts'. Trade-offs or compromises were typical of much of the process, although not without sometimes heated argument. The responses range from substantial public resistance requiring formal public enquiries and conflict between landowners, coast protection agencies and users of the site to compliance with the approach adopted

Table 2
Summary of the application of the tests to the sample sites

Location	Hengistbury Head	Pagham	Birling Gap	Brownsea	Ringstead	Studland
Scenario for next 20 years	Reduced cliff retreat Loss of cliff-top habitat Continued erosion of beach **RSL rise by 100 mm**	Reduced beach Loss of shingle habitat Serious coastal flooding increased **RSL rise by 120 mm**	Continuing erosion Loss of cliff-top property **RSL rise by 100 mm**	Fluctuating erosion with some stability Loss of cliff-top habitat **RSL rise by 80 mm**	Periodic removal of beach and cliff foot retreat Outflanking by erosion alongside protected site **RSL rise 90 mm**	Continuing erosion and accretion Possible breach of beach line Loss of visitor facilities and car park **RSL rise 80 mm**
Scenario for 2020	Cliff top 10 m back	Loss of beach ridge and breached protection	Cliff top 25 m back	Cliff top in places 2 m back	Cliff top 10 m back Serious loss of beach	Extended erosion North dunes advanced by max 10-20 m
Preferred future protection	Maintain present line	Maintain and reconstruct beach	No action	Maintain present line, but monitor over next five years	Maintain present line and monitor	Manage retreat and reconstruct dunes

by the site owner to protection against erosion. The latter approach requires time and patience in explanation and discussion of alternatives. However, it also requires a behaviour change amongst the stakeholders, especially as existing consultation arrangements do not work well. There is a growing emphasis within coastal management research on understanding and developing the processes of coastal management within a cooperative approach rather than an outcome-orientated approach (Davos, 2000). The most successful of the projects had adopted a cooperative approach, even though this had to be implemented within an existing 'command and control' type of approach because of existing legislation and funding arrangements.

Transition may need a combination of hard and soft methods: for example a proposed artificial dune at Studland would restore and maintain a locally important element of the coastal landscape, but the presence of valuable assets along the shore can be protected against the relatively infrequent but damaging event by providing a hardened core to the dune. Ecological appearance and function is largely restored, aesthetic and recreational needs are met, projection against damaging events is provided and breathing space provided for the removal of critical facilities and assets which need a longer transfer time to a safer location. Many solutions are sub-optimal and so communities have to learn not only to adapt to these sub-optimal conditions, but also to recognise the thresholds of safety which such solutions include. Many of the most serious impacts on the sites occurred with extreme events, but protection is rarely designed to protect against the most extreme events or clusters of extreme events.

Table 3. Summary of the transitions

	Hengistbury Head	**Pagham**	**Birling Gap**	**Brownsea Island**	**Ringstead**	**Studland**
Are the design, implementation and evaluation of the softer methods appropriate to the severity of the erosion risk?	Yes, combined with hard methods	No, too dependent upon maintenance and response to events	No	Yes	Yes, in part, but overlook adjacent beach performance	Yes
Do the costs and benefits take account of:						
changes in levels of risk?	Yes	No	No	No, but all under review	Yes	Partially, but under review
frequencies of damaging events?	Partially	No	No		Yes	Ditto
intangible costs and benefits?	Incompletely	No	No		Partially	Yes
Are the coastal communities confident that the softer methods will provide protection?	More confident	Not confident	Not confident	Yes	Yes, but growing concerns about adjacent beaches and cliffs	Yes
Will the systems of softer protection ensure the resilience of						
a. vulnerable human communities?	No	No	No	Partially	Partially	Yes
b. ecosystems?	Partially	Partially	No	Yes	Yes	Yes

Underlying much of the debate are a number of fundamental questions which although often unspoken, interviews show are important. What are the risks if protection fails? Does this imply negligence on the part of the responsible authority/landowner? What action can be taken if the protection fails firstly to deal with the immediate damage and loss and secondly to rehabilitate or relocate the owners in the longer term? Should those who occupy such locations bear the risk? If they are aware of the risk, should they bear the loss and if so, should they bear the costs of protection? What is the role of insurers? The decision process is at least as important as the outcome (Green & Penning-Rowsell 2000), but this needs to be analysed in detail and with due regard to both formal and informal decision developing and making.

5. Discussion

The analysis of the decisions and responses indicates that a transition from resistance (hard) to resilience (soft) requires an understanding of four inter-related transitions which have to occur if the overall transition is to be allowed and will be effective in the long term, that is, sustainable. They concern design, values, trust and adaptability.

First, the design, implementation and evaluation of the softer methods need to be appropriate to the severity of the erosion risk. On some sites, the erosion risk may arise from regular storm wave inputs, whereas in others it may result from infrequent extreme events. The nature of the protection methods thus needs to recognise the intensity, magnitude and frequency of the erosion event. In particular, it may be necessary to develop and use soft protection systems which combine hard components as well as an overall softer structure. One example is the use of walls constructed of gabions and buried within dunes. Although the dunes provide soft protection under most conditions, their hardened interior provides protection against the more extreme events.

Second, the costs and benefits need to be assessed in ways that take account of changes in levels of risk, frequencies of damaging events and intangible costs and benefits. The costs and benefits and the trade-offs inherent within cost-benefits applications need to take account of the nature of these site attributes. Taking a wider view, however, this raises fundamental questions about the value of the coast for particular uses and the coastal communities' overall values. To what extent, for example, can a community accept less than optimal solutions? For example, if a site contains the main breeding site for an endangered species, the most important issue will be the probability of an event that removes the breeding site during the breeding season. On the British coast, a similar problem applies to property on the coast: if the property is lost into the sea, what recompense is available to the landowner or tenant? If the site is of archaeological importance, its destruction is an irretrievable loss. The concept of moving the site might allow the main artefacts to be preserved, but in the future they will be out of context. Insurers play a role in that they provide a level of assurance or otherwise depending upon their acceptance of the risk and the level at which they set insurance premiums. On sites which are of ecological importance or which are used for low levels of recreational settlement, it may be possible to move properties, as for example at Studland, charging £100 ($140: € 160) for the move, but at Mudeford, beach huts sell currently at up to £60 000 (US$ 84 000: € 97 000) in an area where house prices range between £75 000 (US$ 107 000 : € 120 000) to £250 000 (US$ 355 000: € 400 000). These will not be moved! However, the owners do not bear the direct costs of protection of their properties – it is provided from general taxation.

Third, coastal communities must be confident that the softer methods will provide protection. The human communities who occupy the coast have lost the ability or opportunity in many locations to migrate to unoccupied areas. Increasingly there has also been a deterioration in peoples' trust in politicians, 'experts' and scientists (Heeps & May 1997). Land in Western societies has economic value and for many first peoples, the land may carry spiritual values. If communities have been allowed to occupy land or have occupied it for centuries, then continuity of occupation typically carries a very high significance for these communities. However, where lands at risk from flooding have been occupied, there are often strategies in place to cope with the erosion risk. This is well-documented for some locations. Often it has been a tax or a payment or a communal activity that has ensured that seawalls or flood protection have been emplaced. The pattern in Europe of developing low-lying areas usually involved communal systems of flood protection. The Courts of Sewers on English coastal wetlands is an example of this. These had a twofold role. They managed the flood management of the area by controlling water levels and building protective walls and levied local taxes that paid for the works. Much of the work was carried out as communal activity for which there was no payment since it was a duty upon each member of the community to serve the community as a whole.

With the expansion of coastal communities into low-lying and cliffed areas during the nineteenth century and the later expansion of resorts and retirement areas, communities were built in areas which had been avoided in the past. They had generally been agricultural land. The risks of flooding or erosion were often ignored until the events occurred. Seawalls and promenades provided protection. Settlements were often set back to allow for walking areas on cliff tops or on dunes. Communities need either to have the property protected (although they will not bear the total cost themselves) or to be reassured that the steps which are taken will provide protection against the erosive events. Pagham is an example where shingle ridges have traditionally provided protection. Communities have built behind the natural protective barrier. However, coast protection works up-drift have gradually reduced shingle transport, and with a shift in storm events, overtopping and breaches have taken place. The community cannot or will not move and so requires protection which will prevent recurrences of the flooding. They want protection which is assured. How do you provide that reassurance? Managed retreat is not acceptable. Set-back lines cannot be implemented without removal of properties.

Fourth, the systems of softer protection need to ensure the resilience of vulnerable human communities and ecosystems. Some coastal ecosystems are resilient. For example dunes typically migrate in response to wind and wave erosion. They need to be able to move. Dunes need to be allowed the space to adjust. Human communities may be able to adjust to the risk of erosion, Studland being a very good example. Birling Gap is an example where physical adjustment other than abandonment of the site is very difficult. Just as human communities may need to accept sub-optimal solutions, there is a requirement to understand the sensitivity of the coastal ecosystem (*sensu lato*) to sub-optimal conditions. Complex but adaptive systems, which typify the coast, are constantly adjusting to stress and a sub-optimal availability of resources, but coast protection schemes rarely have this information as a boundary condition on what is attainable.

Adoption of softer protection depends significantly on the process of decision-making and public participation. The engineering design needs to be embedded within the socio-economic and politico-cultural processes. Davos (2000) focussed on ways in which more cooperative coastal management could be achieved, emphasising sensitivity to pluralistic and systematic agenda setting and control, that is, a shift from top-down to bottom-up approaches, wide-ranging debate about values and their implications, and a requirement for information which takes account of the perceptions, concerns and capacity of diverse groups. Boehmer-Christiansen (1994, 1995) has criticised the assumption that there is an effective flow of information from the scientist to policy-making and wider society. Eden (1998) has argued for more public exposure, criticism and demystification of science and for more interaction between scientists and the users of scientific information. Eden (1998) has also recognised the problem of a perceived public 'lack of expertise' in science-based debate. As a result, little account is taken either of the value of non-scientific expertise and local knowledge, or what Eden (1996) calls contextual and active knowledge. This arises, according to Orbach (1996), from the different subcultures within coastal management. They are reflected in a lack of mutual respect and communication, and more conflict and competition than cooperation. The overall emphasis shifts to cooperation rather than conflict to achieve a transition from conflict–resolution to transdisciplinary (Kohn & Gowdy 2000) and transstakeholding cooperation. This is the focus of a much wider debate than simply one for coastal engineers and managers. The traditional 'hard' engineering approach has been fundamentally a linear one, but it is increasingly accepted that natural systems are very complex, hierarchically nested and governed by nonlinear dynamic processes (Sneddon 2000). At the same time, there is a growing focus on 'sustainable livelihoods' which require communities to be able to provide opportunities for future generations by coping with and recovering from damaging events, maintaining or improving the community's capabilities and assets (Chambers & Conway 1992). The scale of the community is also an issue in bringing about more sustainable conditions.

If softer shore protection is to be used more widely, then coastal engineers and managers need to better understand the ways in which their knowledge and that of other stakeholders can be used in making coasts sustainable. This requires much better understanding of the sub-cultures of coastal management, how knowledge and information are used and how the changes in behaviour which are needed for successful implementation of soft shore protection might be made. McKenzie-Mohr (2000) has argued that information-intensive campaigns often do not bring about the expected changes in behaviour because developers underestimate the difficulties of making behaviour changes and their lack of psychological knowledge. Information is unlikely to bring about sustained behavioural change unless the information is communicated in ways which are both culturally aware and sensitive to drivers of and barriers to fundamental changes in attitudes. Eden (1996) has argued that environmental education may not produce public participation. More adaptive and resilient approaches to coastal management would, if McKenzie-Mohr's critique is accepted, require soft shore-protection schemes to focus as much on the behaviour of the stakeholders as on the technical issues of the most suitable design. McKenzie-Mohr asks four questions:

- *What is the potential of an action to bring about the required change?*
- *What are the barriers (of perception, attitude and culture) associated with each proposed action?*
- *Do the human resources exist which can work to overcome these barriers?*
- *What type or class of behaviour is to be promoted?*

None of the schemes investigated in the test sites asked these questions. Some were implicit within the way in which the engineers or scientists approached the task and some of the site managers were very sensitive to the expected reactions of some of the stakeholders. This was particularly true of the National Trust site at Studland, where considerable work was undertaken to achieve co-operation. The evidence from the sites indicates that adoption of softer approaches to shore protection requires better understanding of behaviour change, especially as there is much rethinking occurring about environmental management generally (Bryant & Wilson 1998; Eden 2000, 2001). Prochaska et al (1997) have identified five stages of behavioural change. At first there is no intention to change (pre-contemplation), then there is an intention to change (contemplation), followed by active planning of change (preparation) and overt making of change (action). Once behavioural change has occurred, it has to be maintained by taking steps to sustain the changes and to resist temptations to relapse. This has occurred on a number of coastal sites where, despite policies for minimal intervention against erosion because cliff erosion feeds beaches, cliff protection structures have been built. In contrast, Birling Gap, the policy of no action has been maintained despite pressures to introduce harder structures. Hildebrand (1997) identified three stages of improved coastal management through public involvement, co-operation and empowerment. With empowerment, all partners in a coastal management project have the knowledge, skills, information and other resources to play their full part in a co-operative management process. Treby (1999), in an analysis of the consultation process, argues for a five-stage model. The first stage of general education involves both the expert and non-expert communities *informing* each other. This is followed by listening (*consultation*). Possible options are suggested (*inform*), the priorities are discussed (*consultation*) and finally the outcome of the consultation is articulated (*inform*). This approach should meet some of Orbach's (1996) concerns about the conflicts between sub-cultures in coastal management. Soft shore protection is more likely to succeed when the behaviours of those most likely to be affected by its effects are better understood.

6. Conclusions

The sensitivity of these shore processes and systems need to be assessed. In particular, it is essential to identify the thresholds at which demands for greater resistance overcome the softer approaches. Resilience was one, but not the only, common way of dealing with coastal risk in the past. It is a way of approach to managing coastal erosion which may need to be re-learnt. On 27 September 2000, the European

Commission published a strategy on Integrated Coastal Zone Management for Europe (European Commission 2000) saying, 'The Strategy aims to promote a collaborative approach to planning and management of the coastal zone, within a philosophy of governance by partnership with civil society'. The evidence from these case studies and the approach which has been developed for managing the transition from resistance to resilience, suggests that this meets the expectation of the European Commission. However, the investigation described here also shows that this transition is far from easy, for although the scientific basis for soft protection and the ability to emplace softer protection both exist, the widespread adoption of softer protection depends upon our ability to develop the processes of coastal governance through partnership with society. Sadly, many human communities have little faith in scientific advice or political decisions (Heeps & May 1997) and we have much work to do to ensure the success of the transition to softer protection. Otherwise, the resistance will be from the human communities, who will demand resistance to the forces of the sea rather than themselves becoming resilient and adaptive to its relentlessness.

In summary, therefore:

- The design, implementation and evaluation of the softer methods need to be appropriate to the severity of the erosion risk. Appropriate design is essential.
- The costs and benefits need to be assessed in ways which take account of changes in levels of risk, frequencies of damaging events and intangible costs and benefits. Values and acceptability of sub-optimal solutions must be understood and evaluated.
- The coastal communities must be confident that the softer methods will provide protection. They must trust the engineers, experts and politicians, who for their part need to improve the ways in which they communicate probabilities to their publics.
- The systems of softer protection need to ensure the resilience of vulnerable human communities and ecosystems. Both need space to adapt to change. Without space, they cannot be resilient. Without resilience, soft protection schemes will not work and will not be credible.

To achieve these transitions, specific tasks need to be undertaken. The sensitivity of the coastal processes and systems, natural and human, needs to be assessed. It is essential to identify the thresholds at which demands for greater resistance overcome the softer approaches. Better information and understanding must be developed in respect of probabilities, magnitudes and frequencies of extreme events, especially clustered extreme events. Adoption of softer protection depends in part on the process of decision-making and public participation, and so these processes must be understood better. For successful implementation of soft shore protection, the engineering design needs to be embedded within the socio-economic and politico-cultural processes.

Acknowledgements

The author wishes to thank Bournemouth University, the National Trust, West Dorset District Council and Bournemouth Borough Council for support of parts of the work which underpins this paper. Thanks are also due to Emma Treby, who critically reviewed an earlier version of the paper. The views expressed in this paper are, however, the author's own.

REFERENCES

Boehmer-Christiansen, S.: Politics and environmental management, *Journal of Environmental Planning and Management*, 37, 1994, 69–86.

Boehmer-Christiansen, S.: Reflections on the politics linking science, environment and innovation, *Innovation: the European Journal of Social Sciences*, 8, 1995, 275–288.

Bryant, R.L. and Wilson, G.A.: Rethinking environmental management, *Progress in Human Geography*, 22, 1998, 321–43.

Chambers, R. and Conway, G.R.: Sustainable rural livelihoods: practical concepts for the 21st century. *Institute of Development Studies Discussion Paper 296.* London: Institute of Development Studies, 1992.

Davos, C.A.: Sustainable cooperation as the challenge for a new coastal management paradigm, *Journal of Coastal Conservation*, 5, 2000, 171–180.

Dronkers, J. and de Vries, I.: Integrated coastal management: the challenge of transdisciplinarity, *Journal of Coastal Conservation,* 5, 2000, 97–102.

Duvivier, J.: The Selsey coast protection scheme. *Proceedings of the Institution of Civil Engineers*, 20, 1961, 481–506.

Eden, S.: Public participation in environmental policy: considering scientific, counter-scientific and non-scientific contributions. *Public Understanding of Science* 5, 1996, 183–204.

Eden, S.: Environmental issues: knowledge, uncertainty and the environment. *Progress in Human Geography,* 22, 1998, 425–32.

Eden, S.: Environmental issues: sustainable progress. *Progress in Human Geography,* 24, 2000, 111–8.

Eden, S.: Environmental issues: nature versus the environment? *Progress in Human Geography,* 25, 2001, 79–85.

Green, C. and Penning-Rowsell, E.: Inherent conflicts at the coast, *Journal of Coastal Conservation,* 5, 2000, 153–162.

Kohn, J. and Gowdy, J.: Coping with complex and dynamic systems. An approach to a transdisciplinary understanding of coastal zone developments, *Journal of Coastal Conservation,* 5, 2000, 163–170.

European Commission: *Integrated Coastal Zone Management Strategy,* 2000.

Heeps, C. and May, V. J.: Building bridges to science: making coastal science better understood, in Vollmer, M. and Grann, H. (eds.) *Large–scale constructions in coastal environments: conflict resolution strategies*, Springer, Berlin, 1997, 55–66.

Hildebrand, L.P.: Participation of local authorities and communities in integrated coastal zone management, *in* Haq, B. U. et al (eds.) *Coastal Zone Management Imperative for Maritime Developing Nations*, Kluwer Academic Publishers, 1997, 43–54.

May, V.J.: Replenishment of resort beaches at Bournemouth and Christchurch, *Journal of Coastal Research,* Special Issue 6, 1990, 11–16.

May, V.J.: Where will our coast be in 2022? Strategies for managing retreat in commercially valuable and intensively used dunes and cliffed coasts of high nature conservation value, *in* Urban Harbors Institute (eds.) *Vision 2020: the people, the coast, the ocean.* Coastal Zone 99 conference, San Diego, 1999, 212–214.

McKenzie-Mohr, D.: Fostering sustainable behaviour through community-based social marketing, *American Psychologist*, 55, 2000, 531–7.

Orbach, M.K.: Science and policy in the coastal zone, *in* Taussik, J. and Mitchell, J. (eds) *Partnership in Coastal Zone Management*, Samara Publishing Ltd., Cardigan, 1996, 7–13.

Prochaska, J.O., Redding, C.A., Harlow, L.L., Rossi, J.S. and Velicer, W.F.: The transtheoretical model of change and HIV prevention: a review, *Health Education Quarterly*, 21, 1995, 471–486.

Robinson, A.H.W.: The harbour entrances of Poole, Christchurch and Pagham, *Geographical Journal*, 121, 1955, 33–50.

Sneddon, C.S.: 'Sustainability' in ecological economics, ecology and livelihoods: a review, *Progress in Human Geography*, 24, 2000, 521–49.

Treby, E. J.: Frames of influence: embracing culture-centric perspectives on public and institutional participation in coastal zone management, Unpublished Ph.D. thesis, University of Southampton, 1999, pp. 327.

Tunstall, S.M. and Penning-Rowsell, E.C.: The English beach: experiences and values. *Geographical Journal*, 164, 1998, 319–32.

Turner, R. K., Lorenzoni, N., Beaumont, I., Langford, I. and McDonald, A.L.: Coastal management for sustainable development: analysing environmental and socio-economic changes on the United Kingdom coast. *Geographical Journal*, 164, 1998, 269–81.

NOURISHING ERODING BEACHES: EXAMPLES FROM THE WEST–CENTRAL COAST OF FLORIDA

RICHARD A. DAVIS, Jr.,

Coastal Research Laboratory, Department of Geology, University of South Florida, Tampa, FL 22620, U.S.A.

Abstract

Beach nourishment has been widespread and successful in Florida, and particularly so on the west coast of the peninsula. This highly developed coast has seen more than a dozen large-scale projects in past 20 years. The biggest problem with such projects along this coast is the limited sediment available. Most projects rely on the ebb-tidal deltas of tidal inlets but offshore sand bars are present along some coastal reaches.

The type of construction along with the distance traveled by the nourishment material from the borrow area dictates the cost of a given project. Three adjacent projects on Sand Key provide an excellent opportunity to assess the performance of different construction styles and borrow areas. Detailed monitoring of numerous profiles across the beach at each of the projects permits determination of shoreline changes and sediment volume lost during a given period of time. These data show that the Redington Beach project performed best in all respects over at least a four-year period whereas the Indian Shores Beach project performed the poorest. This can be attributed to the type of construction at each. It should be noted, however, that all three projects performed much better than predicted by the design specifications.

1. Introduction

Beach erosion has been a natural phenomenon for as long as we have had beaches. Now that increased population with associated industry and development have put further pressure on maintaining our beaches, we must have some solutions for mitigation and restoration. The State of Florida in the southeastern United States is probably the best example of the combination of developmental pressures, very high coastal populations, and successful programs for dealing with coastal erosion.

The fundamental aspect of the Florida program is beach nourishment. This type of erosion control is based on placing large quantities of sand on the eroding coastal area in order to protect property including buildings, people and infrastructure, and to restore the economically important recreational values of healthy beaches. Such a nourishment program has been underway since the 1960s but has been extensive for only the past 20 years (NRC, 1995). The single project that helped to lead the way and justify this type of construction is the Miami Beach nourishment project that was completed in 1980. This was a huge project that extended about 20 km along the coast and had a price tag of more than US$60 million. The project has performed better than expected and is considered by the engineering community as one of the great projects of the world.

This discussion will consider the beach nourishment situation in Florida. The location, procurement and placement of beach material will be discussed, along with the general characteristics of the design and construction of these projects. Monitoring to determine

C. L. Goudas et al. (eds.), Soft Shore Protection, 17-28.

performance is a critical aspect of all nourishment projects because it is fundamental for establishing long-term management decisions. Three case histories will also be considered as examples of the Florida experience.

2. General Setting

The Florida peninsula is a carbonate platform, quite similar to the Great Bahama Banks but now attached to the mainland of the United States. For most of its history, this platform accumulated various types of carbonate sediments that eventually became limestone. About 5-7 million years ago, during the Miocene epoch, this platform became inundated with quartz-rich sediment from the southern Appalachian Mountains as the Suwannee Straits filled with sediment, attaching the platform to the rest of the United States (McKinney, 1984).

These conditions provided the quartz sediment that now dominates most of the coastal zone of this peninsula. As sea level rise slowed markedly during the past 7 000 years, wave action molded our present barrier islands that extend throughout the State beginning about 3 000 years ago. These barrier islands and their beaches constitute the dominant coastal system throughout most of the peninsula. Because of the fragility of such islands, there is always concern about their stability, especially in light of the accelerating rate of sea level rise over the past century. As climate and beaches became very attractive to North Americans, these peninsular barrier island beaches became populated by about 12 million people; more during holiday times. The combination of rising sea level, sea walls, jetties and other structures with a limited supply of sediment has produced serious erosion problems along many sections of the Florida coast.

The two sides of the Florida peninsula are quite different. The east coast is bound by a steep and narrow continental margin that widens and becomes less steep toward the north (Davis et al, 1992). Wave energy there is substantial but the sheltering effect of the Great Bahama Bank keeps it from being high. By contrast, the west Florida continental margin is very broad and gently sloping. Such conditions, along with the fetch-limited nature of the Gulf of Mexico, minimize wave energy. Another important variable that differs between the two coasts is the availability of sediment. The east coast has a fairly abundant supply of sand on the shoreface and inner shelf, but that diminishes to the south where the supply diminishes and reefs become widespread. On the west coast, sediment is basically absent beyond about a kilometer of the shoreline, and the only sediment sinks are the ebb-tidal deltas on the Gulf side of tidal inlets.

Coastal processes along the two coasts of the peninsula are also somewhat different. On both coasts the dominant direction of longshore transport of sediment is from north to south, driven by northerly winds associated with the passage of cold fronts during the winter. Prevailing winds and longshore currents move from south to north but with little energy. The main difference between the two coasts in this regard is that the east coast is quite uniform in its coastal orientation and sediment transport direction whereas the west coast has a varied orientation to its barrier island coasts and there are many local reversals in transport direction (Davis, 1999). The net result is that the annual volume of littoral transport on the east coast can be up to an order of magnitude greater than on the east coast. All of these factors help to determine the nature and quality of beach nourishment.

3. Sources of Nourishment Material

Once a nourishment plan has been devised and designed, it is necessary to find the appropriate amount of beach quality sediment, typically quartz-dominated sand. In Florida, shell debris is also a common constituent of coastal sediments, including potential borrow areas. Location of the necessary volume and quality is typically accomplished by a combination of high-resolution seismic surveys, surface sampling and vibracoring. This data will demonstrate the distribution and character of the sediments in the proposed borrow area. Once selected and approved in the permit, the borrow area is ready to be mined for the nourishment project.

The shoreface of both coasts of the Florida peninsula has a limited sediment supply for beach nourishment. There are generally two environments where appropriate sediment is available in the required amounts: the ebb-tidal deltas of tidal inlets and the offshore sand bars, some of which may be relict beach deposits. The volume of sediment in various of these deposits ranges from as little as 2×10^6 m^3 to nearly 500×10^6 m^3 (Hine et al, 1984). Permits are required for all extractions of sediment from these resources, and applications must demonstrate that removal of significant sediment will not have an adverse affect on the adjacent coast as well as on its benthic communities..

Other aspects of the borrow area that must be considered and addressed are the presence of human artifacts such as shipwrecks, buildings, fortifications and so on. which must be avoided in the sediment removal process. The stratigraphy of the borrow area as interpreted from the vibracores will show the presence of any fine sediment that must also be avoided.

4. Beach Construction

Although there are multiple ways for constructing beach nourishment, most projects follow a typical format. Removal of the sediment from the borrow area is done using a suction dredge that may range in size depending on the size of the borrow area and the volume of the material removed. Typically, large dredges with a cutter head of 75-100 cm in diameter (Figure 1) are used for projects of at least 500 000 m^3. This rotating head is systematically moved across the borrow area to carefully extract only the sediment permitted and needed for the project.

Another style of removal and construction that has been used is removing the sediment with a clam-shell dredge and placing it in a large barge for transport to the site. This fairly dry sediment is then placed, again by clamshell dredge, on a conveyor and placed on the beach where it is graded using the standard techniques (Figure 4).

Not all borrow material is taken from underwater areas. Upland sources have been used as well. Appropriate sand is mined and transported to the project site by trucks or by a combination of train and trucks. This approach is complicated and causes serious problems with traffic, road deterioration, and disturbance to coastal residents. It is typically not used in large projects as a result.

The unit cost of these projects ranges widely, based primarily on the distance of transport of the sediment. Large projects have been constructed for as little as US$3 per m^3 to as much as US$15 per m^3.

Fig. 1. Large suction dredge (A) and cutter head (B) where borrow material enters the pumping system.

Fig. 2. Aerial photograph showing dredge with system of flexible pipe and floats that transports the sediment slurry to the project. Note also the change in tone that indicates the borrow cut in the substrate.

Fig. 3. Nourishment site with slurry emptying onto the beach and large equipment grading it to construction specifics.

Fig. 4. Dragline and conveyor system of placing nourishment material on the beach used to nourish the Indian Shores Beach project.

5. Example Projects and Their Performance

To date, more than 200 km of beach have been nourished on the Florida coast. Many more areas are waiting for permitting and funding. Three related and adjacent projects on the Gulf of Mexico coast of the peninsula will be used to demonstrate variety in the construction style, borrow areas, and performance. These three projects are located on Sand Key, a 15 km long barrier island that straddles a Miocene limestone headland along the west-central coast of Florida (Figure 5).

The three projects, in order of construction, are Redington Beach (1988), Indian Rocks Beach (1990), and Indian Shores Beach (1992). They represent projects that are similar in size and design, but that were constructed quite differently using a combination of different borrow areas. All of the projects were designed to have a berm elevation of 6 ft (1.8 m) and a constructed foreshore gradient of 1:20. The width of the constructed beach was different at each location but averaged about 30-35 m. Much of the following discussion is taken from a recent paper (Davis et al, 2000) that compares the performance of three projects.

5.1 REDINGTON BEACH

This project is the most southern of the three on Sand Key. It is located on the northwest-southeast trending portion of the shoreline. The adjacent shoreface has the lowest gradient of the three project areas: about 1:700 out to a depth of 10 m. Fairly extensive ledges of Miocene limestone crop out at depths of 4-5 m about 300 m from the shoreline.

The borrow area for this project was the ebb delta at Johns Pass, a tidal inlet located about 3 km from the southern limit of the nourishment project. This proximity permitted the direct pumping of the dredged slurry onto the beach. Sediment from the borrow area ranged from 10 to 35 % by weight shell gravel with considerable variation from location to location throughout the project.

This project extended for 2.5 km along the coast and included 700 000 m^3 of borrow material at a cost of US$2.9 million dollars, US$4.1 per m^3 (Davis et al, 2000). This is judged by current standards to be an inexpensive project.

5.2 INDIAN SHORES BEACH

The final phase of Sand Key nourishment was completed in 1992 and was located between the first two (Figure 1). The ebb delta at the mouth of Tampa Bay was used as the borrow site but an atypical construction style was used. The technique described above utilizing a dragline and conveyor system was used because it saved about 25% of the projected cost of the project. This method produced a very nice, soft beach that was greeted with enthusiasm by all of its visitors even though it contained up to 15% shell debris.

This project was almost 5 km along the shoreline and included 900 000 cubic meters of borrow material. The long transport distance of the nourishment material resulted in a high cost: a total of about US$10 million with a unit cost of US$11.1 per cubic meter.

This was about US$3-4 million lower than expected because of the atypical construction techniques.

5.3 INDIAN ROCKS BEACH

This project was constructed two years after the completion of the Redington Beach project, and it is the northernmost of the three being considered here. This area is at and just to the north of the headland area along this coast. The shoreline tends from northeast-southwest to north-south. The shoreface gradient is the steepest of the three at about 1:400 out to a depth of 10 m. Actually, the inner shoreface is fairly steep with an extensive flat area Gulfward due to the exposed limestone surface that is intermittently covered by small sand ridges.

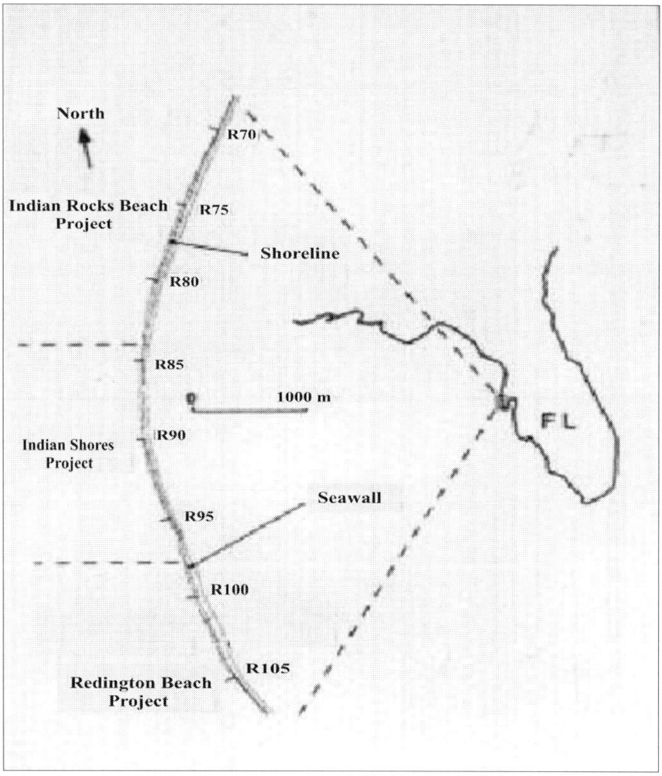

Fig. 5. Location map of Sand Key showing the three nourishment projects discussed in the text.

Because of the absence of sufficient sediment in the vicinity, the borrow area chosen was the huge ebb-tidal delta at the mouth of Tampa Bay. Dredging was on the shallow shoal along the north side of the main ship channel. The 25 km distance between the borrow site and the construction site was covered by barge transport to a staging barge located a few hundred meters offshore. From there, the sediment slurry was pumped onto the beach.

This borrow material contained relatively high concentrations of shell debris: up to 50% as some locations. This caused some concern among recreational users who complained about the shell being uncomfortable for walking. The 4.5 km long project coast a total of US$14.5 million: a unit coast of US$14.5 per cubic meter. This was the most expensive project along this coast of Florida up to that time.

6. Comparison of Nourishment Project Performance

The geographic proximity of these three projects, their different construction styles and sediment sources, and their close times of completion provide an excellent opportunity for assessing their relative performance. The data collected during monitoring of this performance represents one of the most complete data sets for any such projects. A typical monitoring scheme for most nourishment projects is to survey the beach before nourishment, immediately after nourishment, and annually thereafter. A profile is surveyed at each of the monument locations that are spaced at 300 m intervals along the entire coast. By contrast, the Sand Key projects were surveyed at no less than quarterly intervals. The duration of the monitoring was a minimum of four years at each site, with eight years for Redington Beach. In addition, sediment samples were collected and analyzed from three sites along each profile: the backbeach, the foreshore and the surf zone.

Although each of the three nourishment projects was considered to be successful, there were significant differences in their performance. All exceeded their design lifetime. Because of the different completion dates of the projects it was necessary to evaluate their relative performance based on the first year for each one. Another comparison was made during the period of 1992-1996 when all three were monitored under the same scheme.

6.1 SHORELINE CHANGE

One of the primary measurements used to assess beach nourishment performance is the change in shoreline position. There is typically a rapid landward retreat of this datum due to the adjustment of the beach profile from the oversteepened construction configuration to the natural, less steep profile. This has the effect of rapidly changing the shoreline position and generally takes the first 6-8 months, depending on the time of year that the project was completed and the wave energy during that time.

During the first year of performance, the shoreline retreat at each of the projects was remarkably similar (Table 1). All of the values are based on the official datum used in the United States against which all monuments are surveyed.

Table 1 – Shoreline Retreat (m) for Each Project during First Year of Performance

Project	Range	Mean
Redington Beach	10-23	11.35
Indian Shores Beach	8-25	13.04
Indian Rocks Beach	4-22	11.04

A look at the period of time during which each of the projects was monitored simultaneously shows different results. This data is probably a more accurate representation of the long-term performance of the three projects. They show little difference between the projects, but that Indian Shores was the worst (Table 2).

Table 2 – Shoreline Change (m) for Each Project during the Period of 1992-96

Project	Range	Mean
Redington Beach	+12 to -20	-3.22
Indian Shores Beach	-18 to -50	-22.55
Indian Rocks Beach	-5 to -25	-16.29

6.2 VOLUME CHANGE

The most significant parameter for performance of a beach nourishment project is the loss of sediment from the project. Whereas shoreline change can reflect changes in gradient as well as sediment loss, the difference in volume shows the loss and is used to measure the longevity of the project. The volume changes are calculated by taking superimposed profile plots and comparing the differences over the period of time considered. Values are expressed in cubic meters per linear meter of beach.

The performance of the three projects showed great differences over the first year of monitoring. The least amount of sediment was lost by Redington Beach and the most was lost at Indian Shores which was twice as high as the next project (Table 3).

Table 3 – Volume Lost (m^3/m) per Location during the First Year of Performance

Project	Range	Mean
Redington Beach	10-38	15.84
Indian Shores Beach	23-82	43.13
Indian Rocks Beach	10-64	21.92

After stabilization, the long-term changes in volume over the three projects show substantial differences. Redington Beach lost the least sediment and Indian Shores lost the most (Table 4).

Table 4 – Volume Change (m^3/m) per Location during the Period 1992-96

Project	Range	Mean
Redington Beach	+14 to -52	-15.17
Indian Shores Beach	-43 to -103	-65.74
Indian Rocks Beach	-10 to -57	-38.04

7. Summary

It is apparent from the data presented above that a pattern is present for the performance of these nourishment projects. They show that Redington Beach has performed the best by far of the projects and the Indian Shores was the poorest. There are multiple reasons for this. Firstly, the technique of construction: the Indian Shores nourishment was delivered and placed while essentially dry. This produced a very loosely packed beach that was easily eroded. Another factor in performance is that the Redington Beach location receives the least incident wave energy of the three due to its sheltered shoreline orientation and its gentle nearshore gradient.

Another factor in performance is the presence of shell debris in combination with the method of nourishment sediment delivery. Shelly debris when pumped to the nourishment site by slurry forms a platy and well-packed beach that is quite resistant to erosion.

In short, there are many aspects of a beach nourishment project that contribute to its overall performance. Planning of such projects must consider all of these along with the obvious economic factors. The predicted performance and the cost of constructing the project are the most critical aspects of any nourishment project. In order to be successful, there must be a positive benefit/cost relationship.

REFERENCES

Davis, R. A., Reversals in sediment transport along the Gulf coast of peninsular Florida. Amer. Soc. Civil. Engrs, Coastal Sediments, 1999, 761–769.

Davis, R. A., Hine, A. C., and Shinn, E. A., Holocene coastal development on the Florida peninsula. SEPM Spec. Publ. No. 48, 1992, 193–212.

Davis, R. A., Wang, P. and Silverman, B. R., Comparison of the performance of three adjacent and differently constructed beach nourishment projects on the Gulf peninsula of Florida. Jour. Coastal Research, 16: 2000, 396–407.

Hine, A. C., Davis, R. A., Mearns, D. L., and Bland, M. P., *Impact of Florida's Gulf Coast Inlets on the Coastal Sand Budget*, Univ. South Florida, Tampa, Florida, U.S.A., Rept. to Florida Dept. Natural Resources, 1984, 128 p.

McKinney, M. L., Suwannee channel of the Paleogene coastal plain: support for the 'carbonate suppressional model' of basin formation. Geology, 12: 1984, 343–345.

National Research Council, *Beach Nourishment and Protection*. Washington, D. C., National Academy Press, 1995, 334 p.

RELATIVE SIGNIFICANCE OF BACKGROUND EROSION AND SPREADING LOSSES: AN AID FOR BEACH NOURISHMENT DESIGN

ROBERT G. DEAN

Department of Civil and Coastal Engineering, University of Florida, Gainesville, FL 32605, U.S.A.

Abstract

The evolution of beach nourishment projects includes the background (pre nourishment) erosion effects, herein referred to as 'background erosion' and the project evolution in the absence of background erosion, namely the 'spreading out' effects and profile adjustment. This paper examines the relative significance of background erosion and volumetric spreading for idealized initial nourishment planforms. According to approximate beach nourishment theory, it is possible to simply superimpose these two effects. It is useful in the design process to develop estimates of the approximate relative contributions of these two components. It will be shown that in addition to the magnitude of background erosion, the relative effects of background erosion are more significant for the longer projects and increase with time. The interpretation is that the spreading out losses are less for the longer projects and, after evolving for some time, the project behaves as a longer project and thus the relative importance of the background erosion is greater under these two scenarios.

Design aids are presented in the form of graphs to assist in evaluating the relative importance of these two components of beach nourishment project evolution and for preliminary design purposes. These aids provide a basis for determining the significance of background erosion and thus the level of effort warranted in establishing the background erosion rate as compared to the wave forcing which drives the longshore spreading processes. For cases in which the background erosion is secondary, the design engineer can concentrate efforts on the factors (waves) which govern the spreading out losses. Two beach nourishment settings are considered for nourishment with initially rectangular planforms: (1) on a long straight beach, and (2) on a long barrier island with one end of the nourishment adjacent to an inlet. These design aids are illustrated with idealized situations and additional examples are presented for five projects constructed in Florida.

1. Methodology

1.1 SCOPE

Methodology is presented for nourishment: (1) on a long straight beach, and (2) on a long barrier island with one end of the nourishment adjacent to an inlet. In both cases, it is necessary to approximate the background erosion as uniform along the nourishment project and the adjacent beaches of interest. The methodology will be illustrated by application to two idealized cases and to five nourishment projects in a later section of this paper. It will be noted that many of the results in this paper are presented in non-dimensional forms. Any consistent system of units can be used. To convert the longshore diffusivity, G, in ft^2/s to m^2/s, divide by 10.85. To convert G

C. L. Goudas et al. (eds.), Soft Shore Protection, 29-37.

from m^2/s to $km^2/year$, multiply by 31.5.

1.2 GENERAL

The theoretical basis for the linear superposition of 'longshore spreading effects' and background erosion is the so-called Pelnard Considère equation (Pelnard Considère, 1956) which is a result of combining the equation of conservation of sediment and the linearized equation of sediment transport. The conservation of sediment as applied assumes that the profile is translated seaward or landward without change of form in response to an increase or decrease of sediment volume, respectively. The Pelnard Considère equation is linear, thus justifying the linear superposition, which can be expressed as

$$\Delta V_T = \Delta V_S + \Delta V_{BE}$$

in which ΔV_T is the total volume change component, ΔV_S is the volume change component contribution due to spreading and ΔV_{BE} is the volume change component contribution due to background erosion.

1.3 METHODOLOGY DEVELOPMENT FOR NOURISHMENT ON A LONG BEACH

In the absence of background erosion, the evolution of an initially rectangular planform of initially uniform volume density, V_o, and length, ℓ, with longshore diffusivity, G, on a long straight shoreline can be represented as

$$V_S(x,t) = \frac{V_0}{2}\left\{ erf\left[\frac{\ell}{4\sqrt{Gt}}\left(\frac{2x}{\ell}+1\right)\right] - erf\left[\frac{\ell}{4\sqrt{Gt}}\left(\frac{2x}{\ell}-1\right)\right]\right\}, \tag{1}$$

which represents the effects of spreading and is a solution to the Pelnard Considère equation, and 'erf' is the so-called error function. The proportion of volume remaining within the project area for the case of no background erosion is given by Dean and Grant (1989) as

$$M_s(t) = \frac{\sqrt{4Gt}}{l\sqrt{\pi}}\left(e^{-\left(\frac{l^2}{4Gt}\right)} - 1\right) + erf\left(\frac{\ell}{\sqrt{4Gt}}\right), \tag{2}$$

and is presented as Figure 1. Estimates of the longshore diffusivity, G, around the sandy shoreline of Florida have been provided in graphical form by Dean and Grant (1989) and are reproduced as Figure 2 of this paper. The contribution due to background erosion, Δy_{BE} is

$$\Delta M(t)_{BE} = -\, et/Y, \tag{3}$$

where e represents the background erosion rate, t is time and Y is the initial uniform beach width, considering that profile equilibration has occurred. The proportion of volume remaining, M(t), including the contributions of longshore spreading and background erosion is

$$M(t) = \frac{\sqrt{4Gt}}{l\sqrt{\pi}}\left(e^{-\left(\frac{l^2}{4Gt}\right)} - 1\right) + erf\left(\frac{\ell}{\sqrt{4Gt}}\right) - et/Y, \tag{4}$$

which is plotted in Figure 3 for 0<et/Y<1.0 and $0<(Gt/\ell)^{1/2}< 1.0$. The variable on the horizontal axis in Figure 3 determines the spreading losses in the absence of background erosion as can be seen by reference to Eq. (2). The vertical axis represents the proportion of placed material which has been removed from the project area through background erosion. For the upper limit of et/Y = 1.0, the volume of material removed by background erosion equals the total volume placed!

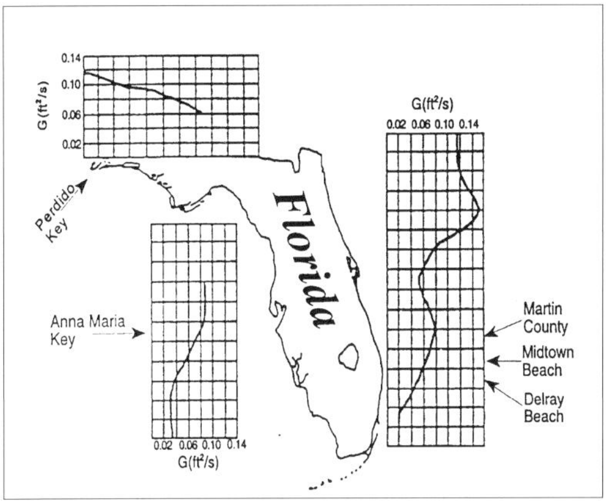

Fig. 1. Approximate Estimates of Longshore Diffusivity G(ft^2/s) around the Sandy Beach Shoreline of the State of Florida. Dean and Grant (1989). Note: Divide by 10.85 to Convert G to m^2/s

There are three sets of isolines present in Figure 3. One set of isolines is the proportion of volume remaining in the project area as a result of both the spreading out losses and background erosion. This set of isolines is labeled 'M(t)'. It will be noted that for zero background erosion (et/Y = 0), the values of M(t) are the same as in Figure 2.

The second set of isolines in Figure 3 is the 'trajectories' along which a project moves during its evolution. Because the abscissa and ordinate of Figure 3 both contain time t, it can be shown that a project will evolve along a line of constant F. The lines defining these trajectories are labeled as constant values of 'F", where

$$F = \frac{e\ell^2}{YG}, \tag{5}$$

For example, a project with a value of F = 0.5 will move along the 'trajectory' defined by that line in Figure 3 during all stages of the evolution of that project.

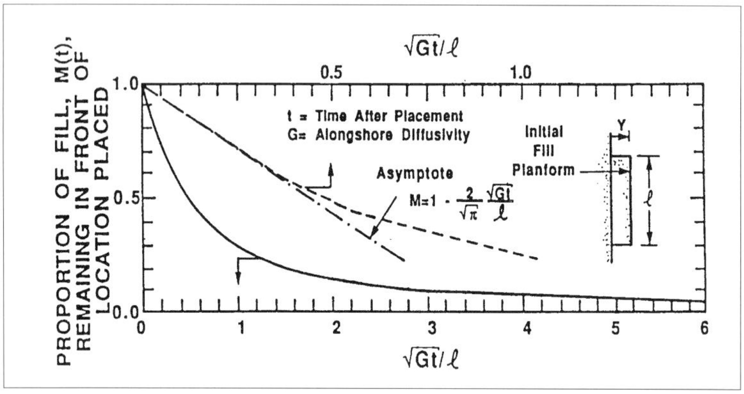

Fig. 2. Percentage of Material Remaining in Region Placed vs. the Parameter $(Gt/\ell)^{1/2}$ for Initially Rectangular Nourishment on a Long Straight Beach. Dean and Grant (1989)

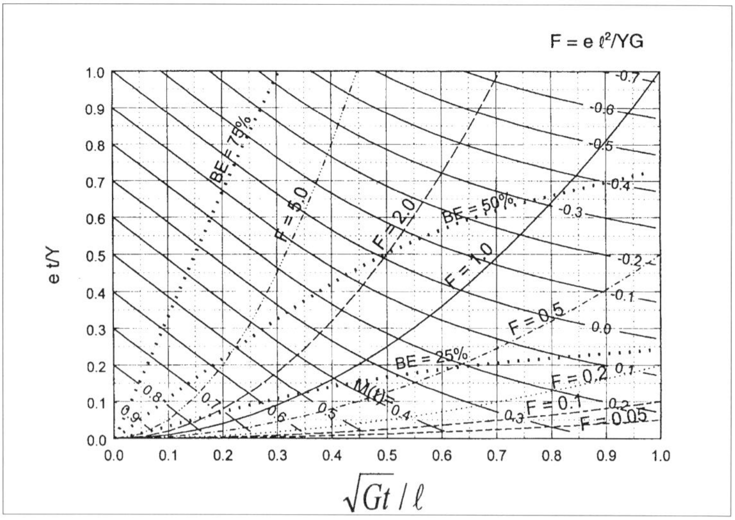

Fig. 3. Isolines of M(t) versus Spreading and Background Erosion Parameters. Long Straight Beach.

The third set of isolines present in Figure 3 defines the proportion of the loss of volume which is attributable to background erosion. These lines are labeled 'BE'. It wil be seen that as the project evolves (following a line of constant F) toward the right in these plots, the proportion of material lost due to background erosion increases for the reasons discussed earlier.

1.4 METHODOLOGY DEVELOPMENT FOR NOURISHMENT ON A BARRIER ISLAND WITH ONE END OF NOURISHMENT ADJACENT TO AN INLET

In this case, the initial nourishment is placed on a long barrier island with one end of the nourishment adjacent to an inlet; Figure 4 shows this diagrammatically. Our understanding of the evolution of this project setting is much less certain than that of nourishment on a long straight beach as discussed earlier. For the current purposes, it will be assumed that the shoreline position at the inlet is maintained at zero displacement, y(0,t) = 0. The solution for spreading losses for the case of an initially rectangular planform is

$$V_S(x,t) = \frac{V_0}{2}\left\{erf\left[\frac{\ell}{\sqrt{4Gt}}\left(\frac{x}{\ell}+1\right)\right] + erf\left[\frac{\ell}{\sqrt{4Gt}}\left(\frac{x}{\ell}-1\right)\right]\right\} - V_0 erf\left(\frac{x}{\sqrt{4Gt}}\right), \tag{6}$$

and the proportion remaining due to spreading losses without background erosion effects is

$$M_S(t) = 2erf\left[\frac{\ell}{\sqrt{4Gt}}\right] - erf\left[\frac{2\ell}{\sqrt{4Gt}}\right] + \frac{\sqrt{4Gt}}{\ell\sqrt{\pi}}\left\{2e^{-\left(\frac{\ell^2}{4Gt}\right)} - \frac{1}{2}e^{-\left(\frac{\ell^2}{Gt}\right)} - \frac{3}{2}\right\}, \tag{7}$$

Thus, the total result for the proportion remaining is Eq. (7) minus the effect due to background erosion

$$M(t) = 2erf\left[\frac{\ell}{\sqrt{4Gt}}\right] - erf\left[\frac{2\ell}{\sqrt{4Gt}}\right] + \frac{\sqrt{4Gt}}{\ell\sqrt{\pi}}\left\{2e^{-\left(\frac{\ell^2}{4Gt}\right)} - \frac{1}{2}e^{-\left(\frac{\ell^2}{Gt}\right)} - \frac{3}{2}\right\} - \frac{et}{Y}, \tag{8}$$

which is plotted in Figure 5 in the same format as Figure 3, which is the result for nourishment on a long straight beach as discussed earlier.

1.5 RANGE OF VARIABLES EXPECTED

Prior to illustrating the application of Figures 3 and 5, it would be informative to examine representative values of the two axes in the figures under discussion for typical projects in Florida.

For this purpose, we shall consider the following ranges of variables:

Project Length, ℓ: 2 km to 8 km

Longshore Diffusivity, G: 0.005 to 0.01 m^2/s (0.16 km^2/yr to 0.32 km^2/yr), h_*: 4 to 6 m

Background Erosion Rate, e: 0.2 to 1.5 m/yr Berm Height, B: 2 to 4 m

Time, t: 1 years to 6 years

Initial Shoreline Displacement (Equilibrated), Y: 10 m to 60 m

For the ranges above, the associated ranges in $(Gt/\ell)^{1/2}$ and et/Y are, approximately:

$(Gt/\ell)^{1/2}$: 0.05 to 0.7 et/Y : 0.003 to 0.9

The ranges of the two axes in Figures 3 and 5 have been selected to encompass these ranges.

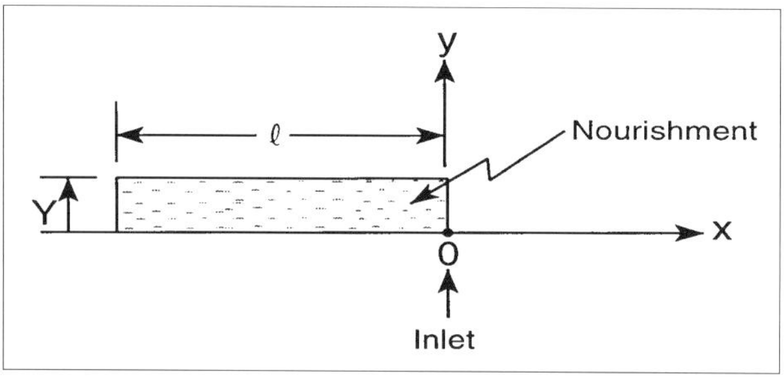

Fig. 4. Plan View of Nourishment Adjacent to an Inlet

2. Examples illustrating applications of methodology

This section presents both hypothetical examples and actual case examples from the State of Florida.

For the latter, reasonably representative values of relevant project characteristics and parameters will be employed as provided in some of the figures presented earlier.

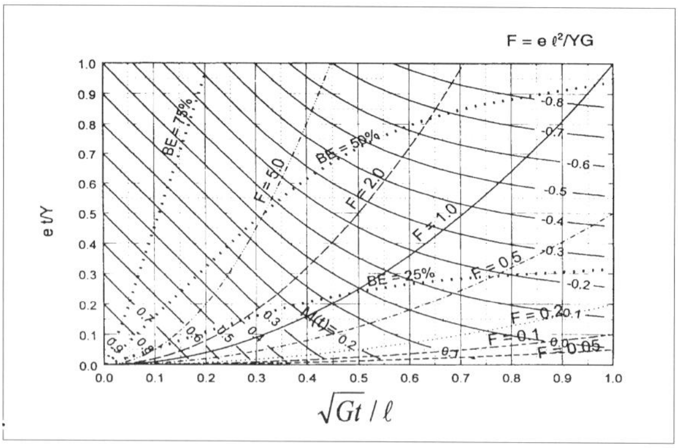

Fig. 5. Isolines of M(t) versus Spreading and Background Erosion Parameters. Nourishment Adjacent to an Inlet

2.1 HYPOTHETICAL EXAMPLES

Two hypothetical examples will be presented with the characteristics and results as shown in Table 1.

Table 1
Characteristics Of and Results For Two Hypothetical Examples

Case		G (m^2/s)	Project Length ℓ (km)	h_*+B (m)	Background Erosion Rate, e (m/year)	$F=\frac{e\ell^2}{YG}$	Time Considered (Years)	$\frac{\sqrt{Gt}}{\ell}$	eT/Y	M	BE
LONG SHORELINE	1	0.007	5.0	6.0	0.7	2.64	0	0	0	1.0	NA
							1	0.094	0.023	0.87	18%
							2	0.133	0.046	0.79	20%
							4	0.188	0.92	0.69	30%
							8	0.266	0.184	0.52	38%
INLET ADJACENT	2	0.007	5.0	6.0	0.7	2.64	0	0	0	1.00	NA
							1	0.094	0.023	0.81	12%
							2	0.133	0.046	0.71	16%
							4	0.188	0.092	0.58	22%
							8	0.266	0.184	0.37	29%

Only the geomorphic settings for these two projects differ from Case 1, located on a long straight beach and Case 2, located adjacent to an inlet. The other characteristics have been chosen to be the same to illustrate the effects of the inlet in reducing the longevity of the project. The calculated value of the trajectory defined by F is approximately 2.64. For Case 1, which is the project constructed on a long straight beach, the proportion remaining after 8 years is 0.52 with the proportion lost due to background erosion being 38%. For comparison, the proportion remaining for Case 2 after 8 years is 0.37, with background erosion accounting for 29% of the reduction. Referring to Table 1, it is clear that the percentage reduction due to background erosion increases with time for both cases. The lines of 'BE' can be used to determine the approximate percentages of reduction due to background erosion. A second approach is to recognize that the ordinate, e t/Y represents the proportion of the volume placed lost due to background erosion. Thus, an alternate and more accurate procedure to obtain BE is: BE (%) = $(et/Y)/(1.0\text{-}M)x$ 100% since the term (1.0 - M) represents the total reduction in volume. As an example, after 8 years, the proportion volume remaining is

0.52, thus the total volume reduction is 0.48 of which the background erosion, et/Y = 0.16 or 0.16/0.48 = 0.33 (or 33%) of the total reduction (see Table 1).

2.2 APPLICATION OF THE METHODOLOGY TO FIVE PROJECTS IN FLORIDA

This section presents the application of the methodology to five projects in Florida. The characteristics of these projects and the associated results of applying the methodology presented here are summarized in Table 2. An attempt has been made to use projects characteristics that are as as representative as possible; however, the values for the uniform background erosion rates have all been taken as 0.7 m/year which is believed to represent an approximate upper limit. More detailed calculations would require correcting for the nourishment grain size (a sediment transport coefficient, K value of 0.77 was used here which is appropriate for a grain size of approximately 0.48 mm). The value of the longshore diffusivity has been obtained from Figure 2, the time considered is the approximate length of time since project construction, the value of h is taken from Dean and Grant (1989) from a figure of the same format as Figure 1 of this paper, and the initial shoreline displacement Y is determined from

$$Y = \frac{V_{TOT}}{(h_1 + B)\ell}, \tag{9}$$

in which V_{TOT} represents the total volume placed and, of course, consistent units are used. The first four projects were constructed on beaches that are reasonably long and thus Figure 3 applies, and the last project (Perdido Key, FL) was constructed adjacent to an entrance (Pensacola Pass) and thus Figure 5 was applied. It will be seen that for long projects, the relative role of background erosion is greater simply because the spreading losses are quite small. For example, the total predicted losses from the Bay County project after two years are only 9%; however, the assumed background erosion represents 41% of this amount. For the shorter projects, the background erosion plays a relatively minor role simply because the spreading losses are relatively rapid. In general, for the five projects examined, the role of background erosion is relatively small.

3. Summary and conclusions

Convenient graphical aids have been presented which allow rapid determination of the proportionate and total volumetric losses due to background erosion and longshore spreading for projects constructed: (1) on a long uninterrupted beach, and (2) on a long barrier island with one end of the nourishment adjacent to an inlet. The background erosion is considered uniform in the alongshore direction. The theoretical basis for this methodology is the so-called Pelnard Considère equation which is linear and thus allows superposition of these two effects. This methodology and associated graphical aids presented in this paper allow the designer to concentrate his or her design efforts on the most significant processes and related variables governing project evolution. Two hypothetical projects are examined and used to illustrate application of the graphical aids. The methodology is also applied to five projects constructed in Florida with lengths ranging from 1.6 to 29.0 km and which have been in place for periods ranging from 2 to 8 years. It is found that the proportional reduction in volume due to background erosion is greater for the longer projects which have small spreading losses. For most beach nourishment projects in Florida, the longshore spreading losses will dominate over background erosion for reasonable evolution times.

Table 2
Characteristics Of and Results For Five Florida Case Studies

Project	Year Constructed	G (m^2/s) (Fig. 1)	Time Considered (Years)	Volume (Millions of m^3)	(h_*+B) (m)	Project Length ℓ (km)	Background Erosion Rate, e (m/year)	$\frac{\sqrt{Gt}}{\ell}$	Y (m)	eT/Y	M	BE
Martin County	1996	0.007	4.0	1.15	6.7	6.0	0.7	0.155	28.4	0.086	0.74	33%
Anna Maria Key	1993	0.0074	7.0	1.69	6.3	7.9*	0.7	0.162	34.0	0.126	0.69	41%
Bay County	1998	0.0088	2.0	6.12	6.4	29.0	0.7	0.026	33.0	0.037	0.93	53%
Midtown Beach**	1995	0.0656	5.0	0.67	6.4	1.64	0.7	0.570	64.4	0.047	0.40	8%
Perdido Key***	1989-1990	0.0106	8.0	4.10	7.0	7.50	0.7	0.219	78.0	0.063	0.58	15%

* Including a 1.13 km transition at the south end of the project

** This project included 11 groins for stabilization. The stabilizing effects of these groins are not accounted for by the methodology presented here.

*** This project is adjacent to Pensacola Pass at the east end and thus Figure 5 is used whereas Figure 3 is used for all other projects.

REFERENCES

Dean, R. G. and Grant, J.: 'Development of Methodology for Thirty-Year Shoreline Projections in the Vicinity of Beach Nourishment Projects', UFL/COEL-89/026, Coastal and Oceanographic Engineering, University of Florida, Gainesville, FL, 1989.

Pelnard Considère, R.: 'Essai de Theorie de l' Evolution des Formes de Rivate en Plages de Sable et de Galets,' 4th Journees de l'Hydraulique, Les Energies de la Mar, Question III, Rapport No. 1, 1956, (in French).

THE USE OF RELICT SAND LYING ON THE CONTINENTAL SHELF FOR UNPROTECTED BEACH NOURISHMENT

FRANCESCO L. CHIOCCI, GIOVANNI B. LA MONICA

Department of Earth Sciences, University of Rome 'La Sapienza", Roma, Italy.

Abstract

The best possible policy in relation to contrast beach erosion and coastal retreat in coasts heavily exploited by tourists is artificial beach nourishment. The success of such a policy strongly depends on the availability of large quantity of sand (in the order of millions of cubic metres). The exploitation of marine sand deposits buried in the continental shelf may meet such a large need. The prospection of sand is mainly realised with reflection seismic profiles, first shot at a regional scale then concentrated in the most promising areas. Seafloor sampling is then used to verify the geophysical data and characterise the sand. The identification of sand-prone deposit is based on a detailed geological interpretation of the seismic data that allows definition of the evolution of the shelf during the last glacial cycle (20 000 years). In fact, the sea level rose from –120 m to 0 in less than 20 000 years and caused the coastline to migrate across the shelf, possibly leaving relict sand prone deposits at various depths. Unfortunately, such deposits are rare, relatively difficult to image and usually buried below the present-day shelf mud. A typical stratigraphic situation, taken from the Regione Lazio, is described as reference.

1. Definition of the problem

Unprotected beach nourishment is usually very effective when a large span of coast is handled and a large amount of sandy material is available for repetitive operations. The latter cannot usually be found in terrestrial quarries because use of such quarries causes serious problems at the regional scale relating to overuse and damage of the transport networks, environmental impact on the quarry site, and rising costs of sand for other civil engineering purposes.

A possible solution to the need for such a large amount of sand is the exploitation of marine deposits. In fact, marine sand usually: 1) is found as large-volume deposits (in the order of 10 to 1 000 millions of cubic metres), 2) rests quite close to the disposal site, 3) is not easy to use for other purposes (given its salt content), 4) is similar to present-day beach sand in texture and mineralogy. Last (but first among the advantages), whilst the cost of the sand from terrestrial quarries is in the order of €12-€15/cubic metre, in marine quarries, the cost may be lowered to €7-€8/cubic metre, possibly even US$ 5/cubic metre for ship-performed nourishment.

Despite the lower cost of the exploitation, marine sand search requires high funding and potentially generates conflicts among different municipalities along the coast if the quarrying site does not coincide with the nourishment site. Consequently, large-scale, long-lasting and possibly multi-site nourishment projects using marine sand need to be organised at the regional level.

C. L. Goudas et al. (eds.), Soft Shore Protection, 39-48.

In the Regione Lazio, the local government and the University of Rome collaborated to find answers to the economic/social/ environmental problems caused by the constant coastal retreat of the highly tourist-exploited beaches. The very first result was a one-million-cubic-metres nourishment of the beach of Rome (Ostia) using sand from a transgressive beach located at –45 m water depth, and the identification of a number of other possible sand quarries accounting for several hundreds of millions of cubic metres of sand.

2. Occurrence of offshore sand

In middle-latitudes micro-tidal seas sand is usually restricted to the immediate coastal area and subject to littoral dynamics. The so-called mud-line (Reineck & Singh, 1980), is located at a depth ranging from a few metres to some 20 metres that is a quite abrupt transition to shelf mud, whose sedimentation is mainly subject to coastal currents, driven by geostrophic forces.

Exploitation of the sand below the closure depth but further ashore in relation to the mud line can be considered but is not very reliable. In fact, even if the closure depth (which has been defined in very different ways by various authors and though mainly in relation to the time recurrence interval) is thought to be the limit of the active littoral sedimentation, the presence of a sandy seafloor indicates processes active in a geological perspective, that is, at a long temporal scale.

The target of the exploration should therefore be the continental shelf, and deposits that play no role in the present-day littoral dynamics and can therefore be removed without affecting the sedimentary budget of the coast.

In meso- or macrotidal coasts, sand can be found offshore because of the tidal current activity that concentrates sand in large-scale bedforms, with sand waves or re-work relict sediments creating a palimpsest blanket that can be successfully exploited.

In microtidal coasts, the main target for sand prospection offshore is sand-prone relict deposits, possibly of littoral origin, that formed on the shelf during glacial times when the sea level was lower than present. If such deposits are found, their thickness, properties and stratigraphic position can be favourable for exploitation in beach nourishment projects.

3. Geological evolution of the continental shelf

The formation and preservation of sand deposit on the shelf has been controlled by the evolution of the shelf or at least since the last glacial-interglacial climatic/eustatic cycle. For this reason, any sand prospection should be based on a geological interpretation of the subsurface geophysical and sedimentological data.

For 20 000 years, continental shelves throughout the world were submerged by rising seas caused by the deglaciation that raised the sea level from ~120 m below the present level to its present position some 10 000 years ago, with an average rising rate of 1cm/year (Fairbanks, 1989; Chappell & Shackleton, 1986).

Due to the fast sea-level change, the transgression (shoreline landward migration) was usually non-depositional; in fact the rapid rise in base level has caused a marked decrease in the erosion/transport capability of rivers and streams and the creation of subaerial basins (river valleys, lakes, coastal plains and lagoons) trapping the small amount of sediment produced. Thus the littoral systems formed during the rise of the sea level were probably extremely underfed and poorly developed.

As mentioned before, there is evidence that the landward migration of the shoreline across the whole shelf did not usually leave any deposits behind due to sediment starvation, summarised as follows: because of the underfeeding of the system, the erosion/transport processes active within the littoral environment (downwelling and rip current, longshore littoral drift, bar migration) were able to keep all the sediments within the system during the relatively slow landward migration of the shoreline. In the meanwhile, the action of waves at the shoreface eroded the sediment on the subaerial part of the beach and backshore environment and hindered the deposition of the shoreface. In this way a 'transgressive surface of erosion' or 'ravinement surface' (Nummendal & Swift, 1987) was created. This surface has been described in the literature as a diachronous erosional unconformity marking the removal of all (or part) of the deposits formed during the former sea level fall, lowstand and during early transgression. Its facies are usually quite subaerial, being buried below the shelf; shoreface and nearshore facies usually lack the stratigraphic column.

Where there was non-depositional transgression, which was the most common situation for continental shelves during the Quaternary, the ravinement surface usually coincides with the subaerial unconformity formed during sea-level lowstand. During the course of time, an extremely flat surface is created that is the product of multiple erosional events (shoreline erosion during sea level fall, subaerial erosion due to shelf emersion during lowstand, shoreline erosion again during transgression). The transgressive deposits, if present, are usually very thin (from a few centimetres to a few decimetres), made up of shell debris and very occasionally pebbles that were left behind by the fast-moving transgression. Only after sea level stabilisation are sediments able to escape from their continental/coastal environments and reach the shelf, giving rise to shelf mud that is typically very variable in thickness, depending on the distance in relation to feeding sources (usually river mouths).

From a sand prospection perspective, the stratigraphy of the shelf is therefore very simple and unfavourable: an erosional unconformity truncates the bedrock and is overlain by a variable thickness of unserviceable shelf mud.

Some exceptions to this standard stratigraphy may occur in very specific geomorphologic/sedimentary conditions. In some cases, part of the material that was eroded at the shoreface during transgression may become part of the submerged beach and therefore be preserved. Elsewhere, if the subaerial unconformity was scoured by depressions, some continental deposits may escape the ravinement erosion. This is the case, for instance, with paleo-river valleys that scoured the shelf during lowstand and were filled up during transgression. Moreover, the thickness of the highstand shelf mud is extremely dependent on the position of the sediment feeding sources and on environmental energy, so that in favourable conditions, it may allow access to the underlying deposits.

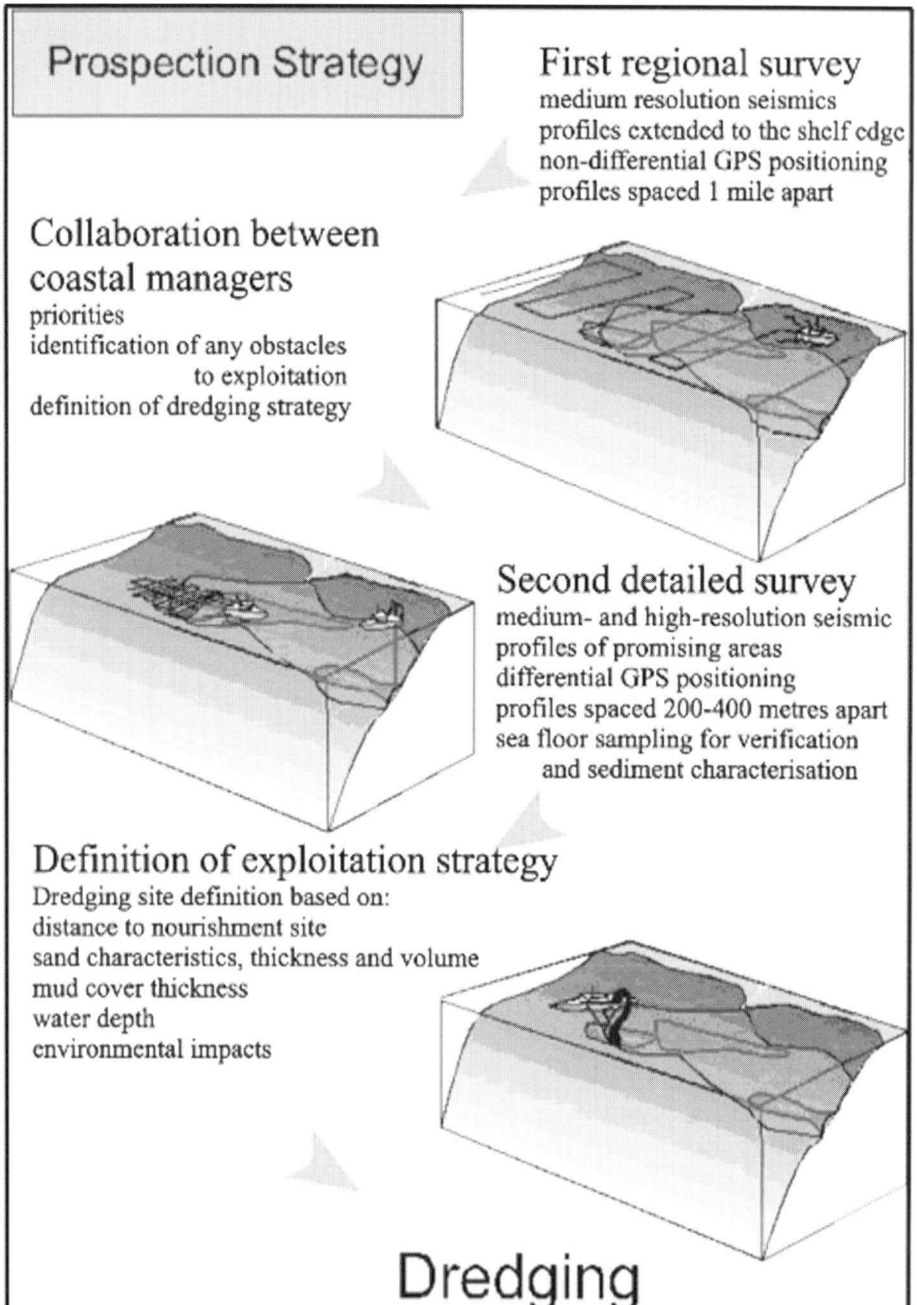

Fig. 1. Search for sand on the continental shelf: prospection and exploitation procedures

4. Prospection strategies

Sand research should be mainly based on high-resolution seismic profiling, a prospection tool able to give a very detailed depiction of the subsurface geometry and some indications of the possible lithology of the seismic units (Fig.1).

While the cost of marine survey is quite high in relation to land survey, the quality and quantity of data is enormous. It is, in fact, possible to cover the whole shelf, obtaining a network of profiles (usually shot parallel to the coast), with a resolution of the shallow subsurface as high as few decimetres.

As the areas to be investigated are usually very large, a *first regional survey* is usually performed to define the general stratigraphy of the shelf, to eliminate areas with unfavourable characteristics and instead identifying areas having an high potential in respect of formation and preservation of sand-prone deposits. In the first regional survey, the seismic source should be strong enough to penetrate a few tens of metres under any possible geological conditions and the profiles should be shot with a spacing of about one km. Given the inverse relationship between resolution and penetration, the source will not have the maximum possible resolution. As for positioning of the seismic profiles, a standard GPS system is required, giving a precision of a few tens of metres. It is recommended that the whole shelf be investigated. Even if the mining target is in the inner and middle shelf, a full geological interpretation of the whole shelf is, in fact, needed for a complete seismo-stratigraphic interpretation.

Determination of the mining potential of the different areas will encompass the following: a rough definition of the mud coverage, usually the main limiting factor in shelf sand exploitation; the depth at which the sand deposits are lying, their thickness and lateral continuity; and the presence of hard bio-constructions, shallow gas or other obstacles to exploitation.

Once the areas have been roughly characterised in terms of their mining potential, geologists/geophysicists and coastal managers, need to collaborate to establish priorities and the feasibility of sand exploitation in the various areas. At this point a *second detailed survey* should be done in the most promising zones. This seismic survey will be characterised by a much narrower spacing of profiles and should encompass both high-penetration/low-resolution and low-penetration/high-resolution seismic sources to define both the volume of the sand-prone deposits and the accurate thickness of the possible mud blanket covering it. If the mud blanket is not too thick, side scan sonar surveys may also be advisable, since sonar backscatter may be influenced by the presence of sand on the seafloor, both that in outcrops and that re-worked by bio-turbation (Fig.2). The seismic/sonar survey could be followed by an extensive vibracoring of the identified targets in order to verifying the geophysical data and defining the textural and mineralogical characteristics of the sandy deposits (Fig.2).

At this stage of the prospection, a regional evaluation of the sand resources should be peformed to define the exploitation strategy.

Considerations such as 1) distance of the quarry and water depth of the borrow site (affecting the cost and dredging technique); 2) sand textural and mineralogical characteristics (affecting quantity to be dredged); 3) environmental impact on the quarry (to be balanced against the environmental impact of other alternative solutions); 4) possible

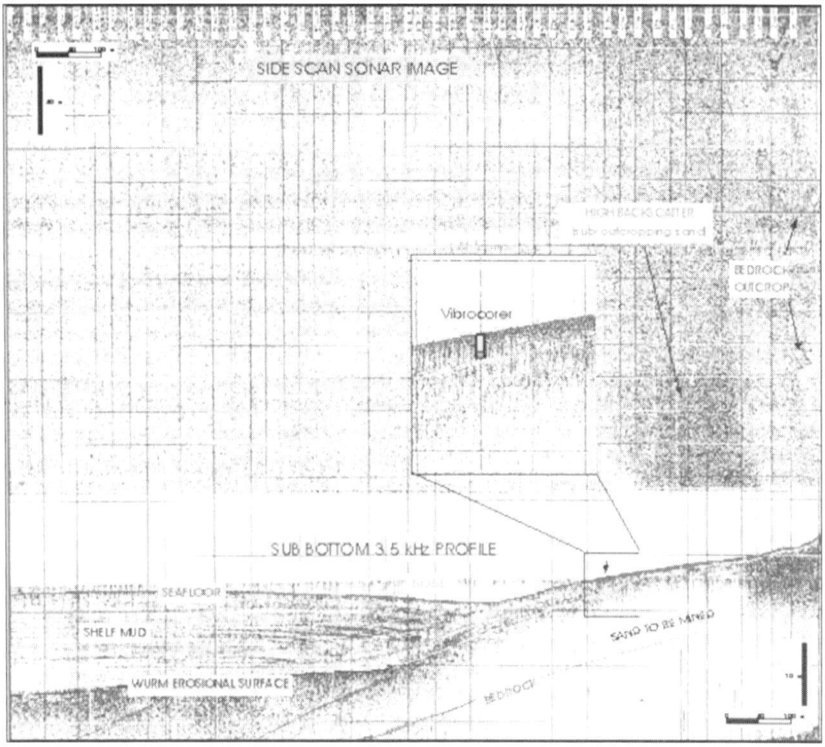

Fig. 2. Side Scan Sonar (upper plan view) and Sub Bottom Profiler 3.5 kHz (lower section) on a possible mining target offshore from Civitavecchia (Central Eastern Thyrrenian Sea). The vibro-corer derived stratigraphy is shown in the lower section of the figure

technical/legal/social obstacles to exploitation (to better attune the dredging techniques) should be defined for each possible quarry site. This will enable the coastal managers to balance all the possible options and find the best possible economic/social/ environmental solution for the span of coast to be nourished and create a long-term strategy for exploitation of the shelf sand and management of the coast.

5. Genesis of sand bodies on the continental shelf

As previously stated, only a detailed geological interpretation of geophysical and coring data is able to define the stratigraphic, paleomorphologic, paleogeographic and paleoenvironmental conditions of the shelf that led to the formation and preservation of sand-prone depositional bodies. However, a typical situation in the Lazio shelf will be described below, as it is a good example of a microtidal siliciclastic sea (Figure 3).

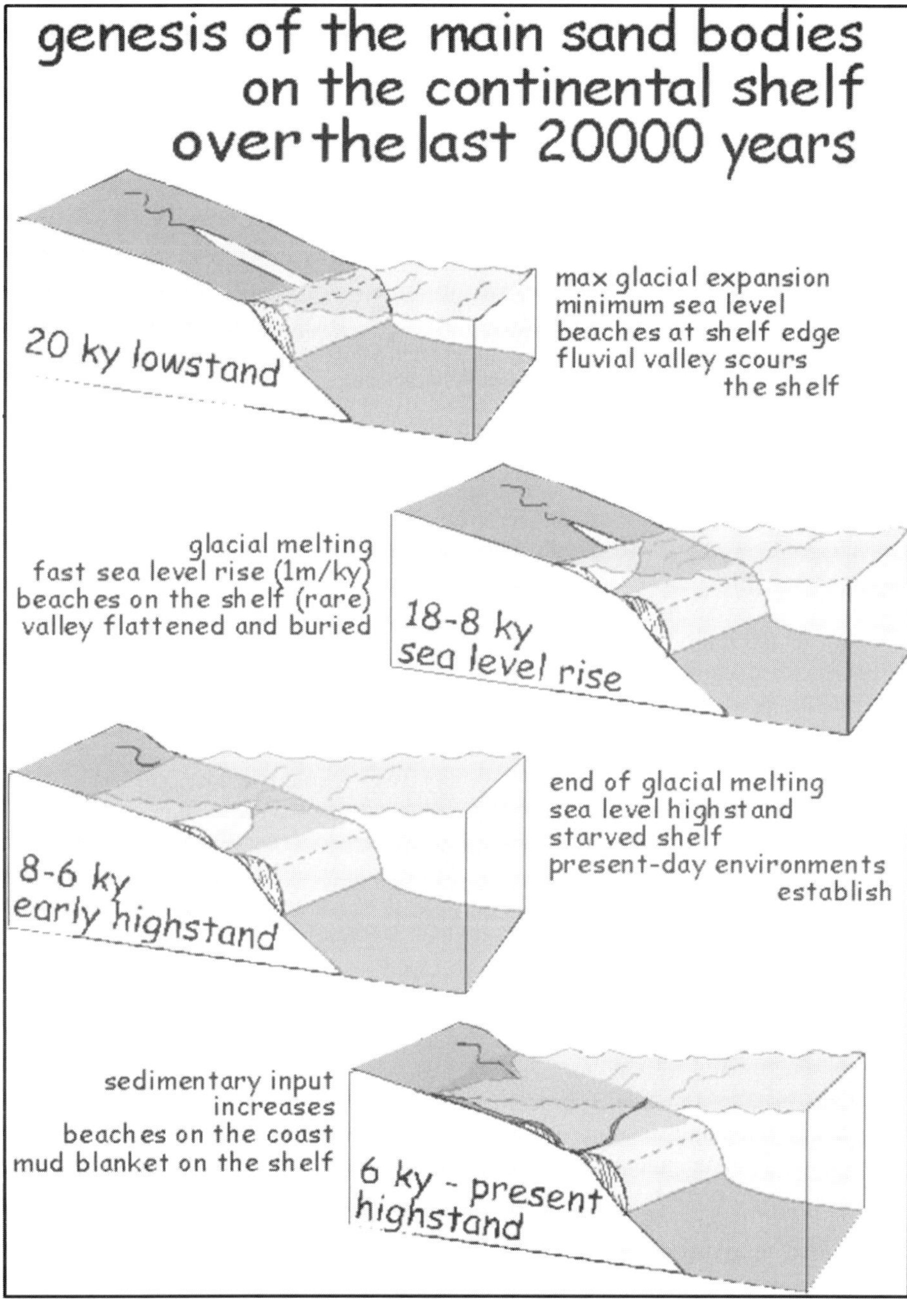

Fig. 3. Evolution of the continental shelf during last glacial cycle and genesis of the main sand-prone deposits

20 000 years ago, the sea level was some 100m lower than at present because a large amount of water was sequestered in continental glaciers. During this period, the rivers scoured the subaerially exposed shelf sometimes depositing fluvial sandy sediment within the incised valley **(a)** as well as on lowstand beaches at or near the shelf edge **(b).** De-glaciation later caused the sea level to rise from the -100m to its present position which it reached about 8 000 years ago. During this period, sandy littoral deposits were constantly formed but also removed by the shoreface retreat process. On rare occasions, littoral deposits were preserved from post-depositional erosion, thus becoming a good target for sand exploitation **(c)**. The present-day depositional processes have taken place during the last 6 000 years with progradation of sandy beaches in the coast and deposition of a very thick mud blanket on the shelf in some places, usually covering the no longer active relics sand-prone deposits.

Each of the deposits will thus have distinct characteristics:

(a) **Incised valley infill** is typically extremely variable in thickness, characteristics and distribution because of the following: 1) the original fluvial sediments may have been heterogeneous, as in typical fluvial facies; 2) the fluvial sediments may be amalgamated with lagoonal/lacustrine facies that usually fill the valleys during the sea-level rise; 3) the fluvial sediments may have been deeply reworked during the shoreline retreat, i.e. ravinement surface formation. The incised valley sediments are therefore usually poorly sorted and coarse (gravel is not rare). It is hard to ascertain the nature of the deposits to be exploited and a high number of corings are required, as seismics usually do not have good resolution due to lateral heterotropies.

(b) **Lowstand beaches** are usually very well preserved and seismically easy to detect. Their textural characteristics are usually very favourable for sand exploitation (well sorted medium sand) even though there is usually quite high percentage of bioclasts that may prove problematical for the siliciclastic beaches to be nourished. Typically, they are only covered by a thin veneer of highstand mud or outcrop directly on the seafloor. The main limit for their exploitation is the water depth at which they are found, typically greater than 100 m.

(c) **Transgressive littoral deposits** are the best mining target for sand exploitation as their textural and mineralogical characteristics usually fit those of the facing beaches. They do not usually occur at very great depths and their volume is in the order of several millions of cubic metres. They are rare on the shelf and often found at the foot of paleo-headlands, whose presence favoured the formation of thick littoral spits and/or hindered their post-depositional erosion. The main obstacle to exploitation is the shelf mud cover that often exceeds several metres or even tens of metres. However, as the shelf mud thickness distribution is strongly controlled by the present-day coastal current and by the position of the feeding sources, it may vary from a few metres (that may be removed during dredging operations) to several tens of metres. Thus, with good seismostratigraphic control, suitable places for accessing these deposits may be found.

6. Conclusions

During the last decade it become evident that beach nourishment is the most convenient strategy to counter beach erosion on highly tourist-exploited coasts. Even though a better strategy would be to re-establish the natural conditions (reconstructing river sediment discharge, removing infrastructures from the coast) the most realistic option if the area is to be restored to anything like a 'natural' state is to artificially nourish the beach. Other forms of coastal protection, such as construction of jetties and breakwaters, only postpones the problem and cause deterioration of the appearance of the coastal environment, thus compromising tourist use of the coast.

If, on the other hand, the coast is highly exploited, the economic value of the beach, as total investment for the local community, may reach € 10 000/square metre, a relatively high outlay for the local authorities. The main problem in nourishment projects is the large amount of sand required for an effective impact on the sedimentary budget of the coast and in the repetition nature of the procedure. In fact, the success of a nourishment strategy mainly depends on the availability of extremely large quantity of sand (in the order of millions of cubic metres per beach kilometre) at a reasonably low cost. The only realistic solution to such a large need is the exploitation of marine quarries, thus avoiding overexploitation of land resources, rising costs and deterioration of the transport network.

Sand prospection at sea is classically based on high-resolution seismics, coupled with seafloor vibro- and gravity-coring. However, very detailed geologic interpretation is needed to reconstruct the depositional/paleogeographical/paleomorphological evolution of the continental shelf which will determine the characteristics of the shelf subsurface stratigraphy. Sand-prone depositional bodies occur only rarely, the situations and processes under which they are created being relatively infrequent in the past 20 k.yrs. Incised valleys that have filled up, lowstand beaches, transgressive littoral wedges are among the main targets for sand prospection; each of them with specific characteristics and potentiality.

A strong increase in the use of marine sand for beach nourishment can therefore be expected in the coming decades.

Acknowledgements

The search for marine sand for beach nourishment in the Latium region involved, among others, Francesco Falese, Roberta Fiorini, Vanessa Signoretto and Eleonora Martorelli. The article refers to IGCP 464 'Continental Shelves In The Last Glacial Cycle' and to 'Coastal Vulnerability and Their Protection', projects financed by GNDCI/National Council of Researches of Italy.

REFERENCES

Fairbanks, R.G.: A 17 000-year glacio-eustatic sea level record: influence of glacial melting rates on the Younger Dryas event and deep-ocean circulation. *Nature*, 1989, vol. 342.

Chappel, J. & Shackleton, N.J.: Oxygen isotopes and sea level. *Nature*, 1986, vol. 324.

Reineck, H.E. & Singh, I.B.: Depositional Sedimentary Environments, (see chapter Examples of Beach-Shelf Profiles from Modern Environments), Springer-Verlag, Berlin-Heidelberg, New York, 1980, 382–386.

Nummendal, D. & Swift, D.J.P.: Transgressive stratigraphy at sequence-bounding unconformities. Some principles derived from Holocene and Cretaceous example. in: Nummendal, Pilkey and Howard Eds. *Sea Level Fluctuations and Coastal Evolution*, Society of Economic Palaeontologists and Mineralogists Special Publication, 1987, n.41.

PREDICTING MORPHOLOGICAL CHANGES ON A COMPLEX 3D SITE USING THE GENESIS MODEL

I.MARIÑO-TAPIA, P.E. RUSSELL

Institute of Marine Studies, Plymouth Environmental Research Centre, University of Plymouth, Drake Circus, Plymouth, UK, PL4 8AA

Abstract

Teignmouth, South Devon, UK is subject to a cycle of bar movement on its nearshore zone. There have been several studies that have tried to elucidate the processes responsible, but there are still conflicting views about the relative efficiency of estuarine flows, tidal currents and waves in controlling the morphology of the area. A numerical investigation was carried out at the site in order to assess the importance of longshore transport processes in the morphological changes of the beach. For this purpose, the one-line model GENESIS (Generalised Model for Simulating Shoreline Change) was used. A period of 4 months is covered, from 11 October 1995 to 5 February 1996.

Three different temporal stages were identified during the period of the simulations. The first stage (October to November) included the presence of a stable middle beach bar, which dispersed and moved southwards during the second stage (December), and suddenly disappeared during the third stage (January to February), when the beach became uniform and with no prominent features. When short-term simulations were carried out with GENESIS, the shoreline trend could be qualitatively reproduced for stages 1 and 2, and for stage 3 the coastline was satisfactorily modelled qualitatively and quantitatively. Long-term simulations failed to reproduce the shoreline changes. The results suggest that during stages 1 and 2 longshore transport of sand was not the governing factor for shoreline change, though it played a role in shifting the middle beach bar southwards and spreading it out. The anticlockwise tidal circulation identified by Robinson (1975) was regarded to be the governing process when the waves were mild and infrequent (stages 1 and 2). During stage 3, longshore transport from ENE and SSE acquired great importance in 'washing off' the bar and shaping the beach. Alongshore gradients in wave height were particularly important for dispersing the middle beach bar when waves approached from the S and SSE quarters.

1. Introduction

One of the most widely applied models in coastal engineering is the Generalised Model for Simulating Shoreline Change or GENESIS (Hanson & Kraus, 1989). It is commonly used to predict volumes of sand for beach nourishment, and often forms the basis for presenting project lifetimes and costs to regulatory agencies, legislative bodies or clients though mainly in the USA. This practice has caused concern in a section of the scientific community, and has given rise to a longstanding debate about the controversial use of numerical models in coastal engineering (e.g. Young, 1995, Thieler, et. al. 2000, etc.). Models such as GENESIS have been categorised as 'poorly founded at best and at worst invalid' (Thieler, et. al. 2000). Criticism of the model has

C. L. Goudas et al. (eds.), Soft Shore Protection, 49-69.

a good scientific basis, though it has been mainly conceptual and theoretical. Careful and objective analysis of model results in real-world situations are needed to assess to what extent the alleged 'poor assumptions and important omissions' (Young 1995) are valid in the simulation of coastal morphology where there is longshore sediment transport. There are a number of cases in the existing literature in which the GENESIS model has been compared with field data (e.g. Rogers & Work 1996, Szmytkiewicz, et. al 2000), with reasonably good results. In this paper, a combination of data analysis and numerical simulations with GENESIS are used to investigate the extent to which longshore transport processes affect beach morphodynamics in a complex 3D site. Emphasis is given to the observed strengths and weaknesses of the model.

2. Description of the GENESIS model

The actual structure of GENESIS was developed in a joint research project between the University of Lund, Sweden, and the Coastal Engineering Research Centre (CERC), USA. Hanson and Kraus (1989) provide the full technical characteristics of the model.

GENESIS is a 'one-line model', which means that the morphological changes of a single elevation contour, the shoreline, is simulated. Shoreline changes are produced by gradients in longshore sediment transport caused by both waves arriving along the coast at oblique angles and alongshore gradients in wave height. While the fact that it only considers longshore transport is bound to be a major weakness, much of the long-term erosion of a nourished beach does seem to be caused by increasing gradients of littoral drift along the length of the nourished area (Nordstrom, 2000).

The first and more basic assumption of one-line modelling is that the beach profile is always in equilibrium, moving landward if the beach accretes or seaward if it erodes, though retaining the same shape. The sand moves over the profile between a defined shoreward limit, the berm elevation (DB), and a certain limiting depth or closure depth (Dc) beyond which the profile does not change significantly during the course of time. In other words, the model is governed by the sediment continuity equation (1). If we consider a Cartesian co-ordinate system in which the y-axis points offshore and the x-axis is parallel to the trend of the coast, the partial differential equation governing the shoreline position is:

$$\frac{\partial y}{\partial t}+\frac{1}{D_B+D_C}\left(\frac{\partial Q}{\partial x}-q\right)=0, \qquad (1)$$

where y is the shoreline position, x the longshore co-ordinate, t is the time, D_B is the average berm height above mean water, D_C is the depth of closure, Q is the longshore transport and q represents line sources and/or sinks along the coast.

Sediment transport Q, is calculated using a modified CERC formula with an added term to account for longshore sediment transport driven by alongshore gradients in wave height ($\partial H/\partial x$) caused by the presence of structures or irregular bathymetry. The second term was introduced into shoreline change modelling by Ozasa and Brampton (in Horikawa, 1988). The sediment transport equation is:

$$Q = \left(H^2 Cg\right)_b \left(a1 \sin 2\theta_{bs} - a2 \cos\theta_{bs} \frac{\partial H}{\partial x} \right)_b , \tag{2}$$

in which, H = wave height, Cg = group speed given by liner wave theory, b = subscript denoting breaking wave condition, θ_{bs} = angle of breaking waves to the local shoreline. The non-dimensional parameters $a1$ and $a2$ are given by

$$a1 = \frac{K1}{16(S-1)(1-p)(1.416)^{5/2}}, \tag{2.1}$$

$$a2 = \frac{K2}{8(S-1)(1-p)\tan\beta(1.416)^{7/2}}, \tag{2.2}$$

where $K1$ and $K2$ are the calibration coefficients, $S = \rho_s/\rho$, ρ_s = density of sediment (taken to be 2.65 x 10^3kg/m^3), ρ= density of water (1.03 x 10^3 kg/m^3 for seawater.), tan β = average bottom slope.

3. Characteristics of the Site

Teignmouth is a relatively small community located on the coast of South Devon, UK. A narrow beach, 2 km long, fronts it, facing ESE into the English Channel. The beach is backed by a seawall, which protects a railway line and prevents further erosion of the cliffs in the northern part. To the south the beach meets the mouth of an estuary and a prominent headland, the Ness, restricts the estuary's inlet (Figure 3.1). The beach is steep at the high water mark (gradient up to 0.110) and less steep at low water

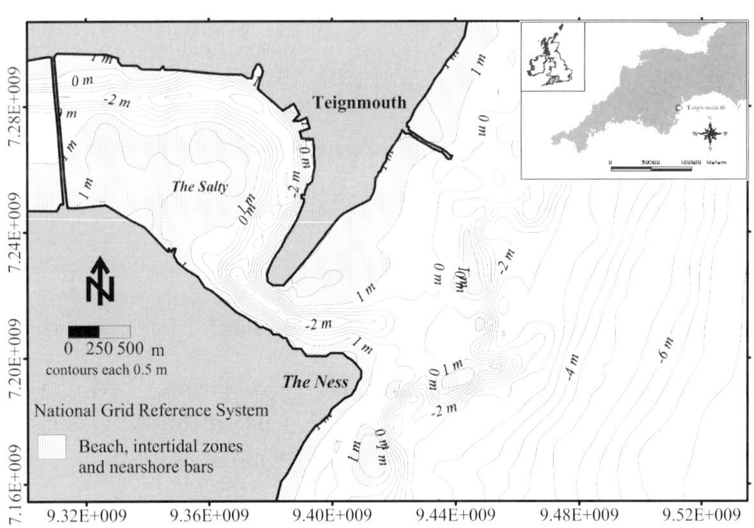

Fig. 3.1. Location of Teignmouth

(gradient = 0.057). It can be classified as a macrotidal reflective beach (Masselink and Short, 1993). The grain size varies widely. It goes from small shingle (5-50mm) to coarse sand (0.4mm) in the upper beach, and fine sand (0.2mm) predominates seawards. The mean grain size is about 0.24mm.

The site is protected from Atlantic swell waves, so only wind driven waves (fetch-limited) affect the site. These waves can drive strong alongshore currents. Measurements of longshore currents on the steeper part of the beach can reach 0.5 m/s in the middle of the surf zone with a flood tide.

Tides are semi-diurnal with ranges of 1.7 m on mean neaps and 4.2 m on mean spring tides, reaching up to 6 m. Tidal current speeds typically reach 2m/s 0.3m above the bed, close to mid-flood and mid-ebb near the estuary's channel. Currents of up to 3m/s have been recorded during large spring tides (Miles, 1997).

4. Problem Definition

The sediment dynamics are very complex at the south end of Teignmouth beach, where the beach meets the estuary's mouth. A sand pole moves from the Ness headland seawards, while an outer pole forms. The outer pole migrates towards the shore and eventually joins the beach forming a middle beach bar (Figure 4.1).

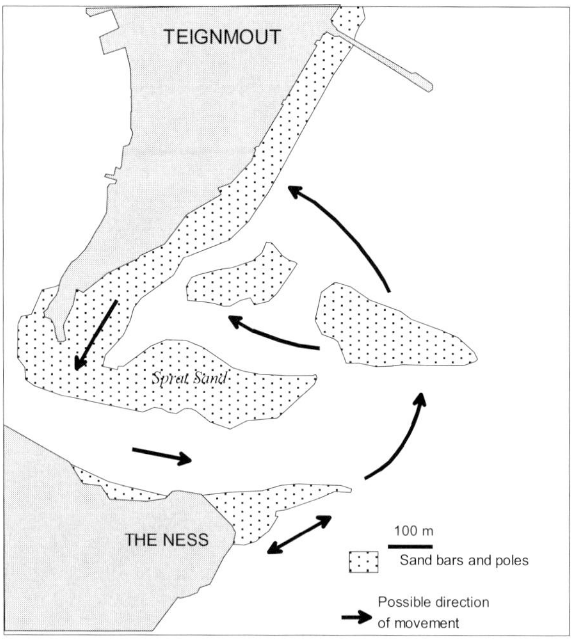

Fig. 4.1. Bar system evolution at Teignmouth (Robinson's conceptual model)

Sprat (1856) and Robinson (1975) suggest that these poles and bank systems move in a cyclic fashion with a period of 40 months, though the changes can be fairly rapid, especially during winter months, when the poles are dispersed because of the storms. Since the mid 19th century, various studies have tried to assess the processes governing the behaviour of this cycle. Probably one of the most important studies is that of Robinson (1975). Using a mostly geomorphological and deductive approach, Robinson (1975) suggested that a flood residual current creates an anticlockwise circulation which must influence the sediment movement in the lower part of the beach. However, along the upper part of the beach, the movements seem to be influenced by wave action, sometimes with a longshore component when waves break obliquely from an easterly or north-easterly quarter.

Other studies and data-gathering programs at the site include those carried out by Hydraulics Research in 1958 and 1966, Wimpol in 1989, the British Beach and Nearshore Dynamics program in 1992, and the COAST 3D project in 1997. In spite of all these efforts, there is still no concensus about the conditions under which estuarine flows, tidal currents, waves, or their interactions produce morphological changes in the area.

This study aims to investigate to what extent longshore sediment transport, as modelled by GENESIS, is responsible for the morphological changes experienced by the southern Teignmouth Beach during the period of October 1995 to February 1996. Results were compared with the observed morphological changes, storm-related parameters, wave sequencing, and tidal influences.

5. Methodology

5.1 DATA

Beach profiles were used to characterise the morphological changes of the beach during the period of the investigation and to define shoreline and related parameters in the GENESIS model. The surveys were carried out along the southern part of the beach, from the pier to the estuary's mouth, on 11 October 1995, 11 November 1995, 11 December 1995, 11 January 1996 and 6 February 1996. Typical survey layout and measurement points are presented in Figure 5.1.

Using this data, the GENESIS co-ordinate system and grid were defined. The alongshore spacing of the GENESIS grid was established as being 17 m.

To determine the location of breaking waves alongshore and to calculate the average nearshore bottom slope used in the longshore transport equation (equation 2.2), an equilibrium profile shape had to be defined. Moore (1982) proposed that the shape of the equilibrium profile is a function of the median beach grain size D50. GENESIS used this approach to determine the equilibrium profile shape. However, in order to ensure that the adopted equilibrium profile was an accurate representation of an average profile for Teignmouth Beach, an 'effective grain size' had to be defined. The equilibrium profile, calculated using Moore's approach, is compared with the measured averaged profile by means of an absolute error defined as Error = |Elevation Measured

– Elevation Calculated|. The value of D50 that better represented the profiles for all stages was 0.75mm.

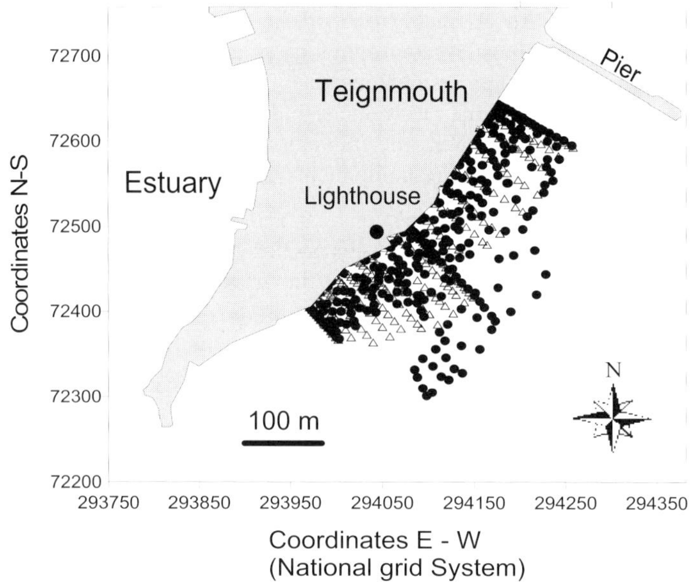

Fig. 5.1. Position of survey points and layout (● October, △ January)

Wave data was obtained from the Meteorological Office at Bracknell, UK. The data came from a continuously recording wave station located 75 km offshore from Teignmouth, at 65m depth in the English Channel (light vessel 62103). The wave station forms part of an international network of wave recording sites locally managed by the Met Office, but administered by the National Oceanographic and Atmospheric Administration (NOAA) through its National Oceanographic Data Centre (NODC) in the USA. The data set consisted of a time series of hourly-recorded data relating to wave height, wave period, wind direction, wind speed and barometric pressure. The time span of the data set coincided with that of the beach surveys (October 1995 to February 1996).

As wave data was adirectional and only wind direction was available, it was assumed that wind direction would be a sufficient indicator of wave direction. There were good reasons for making this assumption, as Teignmouth is quite sheltered from SW swell waves and only locally generated wind waves affect the area. However, there were different conditions at the buoy site, so wave parameters associated with random wind directions were excluded and swell events were filtered by establishing 'maximum allowable' wave periods. This was done using the JONSWAP wave forecasting graphs, the fetch distances from the buoy, and the maximum wind speeds per direction. All wave events with periods above the maximum allowable were treated as swell.

The bathymetric characteristics of the Lyme Bay and nearshore region of Teignmouth are quite complicated, hence an external wave transformation model had to be used to calculate the nearshore wave conditions. For this purpose the numerical wave transformation model RCPWAVE (Ebersole, 1985) was used to calculate pre-breaking wave conditions. RCPWAVE considers the effects of shoaling, refraction and diffraction.

5.2 NUMERICAL EXPERIMENT

No single model prediction could be expected to provide the 'correct answer', so a range of predictions had to be made and judgement exercised to select the most probable or reasonable result (Hanson & Kraus, 1989). The experimental calculations in this study were aimed at investigating the capability of GENESIS to predict morphological changes given a certain wave sequence.

Long-term (3 or 4 months of wave data) and short-term (one month) simulations were carried out for each of the morphological stages at Teignmouth Beach during the period of the investigation. For example, for the short-term simulation of stage 3 (see next section), the shoreline measured on 11 December 1995 was fixed as the original shoreline, and the model was run for January. For a longer-term simulation, October was set as the original shoreline and calculations were made for January.

6. Results

6.1 FIELD DATA

To illustrate the degree to which the equilibrium profile assumption fitted the actual profile characteristics, a logarithmic profile (dotted line) determined with Moore's approach (D50 = 0.75 mm) was plotted against an average measured profile (bold line) in Figure 6.1. It became evident that an equilibrium profile calculated with a single sediment size is a poor representation of a beach like Teignmouth, which has a strong gradient in

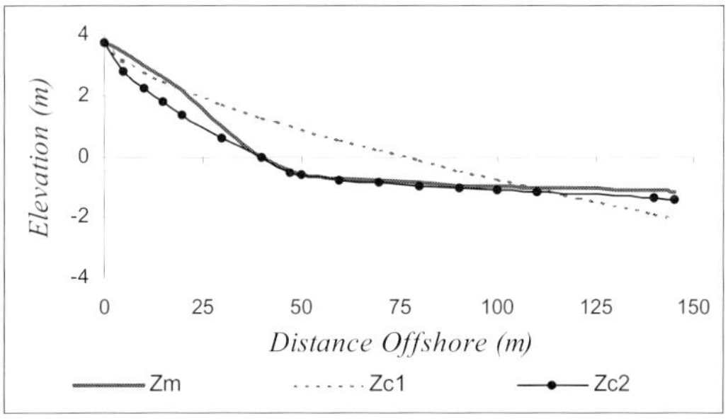

Fig. 6.1 Average vs calculated profiles. Zm = average profile for the northern beach for all stages, Zc1 = equilibrium profile calculated with a single grain size = 0.75 mm (GENESIS approach), Zc2 = equilibrium profile with a coarser grain size (3mm) along the upper beach (0 – 50 m) , and finer sediment (0.04 mm) along the low tide terrace

grain sizes. Coarse sediment predominates on its upper steep face and finer sediments on its flat low tide terrace. If the same approach (Moore's formula) is used but variations in median sediment sizes are allowed, the representation of the profile improves considerably (black circles). The GENESIS model does not include the effect of varying sediment size.

Beach surveys revealed three different stages in the morphological evolution of the beach during the period of investigation (Figure 6.2). During the months of October to November (Stage 1), a stable middle beach bar was present, which had dispersed and moved to the southwest by December (Stage 2). Towards January and February (Stage 3) the bar disappeared rapidly and the beach showed no prominent features and was uniformly straight.

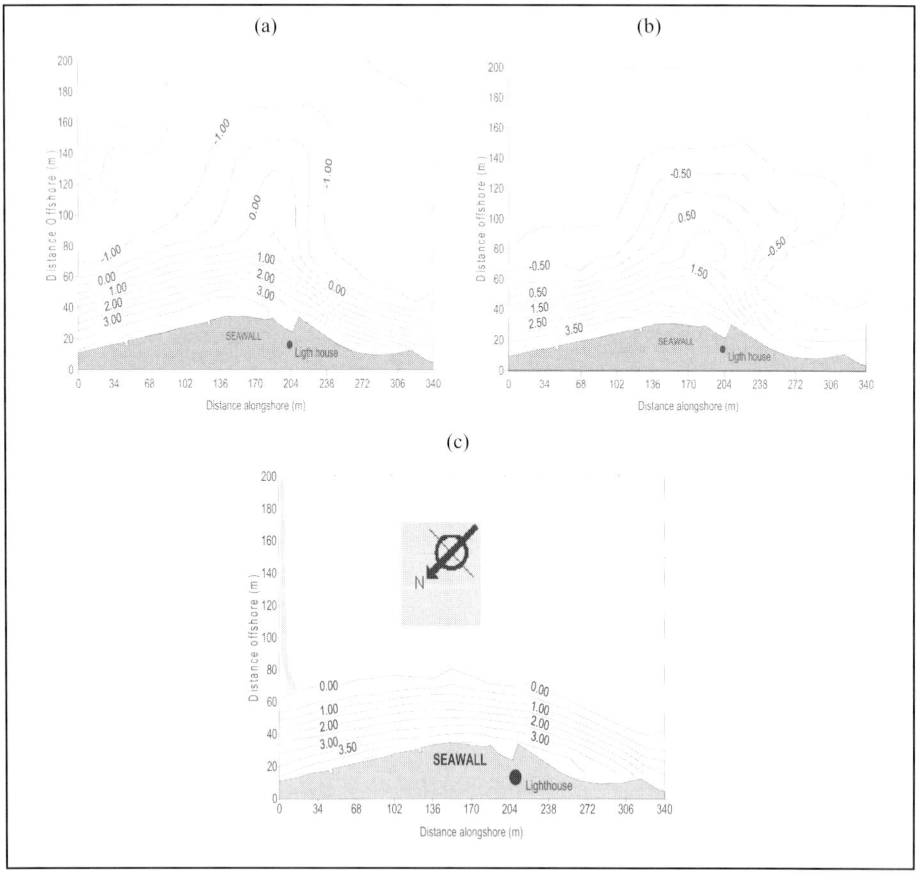

Fig. 6.2. Morphological changes along Teignmouth Beach during the period of study. (a) Stage 1 (October-November), stable bar in the middle of the beach. (b) Stage 2 (December), transition stage with bar shifted towards SW. (c) Stage 3, middle beach bar completely disappears, straight beach

The forcing conditions associated with these three stages are plotted in Figure 6.3, which shows the mean wave energy flux arriving at Teignmouth per stage as an indicator of the wave intensity. This was calculated from the modified offshore time series as $<1/8\rho g H^2>$ where ρ = density, g = gravity force, H= offshore wave height, the brackets indicate time average. The maximum tidal range indicates the likelihood of strong tidal currents. Figure 6.3 shows clearly that the middle beach bar disappeared (Stage 3) during the time of highest wave energy. The tidal range reached a maximum during stages 2 and 3. Tidal currents might also have played a part in dissipating this sand feature.

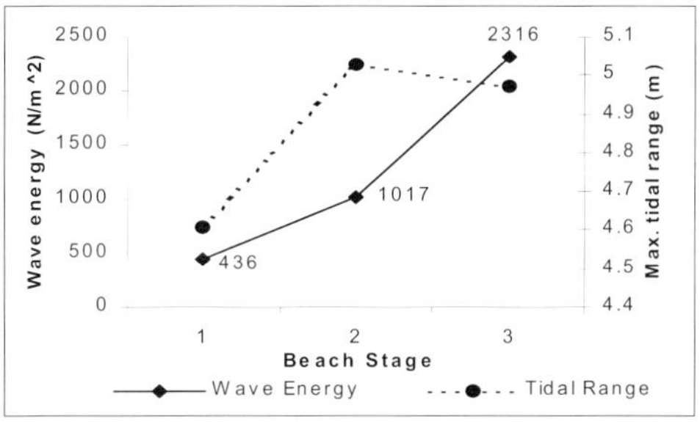

Fig.6.3. Forcing conditions during the observed beach stages

In order to trace the causes of the morphological changes per beach stage in detail, graphic representations of several forcing parameters were made. The figures show some of the forcing conditions preceding a given beach stage and the actual morphological changes caused by them. The indicators used are a time series of tidal elevation, wave height and atmospheric pressure as indicators of extreme events (destructive storms = high tide + low pressure + big waves), wave direction roses as indicators of longshore currents, and erosion-accretion diagrams to establish the actual morphological changes.

Beach Stage No. 1 (Figure 6.2a)

The beach morphology showed little change from October to November. The middle beach bar shifted slightly towards the south and diminished in volume.

During this period, high atmospheric pressures predominated, creating mild weather conditions. The resulting wave events were mild and sporadic calm periods with 80 % of the time (offshore Hs < 0.5 m). There was a mild storm from the S-SSE on October 25, and 3m waves during the spring high tides (Figure 6.4a). The sporadic and mild waves were predominantly produced by winds from the S and SSE directions. There were also waves from the E, which are likely to produce southward longshore currents are also relevant (Figure 6.4b).

Figure 6.4b demonstrates that S and SSE waves are likely to be affected (diffracted) by the Ness headland and the complex bathymetry surrounding the estuary's mouth, hence alongshore gradients in wave height could be expected to be important. This was later confirmed by the GENESIS and RCPWAVE results.

Figure 6.4c does not represent the observed morphology but the actual changes in morphology along the beach during this period. Regions of no change are represented by a bold black mesh, the erosion 'holes' occur below the no-change strata, and accretion 'bumps' are represented by a bold grey mesh. During this period, the morphological changes were minor. Slight erosion was evident along most of the beach, but was more significant near the seawall, where changes of up to -1 m were seen. The beach region with no significant changes corresponded to the location of the middle beach bar.

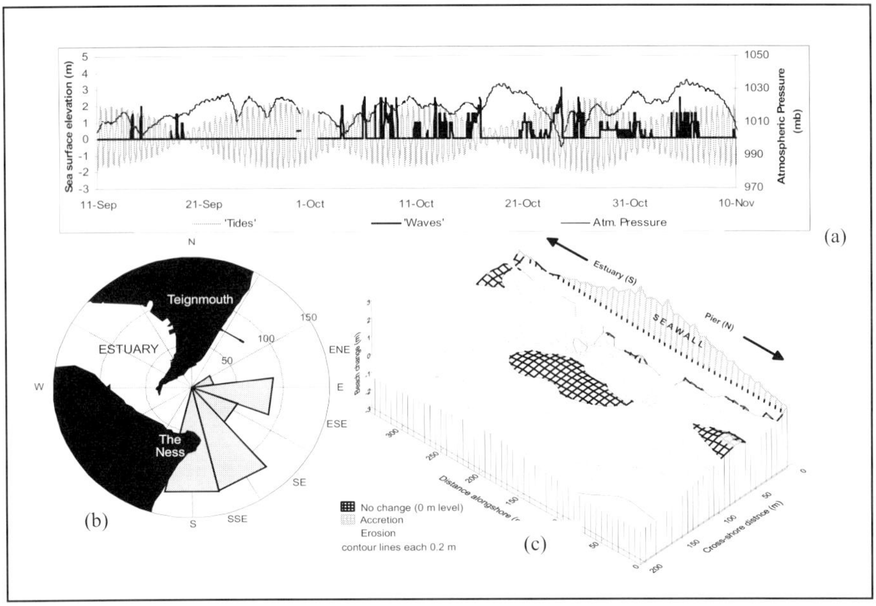

Fig. 6.4. Forcing conditions and morphological changes Stage 1. (a) Tide, wave height and atmospheric pressure time series; (b) wave roses; (c) morphological changes

Beach Stage No. 2 (Figure 6.2b)

The beach morphology still showed a middle beach bar, but this feature had spread out considerably with a net shift to the southwest.

During this stage atmospheric pressures were lower on average, and wave conditions were both more frequent and intense. Calm conditions were present 64% of the time. Two storm conditions occurred, the first one a SSE storm producing 2 m waves on 11 November, and the second a southern storm producing 3 m waves on 27 November. Both occurred very near the spring tides (Figure 6.5a). The waves were predominantly from S and SSE over the whole period, those from other directions being negligible (Figure 6.5b). The probability of strong tidal currents was high, as the largest tidal range occurred during this stage (Figure 6.3).

Beach changes were more pronounced compared with the previous ‘stable’ stage. Erosion ‘holes’ are present in the middle beach (bar site) and again in the upper part of the southern beach, by the seawall (Figure 6.5c). The southern hole was especially deep, with drops of –2 m. In the rest of the beach, accretion is predominant. It is very likely that the sand eroded from the central bar had been spread on the northern and southern regions of the lower beach and had also accumulated to the south. Cross-shore transport of sand might also explain this accumulation because the accretion had the big erosion ‘hole’ near the seawall.

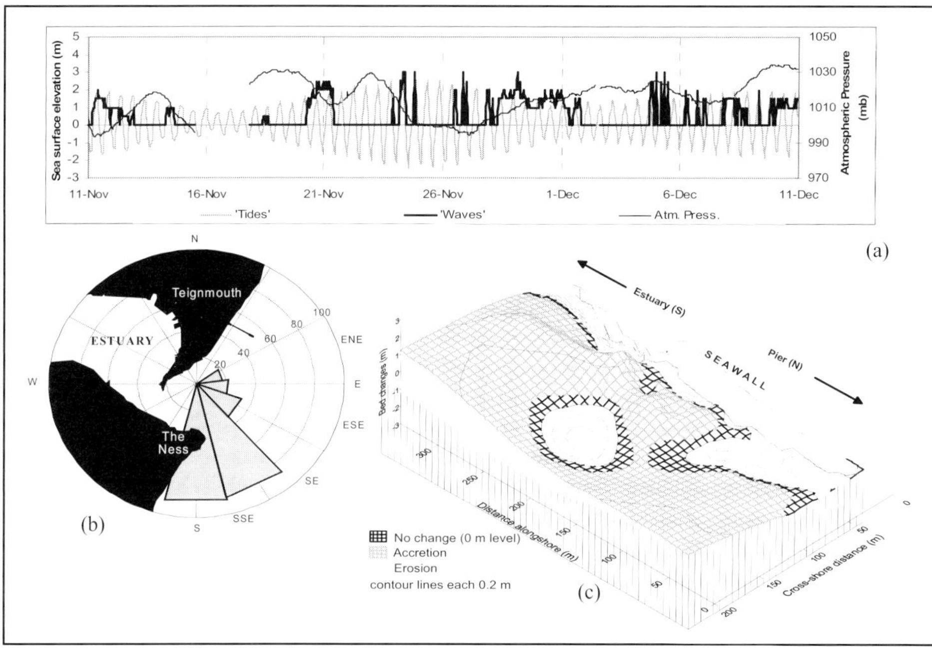

Fig. 6.5. Forcing conditions and morphological changes Stage 2. (a) Tide, wave height and atmospheric pressure time series; (b) wave roses; (c) morphological changes

Table 6.1 Storms during Beach Stage 3

Date	Max. offshore wave height	Approximate direction
22 Dec 95	3m	S-SSE
30 Dec 95	3m	SE
6 Jan 96	3.5m	SSE
9 Jan 96	4.5m	S
12 Jan 96	3.5m	S
5 Feb 96	4m	SSE

Beach Stage No. 3 (Figure 6.2c)

The sand bar had completely washed away. The beach was now smooth and steep with no important prominent features. This period was characterised by extreme storms and very energetic wave activity. A total of six strong storms occurred, with at least three of them coinciding with high spring tides (Figure 6.6a). Most of these storms approached from a southern quarter. Table 6.1 shows the events.

Waves from the east were also quite significant (Figure 6.6b), with offshore wave heights of up to 4 m.

The morphologic changes (Figure 6.6c) were a large erosion 'hole' in the middle beach, representing the disappearance of the middle beach bar. To the south end, large accretions of up to +3 m were found along the upper beach. This suggests that the sand from the spit had been transported alongshore and had accumulated near the seawall in the southern region of the beach.

6.2 GENESIS SIMULATIONS

The importance of wave processes, and specifically the influence of longshore sediment transport on the morphological changes per stage were assessed by means of the GENESIS model.

Calibration/Verification

Calibration and verification was an intrinsic part of this study, and in fact, was the means by which the performance of the GENESIS model at Teignmouth was evaluated.

The main purpose of this procedure was to establish the values of the calibration coefficients K1 and K2 (eq. 2.1 and 2.2). Hanson and Kraus (1989) propose the use of 0.50 for K1 and 0.25 for K2 as initial values. These values have to be modified to obtain the minimum calibration/verification error (C/V error) defined as the absolute value of the difference between observed and calculated shorelines. The values of K1 and K2 used in this experiment are 0.10 and 0.25 for stages 1 and 2, and 0.07 and 0.25 for Stage 3 respectively. The difference between the values of K1 might seem small (only 0.03), but the effects on the resulting shoreline simulations are considerable. Figure 6.7 shows the shoreline simulation for Stage 2 using the values of K1 and K2 adopted for Stage 3.

If Figure 6.7 is compared with the simulation for Stage 2 with the correct calibration coefficients (Figure 6.9b), it will be noticed that the shoreline trend is completely different. Under the conditions presented in Figure 6.7, GENESIS failed to reproduce shoreline changes in spite of having a smaller C/V error. Small changes in the calibration coefficients were not be regarded as negligible, and C/V error was not the only basis for assessing the performance of the model. Visual inspection is required.

Simulation for Stage 1

The shoreline from October was set as the initial condition (dotted line) and the shoreline for November was calculated against the corresponding wave climate. The results are shown in Figure 6.8. The calculated shoreline (diamonds) reproduced the same tendency as the measured shoreline (solid line), the bar spreading and shifting slightly to the south. Although the C/V error was not particularly great, the shape of the bar was not successfully reproduced.

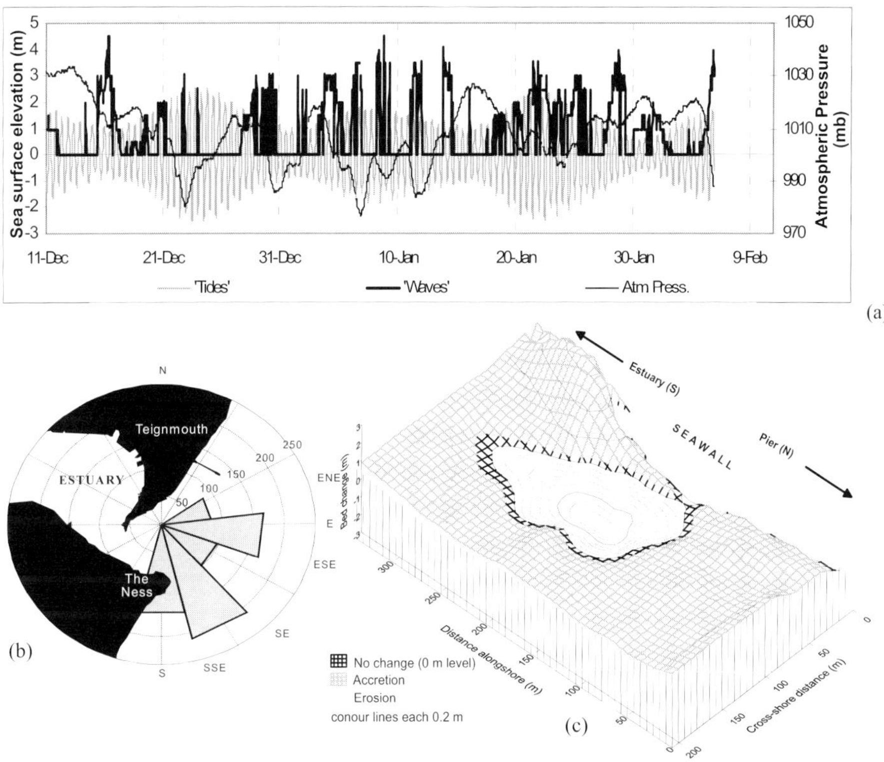

Fig. 6.6. Forcing conditions and morphological changes Stage 3, (a) tide, wave height and atmospheric pressure time series; (b) wave roses; (c) morphological changes

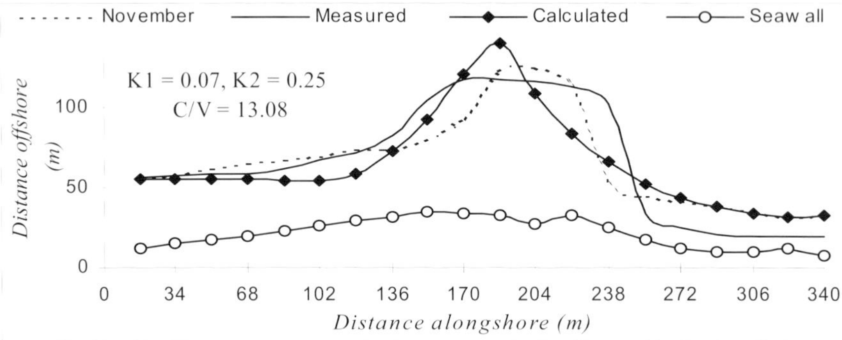

Fig. 6.7 Genesis simulation for Stage 2, with different calibration coefficients

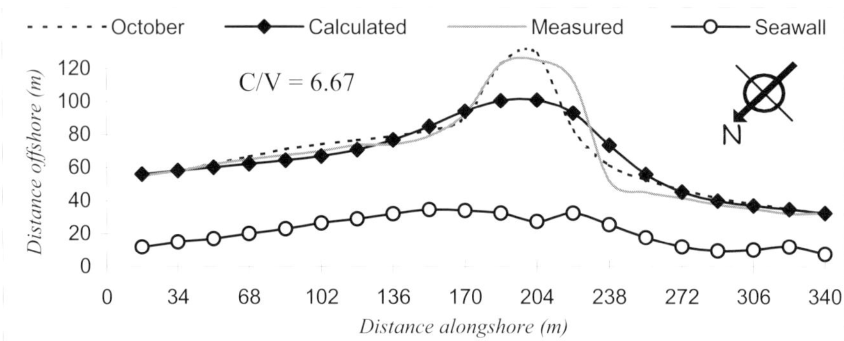

Fig. 6.8. Genesis simulation for Stage 1

Simulation for Stage 2

When a 'long-term' (October-December) simulation was done, GENESIS completely failed to reproduce the middle beach bar and the C/V error was too high (Figure 6.9a). However, when the November shoreline was set as the initial condition, and the shoreline position for December was calculated, the C/V error was reduced and the spreading of the bar was successfully reproduced. Notwithstanding this erosion was overestimated at the bar and underestimated at the southern end of the beach (Figure 6.9b).

Simulation for Stage 3

Again, when a simulation longer that one month was attempted ('long-term') the model failed to reproduce the shoreline changes. Figure 6.10a shows the simulation from October (Stage 1) to January (Stage 3). The shoreline trend was not reproduced at all and the C/V error was high. When the wave climate from stages one and two was excluded, and only the evolution of Stage 3 was attempted, the shoreline trend was simulated very well and the C/V error diminished remarkably (Figure 6.10b).

7. Discussion

7.1 PROCESSES

The numerical simulations showed a strong dependence on initial conditions, and consequently it was difficult to achieve good results when a long-term prediction was attempted (Figures 6.9a and 6.10a). Howevr, when only one month wave conditions were used to simulate shoreline change, GENESIS reproduced beach Stages 1 and 2 (Figures 6.8 and 6.9b) qualitatively well and Stage 3 was satisfactorily simulated quantitatively and qualitatively (Figure 6.10b).

This suggests that the processes included in GENESIS are not the main factors responsible for the morphological changes observed during stages 1 and 2. As a result, when GENESIS performs a long-term prediction, errors accumulate each time step, causing poor performance of the model. Notwithstanding this, longshore transport processes were certainly important during these two morphological stages, producing a southward shift of the bar and spreading it evenly over the beach, as the short-term simulations suggested.

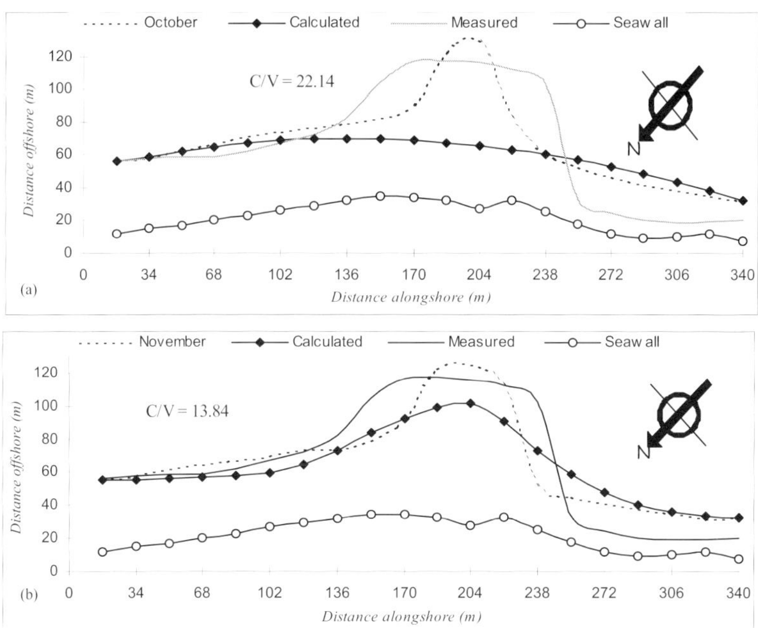

Fig.s 6.9 Simulations for Stage 2. (a) Long-term simulation (October to December), (b) short-term simulation (November- December)

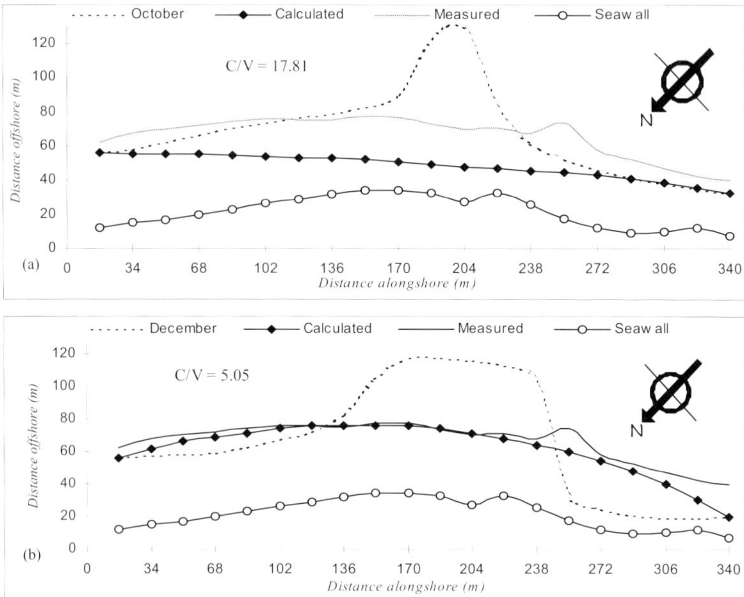

Fig. 6.10 Simulations for Stage 3. (a) Long-term simulation (October to January), (b) short-term simulation (December- January)

Regarding the spreading of the bar, results from the refraction/diffraction analysis (with RCPWAVE) demonstrated that longer period waves (9 seconds) from S and SSE are prone to produce a gradient in breaking wave heights in such a way that longshore currents may have travelled from the middle part of the beach (from the bar) to both north and south, triggering the dissipation of the middle beach bar. Field data shows that the majority of wave events occurring during stages 1 and 2 came from the S and SSE quarters. Due to the angle of approach, waves from this direction also created a northward longshore current on top of the above-mentioned process. The difference in response from one stage to another is due to the intensity of wave energy. Compared to Stage 1, Stage 2 waves were more frequent and energetic, hence dissipation of the bar was much more effective, as seen in the morphological changes at the beach (Figure 6.5c).

The southward shift of the bar during stages 1 and 2 might be partially explained by three different processes. Longshore transport mechanisms include the southward longshore currents produced by waves from the east (θ term in eq. 2), and also by S and SSE waves ($\partial H/\partial x$ term in eq. 2). The flood residual current tending southwest in the vicinity of the beach (Robinson, 1975) is the third mechanism. During Stage 1, longshore currents can be regarded as being responsible for a large part of this southward shift because GENESIS successfully represents the effect (Figure 6.8). Moreover, field data shows that S, SSE and E waves were an important part of the wave climate for the period (Figure 6.4b), and predicted tidal ranges were small

(Figure 6.3). For stage 2, this southward shift is prominent if the overall morphology is examined (Figure 6.2b), but it is not pronounced in the shoreline shape or in the GENESIS predictions. During Stage 2 eastern waves were absent and tidal ranges were at the maximum (Figure 6.3), hence tidal currents and alongshore gradients in wave height were likely to be the main causes for this behaviour.

During stages 1 and 2, there was a consistent scour near the seawall at the southern end of the middle beach bar. GENESIS does not reproduce this scour in either stage and the possible cause is cross-shore sediment transport. During both months, waves arriving from a SE corner could cause cross-shore transport to the beach. Moreover, the morphological changes observed during Stage 2 show a large accumulation of sand just seawards from the southern erosion 'hole' (Figure 6.5c), which might be an indication of cross-shore transport of sand.

During Beach Stage 3, there was clear evidence that longshore transport processes could largely explain the resulting morphology. The results from the GENESIS model are fairly good and field data support them. For this Stage there were five major storm events from the S-SSE quarter. These phenomena could have been very effective in dissipating the bar completely due to longshore transport to the north and the south, the later caused by alongshore gradients in wave height. Figure 6.6c shows the disappearance of the bar (as a hole in the bar's position) and accretion in the vicinity suggesting that the bar evenly dispersed on the beach. Nevertheless, according to the GENESIS simulations and the observed morphological behaviour, sediment was not just dispersed from the bar. Large quantities of sand were transported and accumulated at the southern end of the beach (Figure 6.6c). S-SSE storms and waves from the E are both responsible for the observed behaviour.

One of the aspects that made GENESIS most valuable for this site is the inclusion of the second term in the transport equation (alongshore gradients in wave height, $\partial H/\partial x$). The complex nearshore bathymetry in Teignmouth makes this process very important if waves arrive from S and SSE quarters.

7.2 LIMITATIONS OF THE MODEL

The above results have certain limitations with regard to the degree to which the basic assumptions made by the GENESIS model were met by the beach characteristics of Teignmouth during that period.

An important limitation is the definition of the average beach slope. Teignmouth Beach has a steep foreshore and flat intertidal region, and as a result, the shape of a logarithmic profile calculated with a single sediment size does not correspond to the actual profile shape, and the adopted beach slope is underestimated (Figure 6.1). A gentler beach acts to increase the importance of the second term in the transport equation (eq. 2.2) and erosion is overestimated in the shoreline change simulations. This is evident in figs. 6.6 to 6.8. If the GENESIS model allowed for variation in the median sediment size along the profile, an equilibrium profile produced with Moore's approach would describe the average profile for the site fairly well. Figure 6.1 makes this evident.

Another relatively important constraint is that the refraction analysis could not include all the complicated bathymetry closer to the beach because GENESIS needs pre-breaking wave conditions. Hence wave transformations near the beach underestimate the importance of the alongshore gradients to wave height, which are important in Teignmouth.

Some results found in this study raise uncertainties about the validity of model results when long-term predictions are attempted for practical applications. In fully three-dimensional sites, such as at Teignmouth, the morphological response can be expected to be a function of many factors acting together. Hence, if a model that only considers longshore sediment transport is used to predict shoreline change, the missing processes will add errors to the shoreline calculation at each time step and the resulting shoreline will be inconsistent with reality (see Figure 6.10a). The model only gave reasonably good results when one month wave data was used to simulate shoreline changes in Teignmouth.

Another issue is the calibration/verification procedure. As each site is unique in morphologic and hydrodynamic characteristics, there is general agreement within the modelling community that the calibration coefficients K1 and K2 should be considered site-specific parameters (Horikawa, 1989). Once the coefficients are determined by the calibration procedure, they can be applied to predict morphological changes at any time. The results from this study show that the calibration coefficients are not only site-specific but also condition-specific. In other words, when the system changes significantly (towards Stage 3), a different set of calibration coefficients must be used to successfully reproduce shoreline changes. Shoreline simulations have shown to be very sensitive to small changes in the calibration coefficients for this site (Figure 6.7). Similar results with calibration/verification procedures have been reported across many other disciplines (Oreskes, 1994).

8. Conclusion

Based on a combination of field data analysis and mathematical experiments using the GENESIS model, it was possible to identify the main processes affecting the morphology at Teignmouth from October 1995 to February 1996. The results can be summarised as follows:

1. The flood current tending south-west near the beach (Robinson, 1975) was the main cause of the shifting of the middle beach bar towards the south. This factor was particularly important during months in which tidal ranges were high and wave climate was mild and episodic (stages 1 and 2).

2. S and SSE waves played a major role in spreading out the middle beach bar. This spreading was sluggish during the months of mild-wave climate (Stage 1), but might become important when high and especially long waves are present during the winter (Stage 3). The dissipation was caused by northward transport due to the angle of approach, and also due to longshore transport to the south caused by alongshore gradients in breaking wave height near the estuary's mouth.

3. Waves from the east, which produced a southward longshore current due to the angle of approach, contributeed to the above processes in shifting the bar to the south.

The intensity of the response was, of course, dependent on wave energy. During stage 1, when wave energy was small, the bar is shifted slightly, but during the winter months, southward longshore transport was an important driving force.

4. Waves from the SE quarter tended to cause a cross-shore movement of sand, which was usually related to the consistent scour at the upper-southern beach near the seawall.

The partial conceptualisation of the system (only longshore transport was considered) and the uncertainty of significant changes in the system's conditions (which affect calibration coefficients), are bound to be aspects which limit the ability of GENESIS to carry out long-term predictions in complex 3D sites.

In spite of all the limitations, simple models like GENESIS are very useful for evaluating the short-term effects and importance of longshore sediment transport processes at Teignmouth Beach, South Devon, UK.

Based on the results of this study, it is advisable that the GENESIS model be used only for evaluating the importance of longshore sediment transport processes when applied to a complex three-dimensional site (e.g. close to inlets) instead of being used as an actual prediction tool. Information relating to longshore transport at a site under a given set of wave conditions is a very useful tool for the planning of coastal engineering works.

Acknowledgements

The authors would like to thank Dr. Andrew Saulter and Dr. Jonathon Miles for providing the beach profile data. This project was sponsored by the Consejo Nacional de Ciencia y Tecnología (CONACYT), México.

REFERENCES

Davidson, M.A., Russell, P.R., Huntley, D.A., Hardisty, J., and Cramp, A., An overview of the British beach and nearshore dynamics programme. *Proceedings of the 23rd. International Conference on Coastal Engineering '92*, 1993, pp. 1987–2000.

Ebersole, B.A. Refraction-diffraction model for linear water waves. *Journal of Waterway, Port, Coastal and Ocean Engineering*. ASCE. 111(6), 1985, pp. 939–956.

Galofre, J, Montoya, F.G., Medina, R., Study of the evolution of a beach nourishment project based on computer models. In: Brebbia, C.A., Traversoni, L. Wrobel, L.C. (eds.) *Computer Modelling of Seas and Coastal Regions II. Second International Conference on Computational Models of Seas and Coastal Regions* (COASTAL '95). Computational Mechanics Publications. UK. 1995, pp. 249–256.

Gravens, M.B., Kraus, N.C & Hanson, H., GENESIS: Generalized Model for Simulating Shoreline Change. Report 2. *Workbook and System User's Manual. US Army Corps of Engineers*. CERC, 1989.

Hanson, H., GENESIS: a generalized shoreline change numerical model, *Journal of Coastal Research*, 5(1), 1989, 1–27.

Hanson, H. & Kraus, N.C., GENESIS: Generalized Model for Simulating Shoreline Change. Report 1. *Technical Reference. US Army Corps of Engineers*. CERC. 1989.

Horikawa, K., *Nearshore dynamics and coastal processes*. University of Tokyo Press. Japan, 1988.

Houston, J.R., Discussion of: Young, R.S., Pilkey, O.H., Bush, D.M., Thieler, E. 1995. A discussion of the Generalized Model for Simulating Shoreline Change (GENESIS). Journal of Coastal Research, 11(3):875–886, *Journal of Coastal Research*. 12(4), 1996, 1038–1043.

Inglis, M.L., A numerical model study using GENESIS v.3 of a macrotidal, shingle beach after the construction of extensive sea defences at Sidmouth, Devon. M.Sc. thesis, University of Plymouth, 1996.

Mariño-Tapia, I., Morphological Changes on Teignmouth beach, South Devon, UK: A comparison between field observations and a numerical model. MSc. thesis. University of Plymouth, 1998.

Masselink, G. & Short,A.D. The effect of tide range on beach morphodynamics and morphology: a conceptual model. *Journal of Coastal Research* 9(3), 1993, 785–800.

Miles, J.R., Russell, P.R. & Huntley, D.A., Introduction to the COAST3D field study site of Teignmouth. Unpublished. Institute of Marine Studies, University of Plymouth, 1997.

Moore, B.D., Beach Profile Evolution in Response to Changes in Water Level and Wave Height. MCE thesis Department of Civil Engineering, University of Delaware, Newark, DE. 1982, 164 pp.

Nordstrom, K.F., *Beaches and Dunes of Developed Coasts*. Cambridge University Press, 2000.

Oreskes, N., Shrader-Frechette, K., & Belitz, K., Verification, validation and confirmation of numerical models in earth science. *Science*. 263, 1994, 641–646.

Robinson, A.W.H., Cyclical changes in shoreline development at the entrance of Teignmouth Harbour, Devon, England, in: Hails & Carr, A. (Eds.) *Nearshore Sediment Dynamics and Sedimentation*. John Wiley and Sons. UK. 1975, pp:181–200.

Rogers, W.E. & Work, P.A., Mathematical and physical modelling of beach nourishment projects. *Proceedings of the International Conference on Coastal Engineering*. 1996, pp 2941– 2954.

Silvester, R. & Hsu, J.R., *Coastal Stabilization: Innovative Concepts*. Prentice Hall. 1993.

Spratt, T., Investigation of the movements of Teignmouth bar, London, 1856.

Szmytkiewicz, M., Biegowski, J., & Kaczmarek, L.M., Okrój, T., Ostrowski, R., Pruszac, Z., Rózynsky, G., and Skaja, M., Coastline changes nearby harbour structures: comparative analysis of one-line models versus field data, 2000.

Thieler, E., Young, R.S., & Pilkey, O.H., Bush, D.M., The use of mathematical models to predict beach behaviour for U.S. coastal engineering: A critical review. *Journal of Coastal Research*. 16(1), 2000, 48–70.

Young, R.S., Pilkey, O.H., Bush, D.M., & Thieler, E., A discussion of the Generalised Model for Simulating Shoreline Change (GENESIS) *Journal of Coastal Research*. 11(3), 1995, 875–886.

METHODOLOGY OF SANDY BEACH STABILIZATION BY NOURISHMENT: A LONG–TERM MORPHODYNAMIC MODELLING APPROACH

B. ONTOWIRJO[1] AND C.D. ISTIYANTO[2]

[1]*Research Engineer Coastal Engineering Laboratory at the Agency for Assessment and Application of Technology (LPTP-BPPT) Jakarta - Indonesia.*
[2]*Research Engineer Department of Civil Engineering Kagoshima University, Japan, On leave from Coastal Engineering Laboratory Coastal Engineering Laboratory at the Agency for Assessment and Application of Technology (LPTP-BPPT) Jakarta – Indonesia.*

Abstract

Many attempts have been made to control beach erosion, but in the long term none have succeeded in stabilizing sandy beaches that are being eroded. Coastal structures are of little use for stabilization as they cause beach profiles to gradually become steep. The equation for a stable sandy beach is derived from the set of the equations of continuity for beach change and longshore sediment transport under unsteady and non-uniform conditions, so that theoretical shoreline configurations can be obtained for both static and dynamic stable sandy beaches.

One methodology for beach erosion control is beach nourishment; in which a strip of borrowed offshore sand is reconstructed along the eroded coastline and subject to wave climate to establish new dynamically stable sandy beaches. Most bed evolution models neglect slope-related sediment transport, which though generally weak compared to other mechanisms, is extremely important to the stability of the system. In fact, it is the downslope gravitational transport that enables the beach profile to reach a state of equilibrium. A Lagrangean particle displacement approach has been adopted for bed morphology calculation in the present model since it include gravitational and sheltering effects.

Simplification of equilibrium monotonic profiles (One-line N-line, Genesis) has been shown to be able to match for long-term shoreline change well. In a long-term simulation, the inclusion of downslope gravitational transport by itself may not ensure sensible equilibrium topography, especially when cross-shore profile evolution is important. This is because cross-shore profile evolution depends on the vertical velocity structure comprising of undertow, mass flux and bottom boundary drift. In long-term simulation of a situation where the cross-shore mechanisms are important, a 3D description of the flow structure is required. To improve this aspect the present model includes the role played by undertow, wave asymmetry and bottom boundary drift in sediment transport. Laboratory measurement comparison will also be discussed in this study.

1. Introduction

An adaptive beach nourishment model which represents a rectangular planform of borrowed offshore sand with a strip of small crushed stone placed seaward is reconstructed along the eroded coastline and subject to wave climate. Crushed stone

C. L. Goudas et al. (eds.), Soft Shore Protection, 71-79.

protection must be assumed due to the limitations of the unit displacement model. Calculations relating to the response of particle units to wave induced forces can be expected to give information about the movement conditions of particle units under wave action. The approach selected for this purpose is deterministic one, and although simple, is powerful enough to describe some basic interactions between wave-induced forces and particle units. The assumptions, theoretical considerations, the concepts on which the model is based, and how to determine any related phenomena are described.

A numerical model for reshaping simulation was also developed. The reshaping simulation involves the application of two separate models, which are linked by an iterative procedure. These two separate models are the wave simulation model and the displacement model of particle units. The wave model algorithm is adopted from the Volume of Fluid (SOLA–VOF) code of Nichols and Hirt (1980), while the particle unit displacement model, which will be explained briefly, was developed with reference to the previous work of Norton and Holmes (1992).

Once a stable cross shore downslope profile is reached, a long shore transport model is employed to calculate long-term morphological changes to a strip of armor units.

2. Numerical methods

2.1 UNIT DISPLACEMENT MODEL

Initially, the surface of the bed layer is numerically profiled. This profile will be the input for the wave model. Once a converged solution has been obtained from the wave model, temporal variations of water depth, velocity, acceleration, and gradient of the free surface during a complete wave cycle are stored for as many sampling stations as required along the slope.

Only surface particle units exposed to the external flow are considered to be potentially mobile. These particle units have contact with a maximum of only three other supporting particle units. The stability of each surface particle unit is considered in turn commencing with the particle unit nearest to the toe of the slope. The wave-induced horizontal and vertical components of forces are calculated using the data from the wave model. Velocities and accelerations at the precise particle unit location are obtained by interpolation of data stored for the two nearest sampling stations.

Under threshold conditions, the particle unit is given a discrete displacement of length d_{min} in the direction selected as the path of least resistance. This is the direction associated with the maximum ratio of disturbing to restoring moments. In the case of a two-dimensional situation, the only possible directions are rolling-up or rolling-down according the direction of the least resistance. The magnitude of the discrete displacement must be small enough to provide good resolution of particle unit displacement but sufficiently large to minimize computational times.

After displacement, a rolling sequence is used until the particle unit is relocated in a statically stable position.

After all surface particle units are considered and displaced, if hydrodynamically unstable, the next section step is embarked upon. The adjusted slope is then re-profiled for new input to the wave model followed by further displacement of surface particle units. The entire iterative procedure described is repeated until the required time is reached.

2.1.1 Forces on unit displacement model

The disturbing forces include hydrodynamic and buoyancy forces, while weight of the body, friction force, and interlocking are restorative forces. Disturbing forces resulting from the hydrodynamic waves loading (or as usually called wave-induced forces) are determined using a Morison-type equation including drag, inertia and lift force components. The force work on a single particle unit is given in Figure 1.

The drag force is simply due to the effects of viscosity in the fluid, and is formulated as follows:

$$F_D = 0.5\, C_d\, \rho_w \left(\frac{\pi}{4} D^2 \right) u\, |u|, \tag{1.1}$$

where p$D^2/4$ is the area of spherical cross section across which the force works.

The inertia force is a result of acceleration and can be written as follows:

$$F_I = C_I \frac{\rho \pi D^3}{6} \frac{du}{dt}, \tag{1.2}$$

where pr$D^3/6$ is the mass of the spherical unit, and du/dt is the acceleration.

The force component due to lift F_L would be of the form:

$$F_L = 0.5 \rho \frac{\pi D^2}{4} C_L\, u_x^2, \tag{1.3}$$

where F_L is perpendicular to the direction of wave advance (the x-direction), and p$D^2/4$ is the area of spherical cross section across which the force works.

The buoyancy force will be given as:

$$F_B = C_{SI}\, \rho_w\, g \frac{\pi}{6} D^3. \tag{1.4}$$

Special attention must be paid to the direction of the force. In still water flow, the buoyancy force is an upward force. This force is equal to its weight and will act vertically upward through its centre of gravity. When the wave passes over the fluid, the free surface of water will fluctuate and generate an instantaneous sloping surface. Consequently, there will be a resultant horizontal pressure working on the immersed body.

2.2 THE WAVE SIMULATION METHOD

The Solution Algorithm (SOLA)–Volume Of Fluid (VOF) method has been chosen for the present wave simulation as well as flow inside porous media. This method has been proved to be applicable to problems involving a single fluid with a number of free surfaces, breaking waves for example. By adapting of its velocity component in the governing equation, it is also applicable to the flow inside porous media. Furthermore, application of this method for solving the interaction between the wave and the flow inside porous media may be conducted without any coupling problems at the interface. The original development of the program was carried out in the Computational Fluid Dynamics Laboratory at the Los Alamos National Laboratory, New Mexico, in the early 1960s. The numerical algorithm used was called 'Solution Algorithm–Volume of Fluid (SOLA–VOF)', an extension of the SOLA method (Hirt et al., 1975, 1980). The obstacle geometries in the software were placed in the grids using Fractional Area Volume Obstacle Representation (FAVOR) methods. This method made the geometry and grids completely independent of each other. These techniques generate smoothly embedded geometric features that were constructed from the program's pre-processor or imported from other CAD programs. This eliminated the 'stair-step' boundaries, which were common in the finite difference models.

SOLA VOF employed staggered grid arrangements. All variables are located at the centre of the cell volume, except for velocities, which are located at the cell faces. The flow region is divided into a mesh of fixed rectangular cells. When there are free fluid interfaces surfaces, it is necessary to identify whether those cells are empty, contain a surface, or are full of water. The software considered a cell with an F value less than a unity, but with no empty neighbour, as a full cell.

At the mesh boundaries, a variety of conditions may be set using the layer of fictitious cells surrounding the mesh. There are three types of boundary conditions: symmetry plane (default), rigid wall and specified velocity. As a rule, all rigid and free boundary surfaces are treated as free-slip boundaries (no tangential stresses on the surfaces) and referred to as symmetry plane boundary conditions. For rigid wall boundaries, the normal velocity is set to zero, and the tangential velocities can be set to any value by the wall shear stress model provided by the software for the free-slip type of wall. However, in the no-slip boundary conditions, the tangential velocities in the fictitious cells are set to zero. Furthermore, in the specified velocity conditions, the tangential velocities and normal velocities must be specified.

2.2.1 Governing equations

A characteristic of the SOLA–VOF method is the use of fractional fluid volume for tracking free boundaries. Instead of defining a free surface directly, it works by following fluid regions through an Eulerian mesh of stationary cells.

In their original description Hirt et al (1980) noted that in this method, a function F(x,z,t) is defined whose value is unity at any point occupied by fluid and zero elsewhere. When averaged over the cells of a computational mesh, the average value of F in a cell is equal to the fractional volume of the cell occupied by fluid. In particular, a unit value of F corresponds to a cell full of fluid, whereas a zero value indicates that the cell contains no fluid. Cells with F values between zero and one contain free surface. E means an empty cell with zero value of F, F identifies a full of fluid cell with F value equal one, while S is a surface cell has an F value between zero and one.

The time dependence of F is governed by the equation

$$\frac{\partial F}{\partial t} + u\frac{\partial F}{\partial x} + w\frac{\partial F}{\partial z} = 0, \tag{2.1}$$

where u and w are fluid velocities in the coordinate directions of x and z respectively, and F is the fractional volume of fluid function. This equation suggests that F moves with the fluid.

The fluid equations to be solved are the Navier-Stokes equations. For the external wave motion, the Navier-Stokes momentum equations are written as:

$$\frac{\partial u}{\partial t} + \frac{\partial u^2}{\partial x} + \frac{\partial uw}{\partial z} + \frac{1}{\rho_w}\frac{\partial p}{\partial x} - \upsilon_t\left(\frac{\partial^2 u}{\partial x^2} + \frac{\partial^2 u}{\partial z^2}\right) = 0, \tag{2.2}$$

where u and w are fluid velocities in the coordinate direction x and z respectively, ρ_w is water density, p is water pressure, v is the kinematic viscosity coefficient, and g is gravity acceleration in the coordinate z-direction.

For flow inside a porous structure, an adapted Navier-Stokes momentum equation given by Van Gent (1994) is used, and written as:

$$\frac{1+c}{n}\frac{\partial u}{\partial t} + \frac{1}{n^2}\left(\frac{\partial u^2}{\partial x} + \frac{\partial uw}{\partial z}\right) + \frac{1}{\rho_w}\frac{\partial p}{\partial x} + gau + gbu\sqrt{\left(u^2 + w^2\right)} = 0, \tag{2.3}$$

$$\frac{1+c}{n}\frac{\partial w}{\partial t} + \frac{1}{n^2}\left(\frac{\partial uw}{\partial x} + \frac{\partial w^2}{\partial z}\right) + \frac{1}{\rho_w}\frac{\partial p}{\partial z} + gaw + gbw\sqrt{\left(u^2 + w^2\right)} + g = 0, \tag{2.4}$$

where u and w are velocities in the x and z direction respectively. Coefficients of a, b, and c are as used in the Forchheimer equation as follows:

$$I = au + bu\left|u\right| + c\frac{\partial u}{\partial t}, \tag{2.5}$$

$$a = \alpha\frac{(1-n)^2}{n^3}\frac{\upsilon}{gD_{n50}^2}, \tag{2.6}$$

$$b = \beta_C \left(1 + \frac{7.5}{KC}\right) \frac{1-n}{n^3} \frac{1}{gD_{n50}}, \tag{2.7}$$

where

$$KC = \frac{\hat{U}T}{nD_{n50}}, \quad c = \frac{1 + \frac{1-n}{n}\left(0.85 - \frac{0.015}{A_c}\right)}{ng}, \quad A_c = \frac{\hat{U}}{ngT} > \frac{0.015}{\frac{n}{1-n} + 0.85},$$

and where I is the hydraulic gradient, a and b are the dimensional friction coefficients of laminar and turbulent flow respectively, and c is a coefficient to take the phenomenon *added mass* into account; n is fluid viscosity, n is porosity, U/n is maximum pore velocity; α and β_c are coefficients which depend on parameters such as grading, shape, aspect ratio or orientation of the stones.

Because an incompressible fluid flow is assumed for the present calculation, the momentum equations of Eq. (2.2) to (2.4) must satisfy the mass conservation condition that

$$\frac{\partial u}{\partial x} + \frac{\partial w}{\partial z} = 0. \tag{2.8}$$

The above equations (2.1) to (2.4) and (2.8) are together the governing equations used in the present calculation of the wave and porous flow.

3. Numerical modelling result

The results of the reshaping simulation showed an increased degree of erosion from the upper part of the slope with a corresponding downslope near bed accretion. The toe of the reshaped profile does show a relatively smooth transition to the horizontal bed. Downslope angle variation is tested against the numerical model and the results (see Figures 2 and 3) show that for the same wave condition, the same stable profile is reached no matter what the initial slope angle is. If the wave model were improved, flow inside porous structures could be calculated.

4. Conclusions

Using a deterministic approach, a reshaping simulation model was shown to provide a good prediction of profile development. Improvements to the wave model, including breaking wave conditions, have shown good quantitative capability to predict the physical model. Prediction of transport influence on the reshaped profile can also be validated against physical model data. The influence of variability within the random structure of the unit layer can be investigated by changing the material grading and results have shown that the final steepness of the slope varies depending upon the size of materials. Modification towards a more probabilistic model should include variability of particle shape, which may require a distribution of volume, wave forces

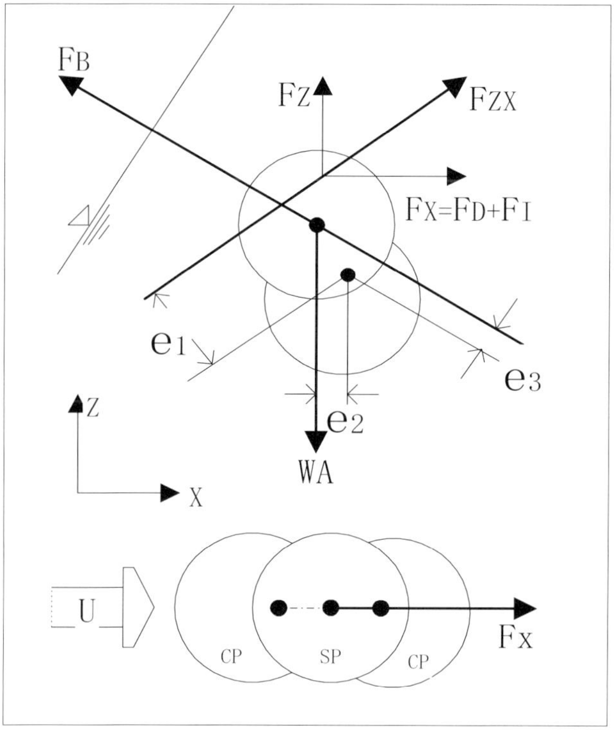

Fig. 1. Diagram of forces onto unit.

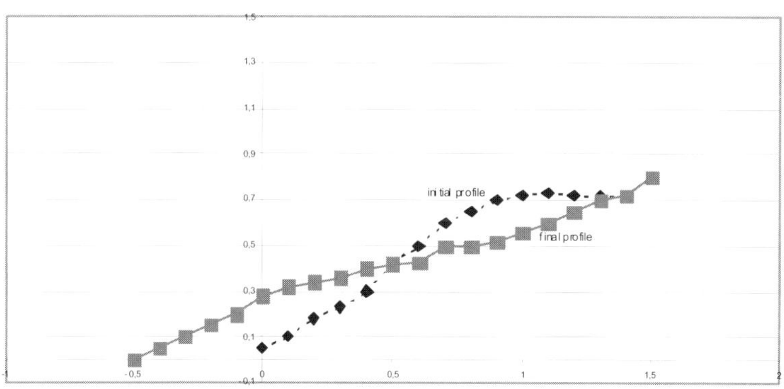

Fig. 2. Slope structure: initial and final profiles (m=1:1.3)

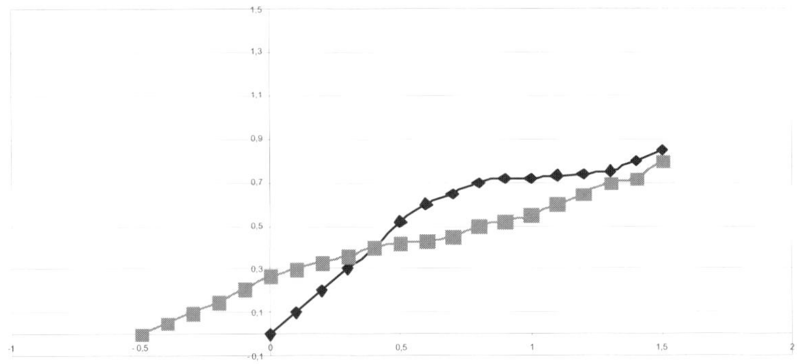

Fig. 3. Slope structure: initial and final profiles (m=1:1)

and interlocking coefficient rather than a single value as used in the present model.

Nomenclature

c : speed of sound in the water — F : fluid function
g_x, g_z: gravity acceleration in x and z direction — H : wave height
k : wave number : $k=2\pi/\lambda$ — p: pressure
T : wave period — u: velocity components in x directions
W : velocity components in z direction — B: beach angle
λ : wave length — υ: coefficient of kinematic viscosity
ω : wave angular frequency — $\omega=2\pi/T$
ξ : surf similarity parameter — $\xi=(H/\lambda)^{-0.5}\tan\beta$

Acknowledgements

This paper is a part of a PhD thesis of the second author at Kagoshima University. The financial support provided by the Ministry of Education of the Government of Japan is gratefully acknowledged. Our thanks are also due to Dr. Hirt and Dr. Nichols for their excellent computational work on SOLA−VOF.

REFERENCES

Flow Science Inc.: FLOW−3D Excellence in Flow Modelling Software, version 7.1, 1997, Los Alamos, NM.

Gross, M.G.: *Oceanography: a View of the Earth*, Prentice-Hall Inc., Engleword Cliffs, N.J., USA, ISBN 0-13-629692-0-01, 1987.

Hirt, C.W., Nichols, V.D. & Romero, N.C.: 'SOLA: A Numerical Solution Algorithm for Transient Fluid Flows', Los Alamos Scientific Laboratory Report, LA-5852, 1975.

Hirt, C.W., Nichols, V.D. & Hotchkiss, R.S.: 'SOLA-VOF: a Solution Algorithm for Transient Fluid Flow with Multiple Free Boundaries', Los Alamos Scientific Laboratory Report, LA-8355, 1981.

Hirt., C.W. & Nichols, B.D.: 'Volume of Fluid (VOF) Method for the Dynamics of Free Boundaries', *Journal of Computational Physics* 39, 1981, pp. 201–225.

Kobayashi, N. & Wurjanto, A.: 'Numerical model for waves on rough permeable slopes', *J. Coastal Research*, SI 7, 1990, pp. 149–166.

Madsen, O.S. & White, S.M.: '*Reflection and transmission characteristics of porous rubble mound*', Technical report no. 207, R.M. Parson Laboratory, MIT Cambridge, Massachusetts, 1975.

Norton, P.A. & Holmes, P.: 'Armour Displacements on Reshaping Breakwaters', J. Coastal Engineering 16, 1992, pp. 1448–1460.

STOCHASTIC ECONOMIC OPTIMISATION MODEL FOR THE COASTAL ZONE

S. VAN VUREN[1], M. KOK[2], R. E. JORISSEN[3]

[1] *Delft University of Technology, Section of Hydraulic Engineering, PO Box 5048, 2600 GA Delft, the Netherlands*

[2] *Twente University, Department of Civil Engineering, PO Box 217, 7500 AE Enschede, the Netherlands*

[3] *Ministry of Transport, Public Works & Water management, National Institute for Coastal and Marine Management/RIKZ, PO Box 20907, 2500 EX The Hague, the Netherlands*

Abstract

This paper describes a stochastic optimisation model for the coastal zone. The model combines stochastic hydraulic and morphological processes in the coastal system with maintenance measures (sand nourishment) and socio-economic and ecological activities. The model optimises maintenance costs in relation to any damage resulting from coastal recession. The expected minimum total costs and the return period on which these minimum costs are based form the output of the model. The model is a useful tool for supporting spatial decisions, assessing various coastal policies and optimising coastal maintenance.

Structural erosion and dune retreat can cause damage to societal activities in the coastal zone. The probability of damage is larger when coastal recession increases or economic development towards the coastline takes place. Coastal maintenance results in a reduction of the probability of damage, but it introduces maintenance costs. Coastal maintenance (return period and magnitude) affects the total costs (the sum of the expected damage to societal activities and maintenance costs). The minimum total costs over a given planning period can be determined by systematically changing the return period of coastal maintenance. The model is influenced by various uncertainties, and these affect the optimisation results. Consequently, uncertainty distributions were assessed and Monte Carlo simulations made with the model. The entire set of model outputs resulting from the Monte Carlo simulations can be used to estimate the expected minimum costs.

1. Introduction

Throughout the world, coasts are subject to physical and societal changes. Coastal policies that address these changes often have political consequences. Physical changes include sea level rise and storm intensification, possibly causing more extensive dune retreat and structural erosion. Societal changes such as the development of socio-economic and ecological activities claim coastal area space. These activities may include enlarging of recreational and tourism facilities, and land reclamation for industry, infrastructure and residence. As a consequence of both physical and societal changes, the pressure upon the coast and spatial claims are increasing.

This might lead to conflicting interests, especially when space is limited. While future physical changes such as sea level rise need to be dealt with, it is also desirable to gain as much profit from the economic opportunities offered by the coastal zone as possible.

C. L. Goudas et al. (eds.), Soft Shore Protection, 81-104.

The landward pressure of societal changes may create irreversible impacts, which may endanger future safety. If societal-economic activities, nature and landscape are prioritised, this will reduce the amount of coast used for future coastal defence purposes. To avoid conflict between these interests and to guide the developments in a responsible manner, spatial decisions relating to the coast have to be taken cautiously. Long-term coastal policies, which enable responsible coastal authorities to organise safety-conditions and spatial development in the dynamic zone, are required. The major aim of this paper is to develop a method to economically assess coastal policies.

The stochastic economic optimisation model which this paper describes, is a useful tool for supporting spatial decisions, assessing various coastal policies and optimising coastal maintenance. The model combines the hydraulic and morphological coastal system processes and maintenance measures with socio-economic and ecological activities. The model optimises maintenance costs in relation to any damage resulting from coastal recession. The expected minimum total costs and the return period on which these minimum costs area based form the output of the model

This paper will focus mainly on sandy coasts (coasts with a flexible character). The forces of nature cause constant moving of this barrier. Coastal zone societal functions and interests may be damaged by structural erosion and dune retreat, and consequently understanding of the dynamics of the coast is of great importance in assessing the probability of coastal zone damage. The probability of damage is greater when coastal recession increases or economic development towards the coastline takes place. While coastal maintenance results in a reduction in the probability of damage, it introduces maintenance costs. Coastal maintenance (return period and magnitude) affects the total costs (the sum of the expected damage to societal activities and the maintenance costs).

The minimum total costs over a given planning period can be determined by systematically changing the coastal maintenance return period. The model is influenced by various uncertainties and these affect the optimisation results. Consequently, uncertainty distributions were assessed and Monte Carlo simulations made with the model. The set of model outputs resulting from the Monte Carlo simulations can be used to estimate the expected minimum costs, and to obtain an insight into the economic and ecological consequences of spatial decisions in the coastal zone.

This paper consists of two parts. Firstly, the stochastic economic optimisation model is discussed. This model is partly based on the model in Peerbolte and Wind (1992). The model contains an input file, three model parts and an optimisation module which combines the model parts. In the second part of the paper, the model is used to do an economic assessment of various coastal policy options in the Netherlands.

2. Stochastic economic optimisation model

2.1 UNCERTAINTIES

Since the model is influenced by various uncertainties, uncertainty distributions were assessed and Monte Carlo simulations made with the model. Monte Carlo simulation involved running a deterministic model many times. Each model run is driven by a different model input set (synthesised on the basis of randomly generated parameters).

The entire set of model outputs resulting from the Monte Carlo simulations are used to determine the statistical properties of the output such as the expected minimum costs.

In the optimisation model, only uncertainties relating to coastal processes and maintenance costs are considered. Coastal processes are stochastic processes: they are not constant, but they change stochastically in space and in time. These processes are characterised by a normal probability distribution function. The maintenance costs are described by a triangular probability density function. The statistical parameters of these distribution functions are based on data (measurements) and model computations. Stochastic model inputs are underlined in the model equations. Future physical changes and societal changes affect the statistical characteristics of stochastic model input.

2.2 STOCHASTIC ECONOMIC OPTIMISATION MODEL

The stochastic optimisation model consists of three models: a coastal process model, which describes the hydraulic and morphological processes, a coastal defence model and an economic model. An optimisation module combines these three models. The coastal measures return period and the expected minimum total costs form the output of the model.

Hydraulic and morphological processes, societal activities, the ecology and coastal maintenance play a central role in this model. It is able to give a clear overview of the damage to societal functions in the coastal zone caused by the conflict between coastal and societal dynamics, as well as find an optimal solution in which the total annual costs are minimised. This solution is based on the relationship between the coastal maintenance costs and damage to societal functions. The model framework is given in Figure 1.

Figures are included in the stochastic optimisation model and coastal recession, sand nourishment costs and the economic value of a coastal area are calculated per linear meter along the coastline. The equations given in the next sections relate to a linear meter of coastal stretch.

2.3. LIMITATIONS AND ASSUMPTIONS

Limitations and assumptions made in this paper include the following:

- The optimisation model was developed for dynamic sandy coasts.
- Safety is the most important function of the dune coast. Coastal defence is also inextricable linked with other coastal zone societal functions. Integrated coastal zone management is aimed at sustainable preservation of these functions in the coastal dune area without harming the water protection function. Awareness of the presence and importance of other functions in the coastal zone will thus increase. In view of the importance of the societal functions and their spatial claims on the coastal zone, these functions need to be specified.
- The only form of coastal maintenance is beach nourishment.

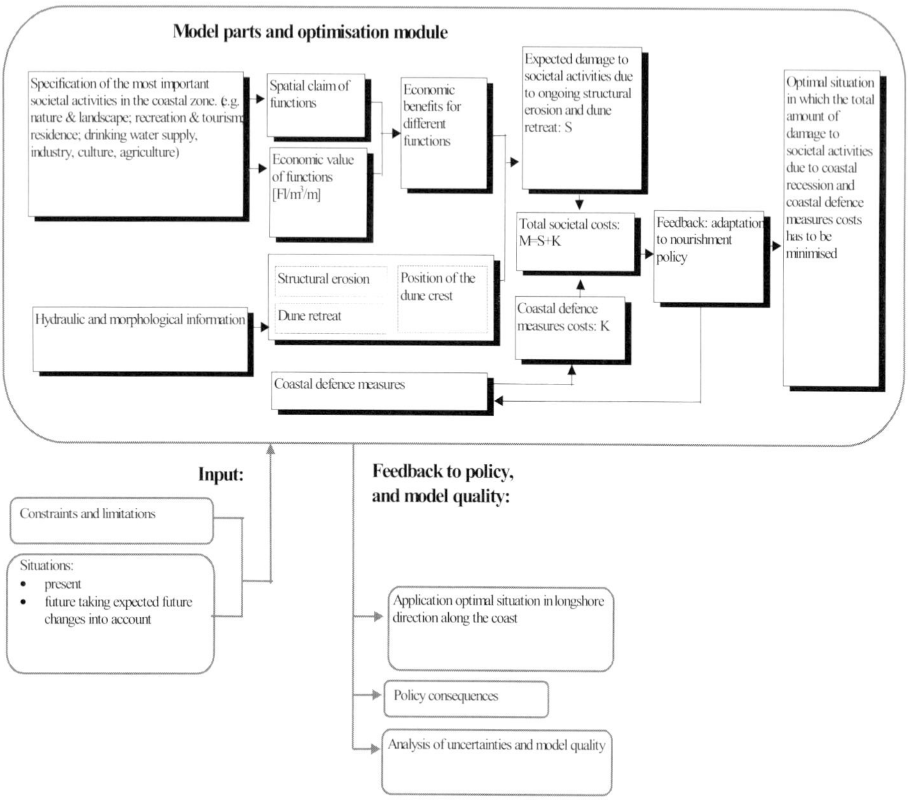

Fig. 1: Stochastic economic optimisation model framework

- The present situation and that in 2100 are taken into consideration.
- In the event of damage to societal functions, reconstruction takes place at a location which is not susceptible to damage. Consequently, damage will only be incurred once.
- Damage to societal functions includes both investment costs and loss of profits. Loss of profits due to damage is calculated for two years.
- Damage can be interpreted according to various economic scales. At the local scale, the entrepreneur will experience damage and loss of profits. At the macro-scale, this damage might be negligible. This paper addresses nourishment costs and damage at the local economic scale.
- The planning period over which the total costs of coastal defence measures and damage to societal functions are optimised is 30 years.
- The stochastic optimisation model is a sequential model. Changes to the input conditions that occur over time are not included in the model. The costs of coastal

defence measures and damage to societal functions are calculated in time steps of one year. The optimisation process is based on constant input conditions and includes two reference points: the present and a future point in time. Feedback between changing hydraulic conditions and their impact on morphological processes and visa versa is omitted in the model.

2.4 BOUNDARY CONDITIONS

The boundary condition related to the coastal dune area. A representative coastal cross-section is defined based on historical data. Societal functions and coastal dynamics are considered within the coastal dune area of this representative coastal cross-section. The coastal dune area stretches out from the dune crest till the landward boundary of the coastal dune area (see Figure 2). The dune crest is a reference point for the coastal dynamics and the position of societal functions in the dune area.

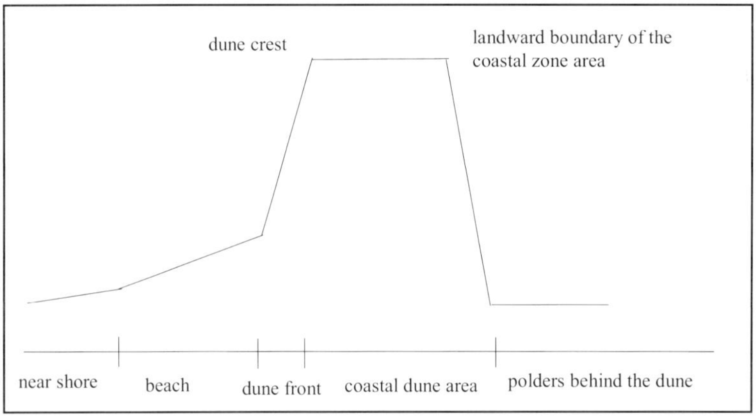

Fig. 2. Coastal zone

2.5 COASTAL PROCESSES MODEL

The morphological development of the coast has been somewhat irregular, and the various different processes and human interference have left their mark. Morphological processes occur at various scales. The coastal processes model focuses on meso-macro scale level. This level includes the principal morphological features and the interaction between these features. Seasonal and interannual variations in tide and weather conditions, human activities (sand mining) and extreme events have prompted the main changes. In this model, two types of coastal development are considered: structural and incidental erosion (Van Rijn, 1998). Structural erosion includes the ongoing sand loss in the coastal zone due to waves and flows. The general effects of structural erosion are predictable within uncertainty bounds, which means that the coastal manager is able to anticipate structural erosion. However, what influence future climate changes (those involving sea level change and storm intensification) are likely to have on structural erosion is uncertain.

Structural erosion is expressed as follows:

$$\underline{x_s}(t) = \underline{b} \cdot t \tag{1}$$

In which:

$\underline{x_s}(t)$	structural erosion after year t	[m]
$\underline{b}$	annual structural erosion rate	[m/yr.]
t	time	[yr.]

The stochastic variable consists of an average annual coastal recession (or coastal extension) trend and a standard deviation. Using a regression analysis, the parameters of the stochastic variable of structural erosion can be determined based on coastal measurements (if available). The coastal recession (extension) due to structural erosion is shown diagrammatically in Figure 3.

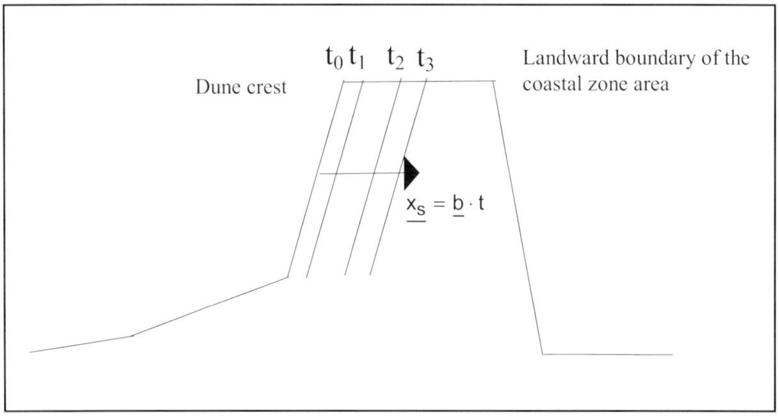

Fig. 3. Structural erosion in the coastal zone

Severe storm surges cause incidental erosion. Incidental erosion is a rapid process, with significant sand loss in the top part of the cross section occurring within a short period of time. Sand that has been removed from the dunes is deposited on the foreshore, [see the diagrammatic representation of incidental erosion in Figure 4 (TAW, 1995)].

To calculate the extent of incidental erosion, dynamic equilibrium is assumed at the annual scale. This means that the dune retreat in a certain year is assumed to be restored the same year in a natural manner. The cumulative effects of coastal recession over several years due to incidental erosion is equal to the annual contribution of dune retreat to coastal recession. Dune retreat is therefore a temporal condition. The deposited sand will be transported back to the dunes in a natural way. Natural recovery takes place under calm weather conditions.

The Bruun rule of erosion is an important factor in calculations involving future dune retreat (Bruun, 1983). It is a rule relating to the long-term effects of onshore/offshore

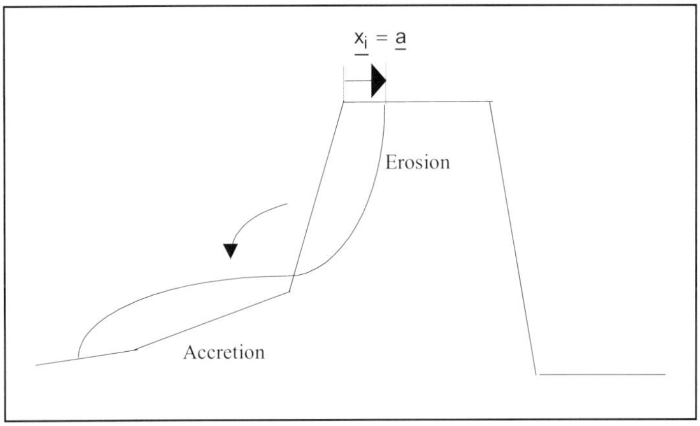

Fig. 4. Schematisation of Incidental erosion in the coastal zone

movement of sediment caused by future rises in the sea level, and includes an interpretation of the beach profile in which distance s following a sea level rise a results in shore erosion and deposition of sediments. The changing cross-section profile will have an impact on the dune retreat computations.

$$\underline{x}_i = \underline{a} \tag{2}$$

In which:

$\underline{x}_i$	incidental erosion in year t	[m]
$\underline{a}$	annual incidental erosion	[m]

Superposition of both structural erosion and dune retreat results in the total coastal recession (see Figure 5):

$$\underline{x}_{tot}(t) = \underline{x}_s(t) + \underline{x}_i = \underline{b} \cdot t + \underline{a} \tag{3}$$

2.6 COASTAL DEFENCE MEASURES

The purpose of coastal protection measures is to provide compensation for coastal recession that has already occurred or prevent structural recession. Incidental dune retreat will be restored in a natural manner. Coastal protection measures are limited and usually only involve sand nourishment (TAW, 1995). Sand nourishment has advantages over other coastal defence measures such as seawalls and groins:

- Sand nourishment is cheaper in comparison to other coastal protection schemes. This is mainly because sand is a relatively cheap material, despite the fact that sand nourishment is a temporary solution and has to be repeated on a regular basis.
- Sand nourishment is comparable with the natural character of the coast. The natural processes along the coast remain virtually undisturbed.

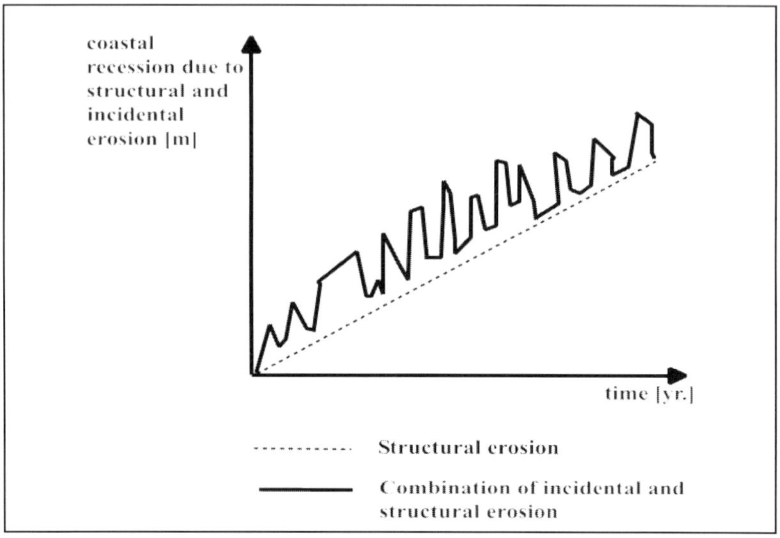

Fig. 5. Coastal recession: structural and incidental erosion

- Sand nourishment is a very flexible approach to combating coastal recession: the method can be utilised virtually everywhere.

The costs of beach nourishment per linear meter coastal stretch are divided into constant starting costs and variable costs per metre of coastal extension. If the return period is increased the sand nourishment volume will need to be increased, and average costs will decrease. The relative high starting costs will be absorbed by the larger sand nourishment volume. However, the probability of damage to societal activities in the coastal zone will increase. The relationship between sand nourishment costs and coastal extension is described by:

$$\underline{C_T} = \underline{c} + \underline{d} \cdot \underline{x_n} \tag{4}$$

in which:

$\underline{C_T}$	total sand nourishment costs	[$]
$\underline{c}$	constant starting costs	[$]
$\underline{d}$	variable costs per meter coastal extension	[$/m]
$\underline{x_n}$	coastal extension	[m]

Coastal extension is related to the erosion rate. Sand nourishment prevents and corrects structural coastal erosion. The range of coastal extension can be determined using the stochastic variable of structural erosion:

$$\underline{x_n} = \underline{b} \cdot T_R \tag{5}$$

in which:

$\underline{x_n}$	coastal extension	[m]
$\underline{b}$	structural erosion rate	[m/yr.]
T_R	return period sand nourishment	[yr.]

Therefore, the relationship between sand nourishment and structural erosion corresponds to the following:

$$\underline{C_T} = \underline{c} + \underline{d} \cdot \underline{b} \cdot T_R \tag{6}$$

2.7 ECONOMIC MODEL

The economic model calculates the values of several societal functions over the cross sections. The investment costs and the profits are included in these functions.

Various coastal types are dealt with in the model: urban coastal areas (coastal resorts focussing on recreation, other recreational zones, harbours and industry), areas of natural landscape and coastal areas where agriculture is the dominant function. The variation in the width of the cross sections needs to be taken into account in determining the damage functions. The direct economic impact of the damage needs to be calculated for all of the selected societal functions, with the exception of the impact on areas of natural landscape. For these areas, indirect factors have to be assessed: for example, the effect on recreation and drinking water supply. To do this, optimal conditions will firstly have to be ascertained, and on that basis, the effects on the various natural landscapes determined.

An example of a damage function is shown in Figure 6. The damage function shows the cumulative damage in the coastal zone due to coastal recession $\underline{x_{tot}}(t)$.

The damage to societal functions depends on the coastal recession measured from the dune crest. The coastal recession is a function of time. The damage function can be used to determine the damage as it contains the cumulative economic value for an arbitrary point at a distance x from the dune crest. The total damage at time t can be described as follows:

$$\underline{D} = f\left(\underline{x_{tot}}(t)\right) = f\left(\underline{b} \cdot t + \underline{a}\right), \tag{7}$$

in which:

$\underline{D}$	damage in the coastal area after year t	[$]
$f(\cdot)$	damage function	[$]
$\underline{x_{tot}}(t)$	total coastal recession after year t	[m]
$\underline{b}$	structural erosion after year t	[m/yr.]
$\underline{a}$	incidental erosion in year t	[m]
t	time	[yr.]

For the future value functions, various development possibilities can be investigated. The scenarios on which future value functions are based are purely hypothetical. The increase of capital value (inflation) or/and economic developments could be taken into account.

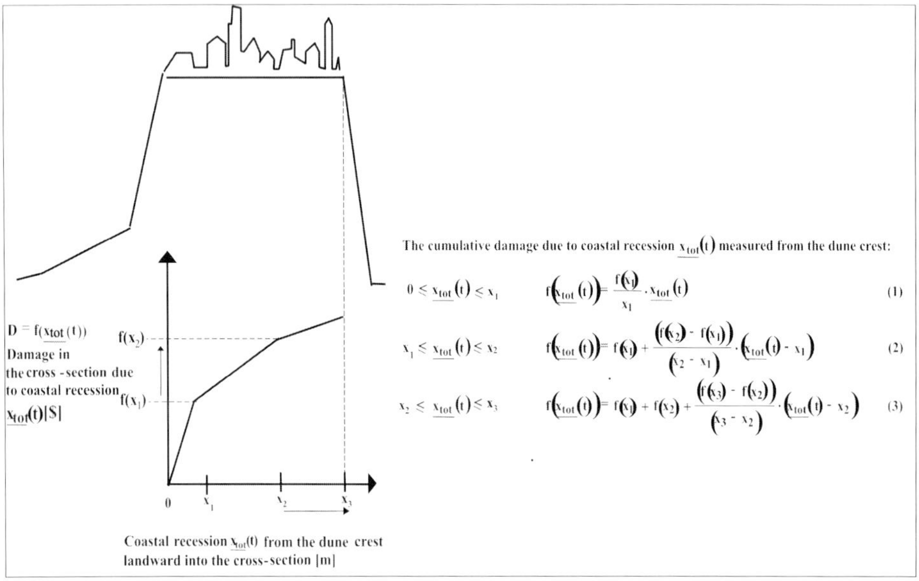

Fig. 6. Damage function

2.8 OPTIMISATION MODULE

The various model parts are linked in the optimisation module. The optimal situation is a function of nourishment costs and societal benefits. While nourishment costs and damage may occur at different points in time, their values can be made comparable by reduction to prices of a specific time. If no coastal defence measures are taken, the damage after the planning period corresponds to:

$$\underline{NPV_D} = \sum_{t=0}^{T} \left[\frac{\left(f\left(\underline{x_{tot}}(t)\right) - f\left(\underline{x_{tot}}(t-1)\right) \right)}{(1+r)^t} \right] \tag{8}$$

In which:

$\underline{NPV_D}$	net present value of the expected damage	[$]
$f(\cdot)$	damage function	[$]
$\underline{x_{tot}}(t)$	total coastal recession after year t	[m]
T	planning period	[yr.]
r	net interest rate	[-]
t	time	[yr.]

Postponing or cancelling coastal defence measures will have a financial impact on the societal functions in the coastal zone. However, taking coastal defence measures involves costs. The cumulative net costs of coastal defence measures and damage after the planning period will depend on the return period of sand nourishment. The first sand nourishment takes place after the first return period.

$$\underline{NPV}_{D+C_T} = \sum_{t=1}^{T_R}\left[\frac{\left(f\left(\underline{x}_{tot}(t)\right) - f\left(\underline{x}_{tot}(t-1)\right)\right)}{(1+r)^t}\right] + \sum_{k=1}^{n_S}\left[\frac{\underline{C}_T(T_R)}{(1+r)^{k\cdot T_R}}\right] \tag{9}$$

In which:

$\underline{NPV}_{D+C_T}$	net present value of the total costs	[$]
$\underline{C}_T(T_R)$	sand nourishment costs	[$]
$f(x_{tot}(t))$	damage function	[$]
$\underline{x}_{tot}(t)$	total coastal recession after year t	[m]
T	planning period	[yr.]
T_R	return period sand nourishment	[yr.]
r	net interest rate	[-]
t	time	[yr.]
n_s	amount of sand nourishment in planning period T: $n_s = \frac{T}{T_R}$	[-]
k	k^{th} sand nourishment from k=1 to n_s	[-]

The total costs of sand nourishment and damage in the coastal zone can be minimised by systematically changing the return period of sand nourishment.

$$\begin{aligned} &\underset{T_R}{MIN}\left\{\underline{NPV}_{D+C_T}\right\} = \\ &= \underset{T_R}{MIN}\left\{\sum_{t=1}^{T_R}\left[\frac{\left(f\left(\underline{x}_{tot}(t)\right) - f\left(\underline{x}_{tot}(t-1)\right)\right)}{(1+r)^t}\right] + \sum_{k=1}^{n_S}\left[\frac{\underline{C}_T(T_R)}{(1+r)^{k\cdot T_R}}\right]\right\} \end{aligned} \tag{10}$$

3. Coastal zone management in the Netherlands

3.1 THE DUTCH COAST

The Dutch coast has a total length of 432 km of which the straight coast from Hook of Holland to Den Helder covers 118 km. The main part of the Dutch North Sea coast is made up of dunes which vary in width from less than a hundred metres up to a few kilometres. This flexible dune coast together with the beach and the coastline protect the western part of the Netherlands against flooding. Failure to take coastal protection measures would have a disastrous impact on the Netherlands.

Along the coast, the forces of nature are causing constant moving of this dune coast barrier and put societal activities and the ecology in the coastal zone under threat. Erosive coasts and coasts with accretion alternate along the coast. From the Hook of Holland to Den Helder, there is quite a lot of variation within both the dynamic characteristics and the coast area types. Alternately urban coastal areas and coasts with nature reserves are situated along the coast. A responsible coastal policy and coastal management must be based on knowledge of the dynamics of the coast and the consequences of spatial decisions.

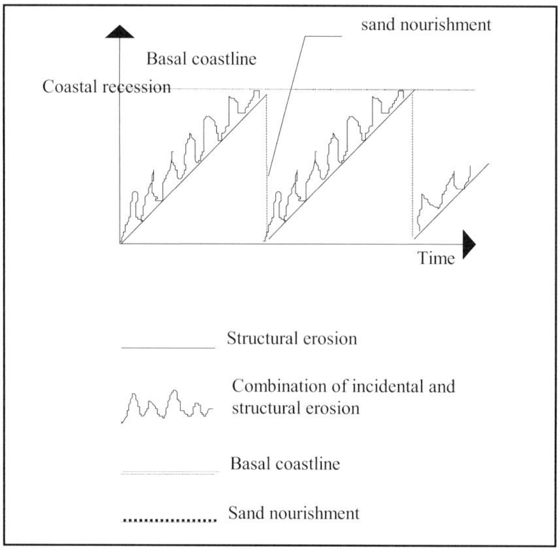

Fig. 7. Basal coastline policy

3.2 COASTAL ZONE POLICY IN THE NETHERLANDS

Dynamic preservation and natural processes play an important role in current Dutch policy. However, if the entire coast were left to natural processes, the safety of the polders behind the dunes would be endangered by erosion. Consequently, a policy of preserving the coastline at its 1990 position in order to stop ongoing structural coastal recession has been adopted. This reference point is known as the basal coastline. If the coastline recedes across the basal coastline, preventive measures have to be taken (see Figure 7). The main instrument in this dynamic preservation policy is sand nourishment (Ebbing, 1996).

Besides protection against flooding, integrated coastal zone management is aimed at sustainable preservation of functions in the dunes area. Attention is focused on maintaining and strengthening the position of the functions in the coastal zone.

The Dutch coast is subject to ongoing changes. Given the expected physical and societal changes, the question arises of whether it is still possible and economically attractive to try to reconciliate coastal defence and the increasing spatial claims of other interests and values under the basal coastline policy. One of the effects of the basal coastline policy is that societal activities in the coastal zone are tending to develop seaward up to the designated basal coastline. Moreover, physical changes will claim additional space along the coastal zone.

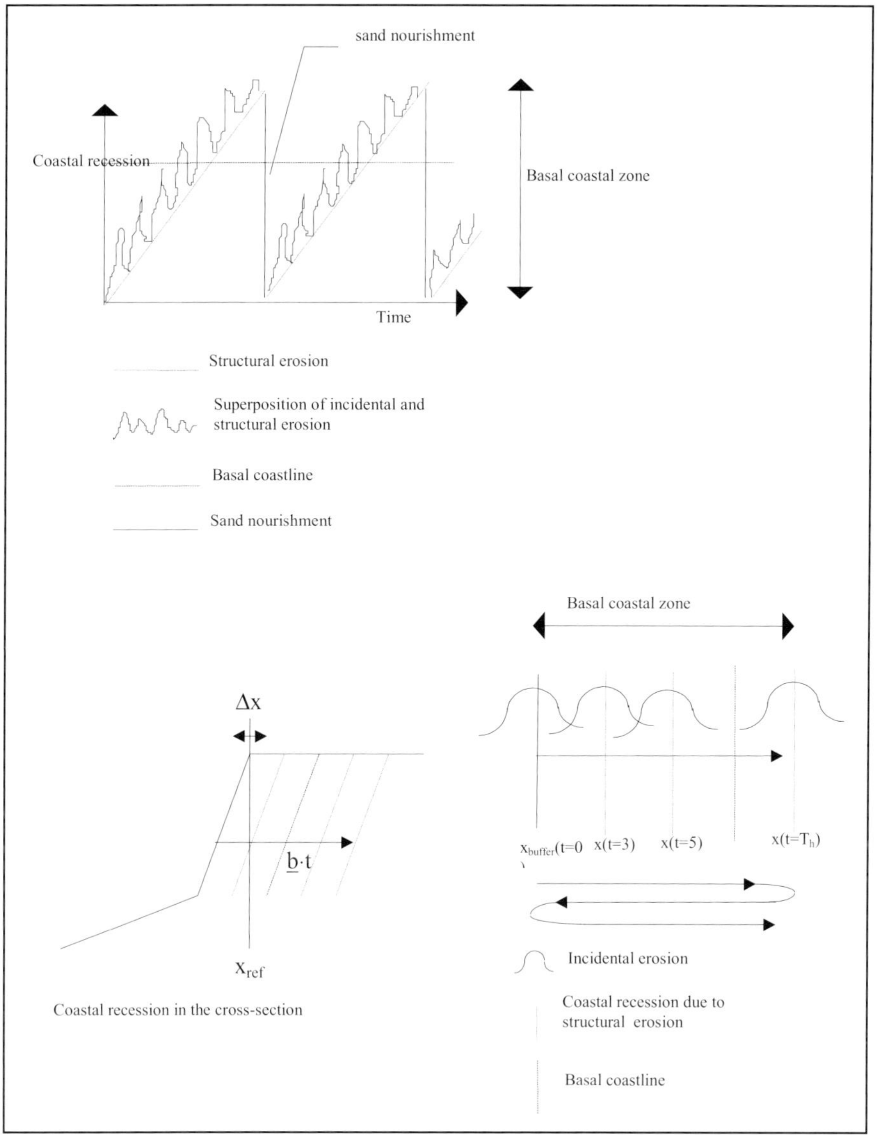

Fig. 8. Basal coastal zone policy — landward strategy

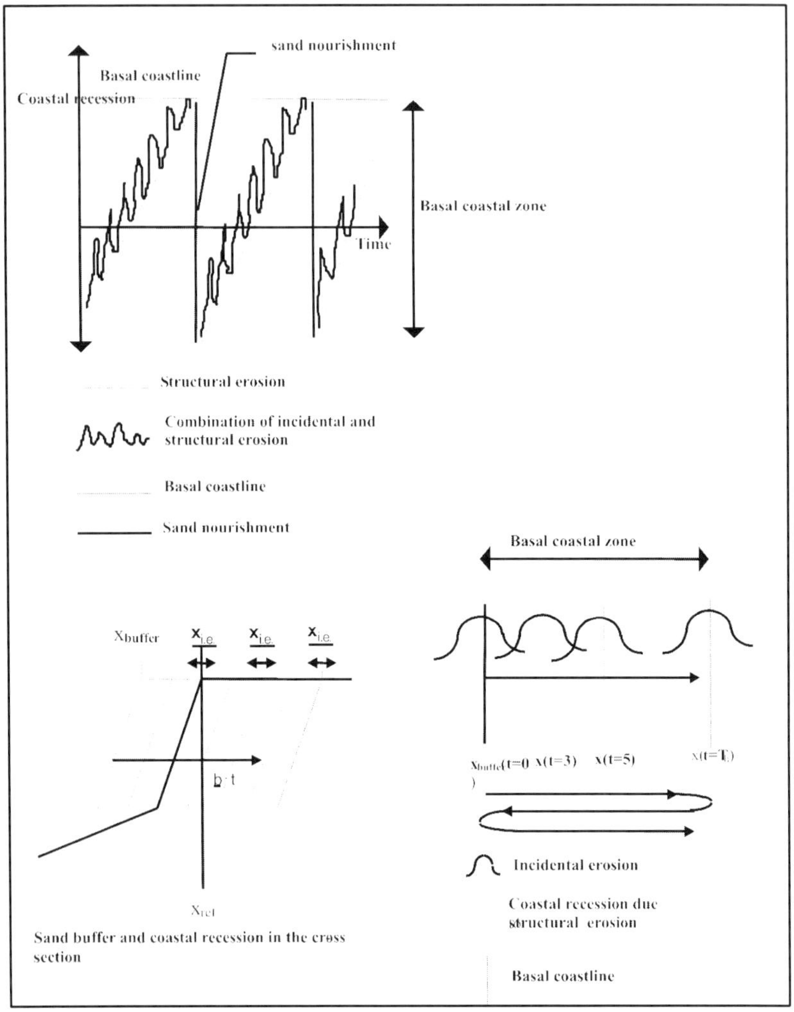

Fig. 9. Basal coastal zone policy — seaward strategy

Experience shows that the present coastal policy has tended to been interpreted as less flexible than desired. Relaxing the basal coastline policy and permitting the coastline to move within a designated zone has much to recommend it. Two other coastal policy options have been drawn up (see Figure 8 and Figure 9): a coastal zone policy with a landward strategy and a coastal zone policy with a seaward strategy. A coastal zone policy with a landward strategy would allow the coastline to recede across the basal coastline in a landward direction. A seaward sand buffer would be created under the coastal zone policy option with a seaward strategy. The economic and ecological consequences of both new policy options and the current basal coastline policy are analysed below using the stochastic optimisation model.

4. Stochastic optimisation model applied to the Dutch coast

There are a great many types of coasts throughout the world, particularly as far as their dynamic behaviour is concerned. Consequently, the stochastic optimisation model might need to be adapted to suit each specified coast it would be applied to. The necessary adaptations to make the stochastic optimisation model applicable to the Dutch coast are explained below.

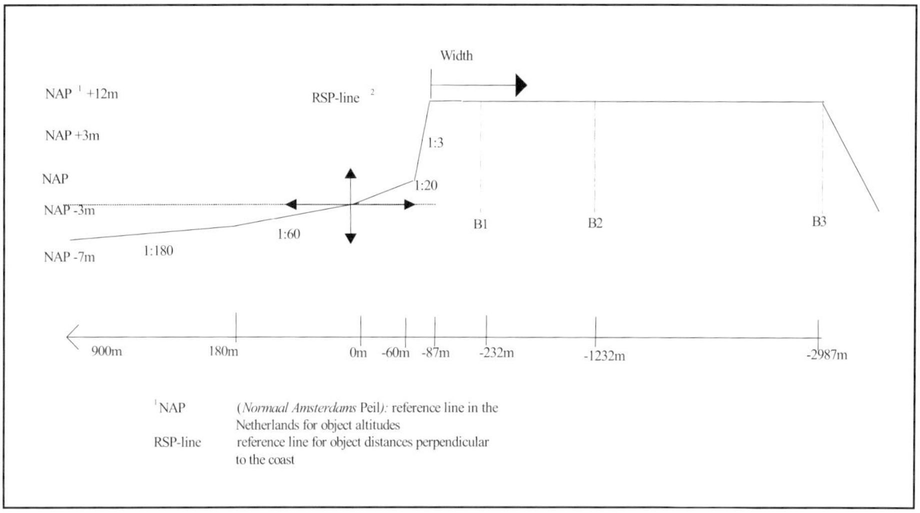

Fig. 10. Representative cross-section: dune area varying in width from B1 till B3

4.1 LIMITATIONS AND ASSUMPTIONS

The Dutch coastal zone is home to a great many human activities. The sea has always exerted an attraction because of its beauty and its economic advantages. Only a limited number of the coastline's functions will be considered here. Those included in the optimisation model as particularly relevant to the Dutch coast are safety, natural landscape, recreation and tourism, residence, industry and water catchments (drinking water supply).

The optimisation is performed at two points in time, the present (2000) and the future (the year 2100).

Many different combinations of hydraulic and morphological characteristics and economic activities are to be found in the coastal zone along the Dutch coast. Representative morphological classes, storm conditions, cross-sections and economic distribution functions show for the varying situations along the Dutch coast. A representative cross-section is shown in Figure 10.

4.2 BOUNDARY CONDITIONS

In addition to the boundary condition in section 2.4 a second restriction has been imposed on the coastal zone. The Dutch coastal policy is aimed at the safety of the low-lying western part of the Netherlands. Following the 1953 disaster, legally imposed safety constraints were introduced. These safety constraints impose an additional boundary condition on the optimisation model.

According to these safety constraints the probability of failure equals 10^{-5}. That means that once in 100 000 years failure may be expected. Under the safety constraints, a limiting zone additional to the 10^{-5}-dune retreat zone must be present in the coastal zone at all times (see Figure 11). This constraint limits the space for dynamics. To meet this constraint, the distance over which the coastline is allowed to recede was calculated and included as a boundary condition in the model (DWW, 1996).

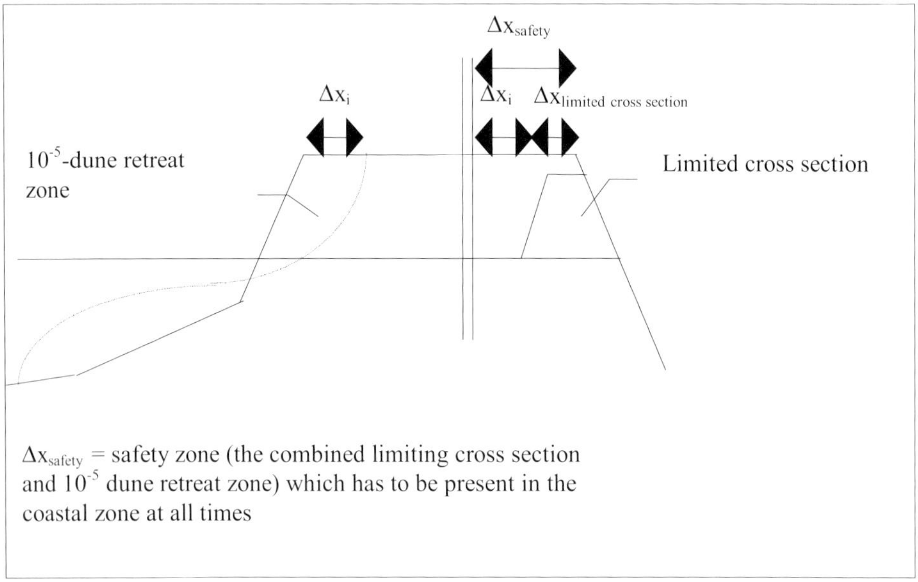

Fig. 11. Safety conditions

4.3 HYDRAULIC MORPHOLOGICAL MODEL

Structural erosion was determined using annual coastline measurement data. The annual erosion or accretion trend was determined by a linear trend analysis (Bolle et al., 1983). There was a normal distribution trend for structural erosion or accretion.

Incidental erosion was calculated for three storm climates along the Dutch coast: Those at Den Helder, IJmuiden and Hook of Holland. A dune-retreat program called Super Dune executed retreat calculations. The calculations were repeated for different storm surge levels and corresponding significant wave heights. The relationship between the dune retreat calculations and the storm surge levels was determined with a curve fit-program. Using this relationship and the probability distribution of the storm surge level, the probability distribution of dune retreat was calculated with a Monte Carlo simulation. The dune retreat was Gumbel-distributed.

Structural and incidental erosion were also predicted for the future (2100). Climate changes might result into a(n) (accelerated) sea level rise and intensification of storms. This will influence the morphological processes. It is possible to have a long-term perspective and to outline a recommendable long-term policy in the future. Table 1, 2 and Table 3 show the values of structural erosion and incidental erosion in the present and in the future.

Table 1. Trend of structural erosion in the present and in the future along the Dutch coast

Coastal area stretches	Trend [m/yr.]	Trend [m/yr.] 100+	Coastal area stretches	Trend [m/yr.]	Trend [m/yr.] 100+
Den Helder	-5.1	-15.2	Zandvoort	-1.1	-3.4
Callantsoog	-0.4	-1.3	Zandvoort Z.	0.7	2.0
Petten	-1.4	-4.1	Noordwijk	0.5	1.6
Bergen a. Zee	-1.4	-4.1	Katwijk	0.3	1.0
Egmond a. Zee	0.9	2.7	Scheveningen. N.	-0.7	-2.1
Wijk aan Zee	-0.8	-2.5	Scheveningen	0.2	0.5
IJmuiden N.	2.8	8.4	Ter Heijde	-0.5	-1.6
IJmuiden Z.	7.4	22.3			

Table 2. Parameters Gumbel-distribution function for incidental erosion at the present time (2000) along the Dutch coast

Gumbel-distribution function, $F(x_i) = e^{-e^{-\frac{x_i - A}{B}}}$, with parameters:

Location	A	B
Den Helder	-15.2	6.5
IJmuiden	-17.7	6.2
Hook van Holland	-19.9	6.1

Table 3. Parameters Gumbel-distribution function for incidental erosion in the future along the Dutch coast

Gumbel-distribution function, $F(x_i) = e^{-e^{-\frac{x_i - A}{B}}}$, with parameters:

Location	A	B
Den Helder	-15.21	7.03
IJmuiden	-17.74	6.71
Hook van Holland	-19.82	6.55

The total coastal recession equalled the combined effects of structural erosion and dune retreat, see equation (3). In case a sand buffer is to be created on the basis of the coastal zone policy with the seaward strategy, then equation (3) has to be modified as follows:

$$\underline{x_{tot}}(t) = -x_{buffer} + \underline{b} \cdot t + \underline{a} \tag{11}$$

4.4 COASTAL DEFENCE MEASURES

The relationship between the costs of beach nourishment per linear metre of coastal stretch and coastal extension due to sand nourishment is given in equation (6). Once information on nourishment cost is available then the constant starting costs and the variable costs are described as triangle distribution functions (See Table 4 and Table 5).

Table 4. Present sand nourishment costs: parameters triangle distribution function

The present	Parameter A – most optimistic cost estimation	Parameter B– most realistic cost estimation	Parameter C – most pessimistic cost estimation	Average μ_K	Standard deviation σ
Constant costs	55	280	505	280	92
Variable costs	66	75	84	75	4

The average and the standard deviation are determined as follows:

$$\mu_K(K) = \frac{A+B+C}{3} \tag{1}$$

$$\sigma^2_K(K) = \frac{1}{18} \cdot \left(A^2 + B^2 + C^2 - A \cdot B - A \cdot C - B \cdot C\right) \tag{2}$$

Table 5. Future sand nourishment costs: parameters triangle distribution function

The future	Parameter A – most optimistic cost estimation	Parameter B– most realistic cost estimation	Parameter C – most pessimistic cost estimation	Average μ_K	Standard deviation σ
Constant costs	1106	5628	10150	5628	1846
Variable costs	1327	1508	1688	1508	74

The costs of sand nourishment measures will increase in time as a function of the growth rate of capital:

$$\frac{dC}{dt} = r \cdot C \tag{12}$$

$$C(t) = C_0 \cdot e^{r \cdot t} \tag{13}$$

In which:

$C(t)$:	costs at t	[$]
C_0:	present sand nourishment costs	[$]
r:	real interest rate, growth rate of capital	[%]

Taking a growth rate of 3% into account, the sand nourishment costs in the future (2100) will increase by a factor 20.

4.5 ECONOMIC MODEL

Economic value functions were determined in respect of the cross sections of the Dutch coast. To do this, a distribution based on an inventory along the Dutch coast had to be made between urban coastal areas and coastal areas with nature reserves. Urban coastal areas were subdivided into three types: coastal resorts with small-scale recreation, coastal resorts with large-scale recreation and areas with large-scale recreation, harbours and industry.

Nature type varies over the dune cross-section (see Figure 12). Nature characteristic of dynamic coastal areas appears in the seaward region of the dune cross-section and nature whose genesis was hundreds of years ago appears in the landward region of the dune cross-section.

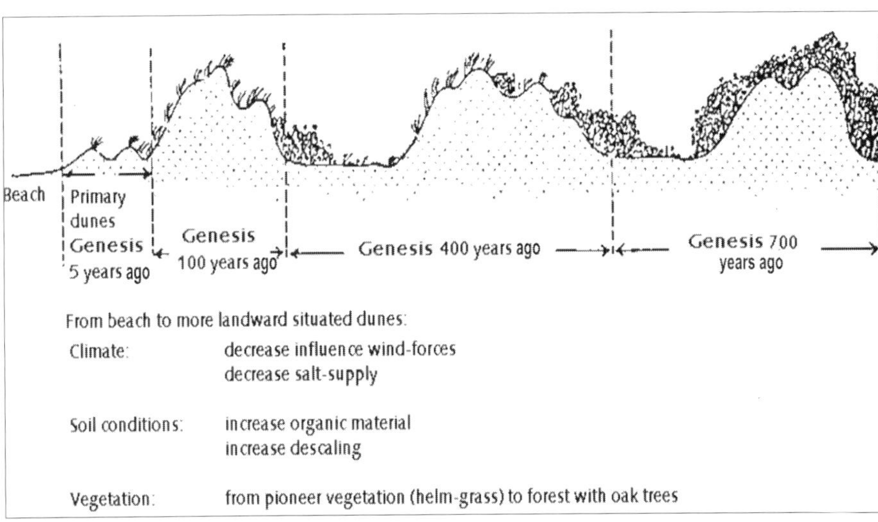

Fig. 12. Nature types over the cross-section

The assumption was made that nature in coastal areas with nature reserves coincides with the functions of extensive recreation and drinking water supply. The economic value of extensive recreation and drinking water supply was taken into account, though the economic value of nature was not included. The economic value of coastal areas with natural landscape is relatively low. After an optimal situation was determined, the effects on the various different types of nature were considered.

The economic value per function in the coastal area was based on standard economic values per unity or square measure for each given function.

There is no doubt that changes in spatial claims and the economic value of societal functions will occur in the future. For the Netherlands, two scenarios for societal changes were drawn. The economic value increases with a factor 20 (see section 4.4) for both scenarios. Based on the development of societal functions in the present Dutch policy being abandoned, the first scenario considers a situation in which there is an increase in capital value and no more economical development takes place than in the present situation. The second scenario considers a situation in which both inflation and economical developments occur. In fact, many different scenarios for future societal development in the coastal zone can be imagined.

4.6 OPTIMISATION MODULE

The optimisation module links the various parts of the model.

With the Dutch coastal situation, the comments made previously in respect of optimisation need to be qualified. When a sand buffer is created the net present value of coastal defence measures costs and damage after the planning period is given by equation (9). However, the index k in the second term starts at zero instead of one and stops at n_s-1 instead of n_s. The first sand nourishment is the creation of the sand buffer.

5. Optimisation

Because of the considerable variety the Dutch coast exhibits, a choice of representative situations had to be made. The following were selected for the optimisation analysis:

- a narrow and a broad urban coastal area
- a broad coastal area with nature reserves

The stochastic optimisation model gave an insight into the economic and ecological consequences of dynamic preservation of the basal coastline and the two basal coastal zone policies.

5.1 URBAN COASTAL AREA

The optimisation results show that in urban coastal areas, relaxation of the basal coastline policy and permitting the coastline to move within a designated zone both in the present and in the future has much to recommend it. This zone should be created by a seaward sand buffer. In order to allow for future circumstances, the buffer should be made larger than it currently needs to be. A situation in which the coastline is allowed to recede across the basal coastline in the landward direction has to be avoided. If this were to happen, the damage would be greater due to the high economic value of urban coastal areas.

The optimisation model does not allow for any additional profits derived from future economic developments, and consequently the question of whether it is more attractive from an economic point of view to allow further economic development in the coastal zone or to maintain the policy of prohibiting economic development cannot be answered.

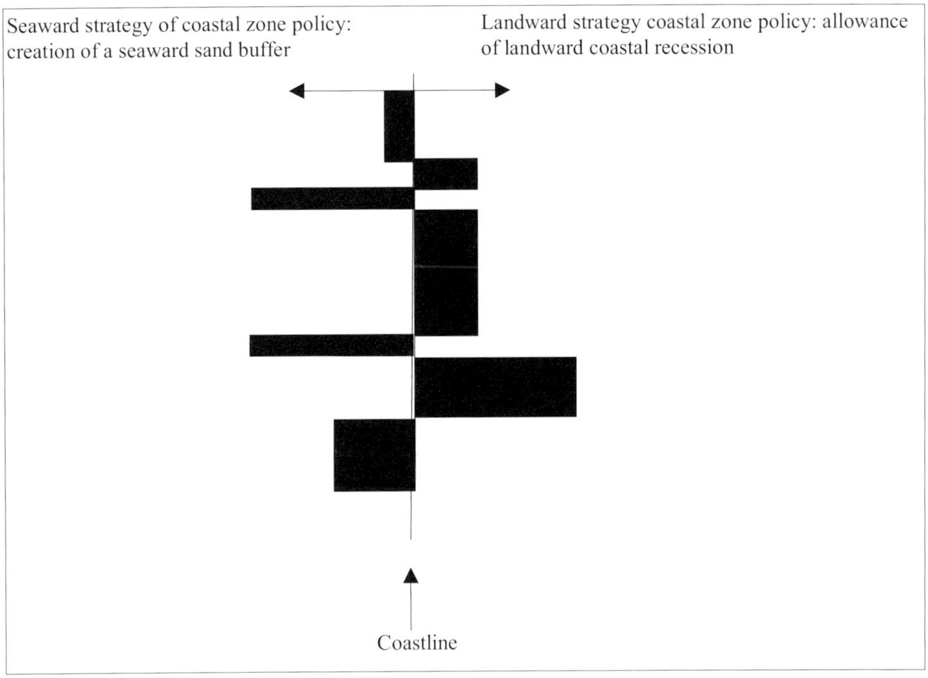

Fig. 13. Results

5.2 COASTAL AREA WITH NATURE RESERVES

Since these areas possess little economic value, consideration could be given to permitting the coastline along such areas to move within a broad designated zone landward. The damage due to coastal recession would be relatively low, and the nourishment costs to maintain the coast at its position would be high in comparison with the damage. A coastal zone policy with a landward strategy is the cheapest maintenance option.

The zone over which the coast is permitted to move landward should be broad (50–110 metres). As the type of nature varies over the cross-section, it might be influenced by coastal recession might influence the type of nature. Where there is extreme coastal recession natural landscape that have evolved over hundreds of years will disappear and be replaced by 'dynamic' landscapes. Although economic considerations are relatively unimportant in such zones, it will be important to ensure that there is broad societal acceptance of the increase of a certain type of nature being at the expense of the overall disappearance of another type of nature. The economic optimisation model cannot be indiscriminately applied in coastal areas dominated by nature reserves and landscape.

5.3 FEEDBACK TO DUTCH COAST: HOOK OF HOLLAND TO DEN HELDER

To guarantee safety and liveability in the Netherlands both for the current and future generations, it is important that there be a long-term strategy for coastal protection against flooding. Anticipation of future changes is essential. The optimisation model is able to estimate the effects of the various policy options as well as to gauge the likelihood of the anticipated changes actually occurring and to calculate the economic consequences of decisions.

In both urban coastal areas and coastal areas with nature reserves it is advisable to relax the basal coastline policy and to introduce a coastal zone policy. In urban coastal areas, the designated zone will need to be located on the seaward side, while in coastal areas with nature reserves, the coast should be allowed to recede landwards. Due to the fact that urban coastal areas and coastal areas with nature reserves alternate along the Dutch coast there will be a variety of optimal situations. The frequent alternation of coastline zones is likely to cause sand losses along those coastal areas which have been regarded as needing to be maintained at their current position. Because of the difficulty of maintaining this position, the chosen solution will have to be one in which allowance is made for the creation of transitional zones.

Societal acceptance of the changes proposed by the optimisation model will also have to be sought. Decisions should not only be based upon cost-benefit analyses. Moreover, since opinions and ideas change during the course of time, new options may arise.

6. Discussion and recommendations

6.1 SIGNIFICANCE OF THE OPTIMISATION MODEL FOR COASTAL AREAS WORLD-WIDE

The optimisation model is a useful instrument for analysing the relationship between societal activities and coastal dynamics and for calculating the economic and ecological consequences of spatial decisions relating to various policy maintenance options for sandy coasts throughout the world.

The stochastic model provides answers to questions such as the following:

- In an arbitrary sandy coast anywhere in the world, what is the most appropriate coastal zone policy from both an economic and an ecological point of view?
- How to deal with the pressure upon coasts and the increase of spatial claims due to both future physical changes and societal changes?
- How to guarantee the future safety of the lowlands behind the coastal dunes while at the other hand deriving maximum profit from the economic opportunities offered by the coast and the sea?
- Is it advisable to adapt the present coastal policy in anticipation of future changes?

6.2 THE MODEL'S SHORTCOMINGS OF THE MODEL AND RECOMMENDATIONS

While this model is able to provide some insight into the nature of the interactions between physical processes and societal activities in the coastal zone and is able to estimate the consequences of future changes, one of its main shortcomings is that it primarily focuses on cost benefits. However, such an approach is of limited applicability for coastal areas with natural landscape and consequently societal acceptance of the change proposed by the model will have to be sought. This also applies where the model recommends extreme coastal recession. Lack of understanding and fears about the safety of the coasts are likely to be the result, and would be quite understandable. It would also be welcome, since it may well lead to increasing awareness of not only the likely effects of policies and proposed civil engineering works, but also of the unforeseen effects (Dubbeldam, 2000). How to cope with societal preferences in relation to proposed changes is beyond the scope of this paper and could form the basis for further research.

The model could be improved in the following ways:

- Because the present optimisation model is a sequential one, it only takes current input variables into account. It is important that as these variables change during the course of time, they be included as new input.
- An additional module in which the additional benefits of economic development alternatives are included is necessary, since it would make it possible to evaluate the economic consequences of future economic development strategies.

Acknowledgements

Valuable contributions and comments have been made by H.J de Vriend and C.M. Dohmen-Janssen. The National Institute for Coastal and Marine Management in the Netherlands provided useful comments and information relating to the structural erosion computations.

REFERENCES

Bolle, E.A.W., Lenoir, J.M.H., Loon, J.N.M. van: 'Wiskundige statistiek', second edition with changes, Deventer, The Netherlands: Van Loghum Slaterus, 1983, pp 169–179 (in Dutch).

Bruun, P.: 'The Bruun Rule of Erosion by Sea-level Rise: a discussion on large-scale Two- and Three-Dimensional Usages', *Journal of Coastal Research*, Vol. 4, No. 4, 1988, pp 627–648.

Bruun, P.: 'Review of conditions for uses of the Bruun Rule of erosion'. *Coastal Engineering*, Vol. 7, 1993, pp 77–89.

Dubbeldam, H.: *Maatschappelijke golven in de waterbouwkunde*, Delft: Delft University Press, 1999, (in Dutch).

DWW, Dienst Weg- en Waterbouwkunde 'Hydraulische randvoorwaarden voor Primaire Waterkering'. Ministerie Verkeer en waterstaat, Directoraat-Generaal Rijkswaterstaat, Delft, 1996, (in Dutch).

Ebbing, H., Heuvel, T. van, & Kruik, H. de.: 'Coastline management, from coastal monitoring to sand nourishment', Ministry of Transport; Public Works and Water Management, Directorate-General of Public Works and Water Management, National Institute for Coastal and Marine Management/RIKZ, second edition, The Hague, 1996.

Peerbolte, E.B. & Wind, H.G.: 'Coastal zone management; tools for initial design', Proceedings of the International Coastal Congress, Kiel, 1992, pp 332–34.

Rijn, L.C. van: 'Principles of coastal morphology', University of Utrecht, Department Physical Geography, 1998.

TAW (Technical advisory committee on water defences): *Guide to the assessment of the safety of dunes as a sea defence*, The Hague, 1989.

TAW (Technical advisory committee on water defences): Basisrapport Zandige Kust, Bijbehorende bij de Leidraad Zandige Kust. (Report on sandy coasts, part of the guide on the assessment of the safety of dunes as a sea defence), Delft, 1995, (in Dutch).

Vrouwenvelder, A.: 'Memorandum, 2000-CON-DYN/M2020, Kansen, onzekerheden en de presentatie daarvan', Sprintproject, TAW-V-vergadering, TNO, Delft, 2000.

SHORELINE CHANGES INDUCED BY A SUBMERGED GROIN SYSTEM

TH. V. KARAMBAS,[1,2] CH. KOUTITAS,[2] S. CHRISTOPOULOS[1]

[1]*HYDROMARE, Ethn. Antistaseos 3A, Kalamaria, Thessaloniki, 551 34, GREECE*
[2]*ARISTOTLE UNIVERSITY OF THESSALONIKI, Dept. of Civil Engineering, Division of Hydraulics and Environmental Engineering, Thessaloniki, 540 06, GREECE*

Abstract

In the present work, a theoretical investigation of the realization of coastal processes in presence of a system of submerged groins used for coastal erosion control, is presented. The approach is based on three numerical models: a linear wave propagation model, a wave-induced circulation and sediment transport model, and a one-line model for the prediction of shoreline evolution. The wave model is based on the hyperbolic-type mild -slope equation and is valid for a compound wave field near coastal structures where the waves are subjected to the combined effects of shoaling, refraction, diffraction, reflection (total and partial) and breaking. The estimated radiation stress components drive the depth-averaged circulation model, which describes the nearshore currents and sediment transport in the surf zone. The model is coupled with a one-line model to provide the shoreline changes. Some important notions concerning the restructuring of the known longshore current in the breaker zone become evident. Circulation cells bifurcate from the mainstream of the longshore current, causing it to transport sediment inshore and trap a proportion of it inshore, contributing to shore accretion.

1. Introduction

Modification of the wave field in the inshore zone in the presence of a pattern of submerged structures, such as a system of submerged breakwaters or groins, by induced refraction, diffraction, breaking of the waves and near-bottom turbulence, and the consequent restructuring of the coastal currents patterns and the sediment transport, are some of the processes of physical interaction integrally responsible for coastal morphology evolution.

Submerged groins as a means for coastal accretion and erosion control constitute a technological advancement due to the method of their construction, their effectiveness, their aesthetic integration with the environment and their low cost. From the scientific point of view, little is known of how they act on the existing wave climate and sediment transport mechanisms. The present study deals with a theoretical investigation into the effects of coastal processes in the presence of such structures, and provides a quantitative description of wave-induced currents and sediment transport patterns. The final goal is to rationalize the range of operability and the optimal layout of such structures in terms of beach geometry, sediment properties and the wave climate of a coast.

Many models exist for the evaluation of wave deformation in the coastal region but most of them are based on the progressive wave assumption (period-averaged refraction wave models) and employ elliptic or parabolic-type differential equations which are in general difficult to numerically solve. Moreover, they are not valid for a

C. L. Goudas et al. (eds.), Soft Shore Protection, 105-114.

compound wave field near coastal structures where the waves give rise to the combined effects of shoaling, refraction, diffraction, reflection and breaking.

In the present work, a wave model based on the hyperbolic-type mild-slope equation valid for a compound wave, is used. The incorporation of breaking and the evaluation of the radiation stress, the model drives the depth-averaged circulation and sediment transport model which describes the nearshore currents and beach deformation. In the case of the submerged groins, the above models are applied to describe the coastal processes in presence of such structures.

2. Wave model

The breaking and non-breaking wave model is based on the hyperbolic-type mild-slope equation without using the progressive wave assumption. The model consists of the following pair of equations (Copeland 1985a, Karambas 1999):

$$\frac{\partial \zeta_w}{\partial t} + \frac{c}{c_g} \nabla \frac{c_g}{c} \mathbf{Q}_w = 0,$$

$$\frac{\partial \mathbf{U}_w}{\partial t} + \frac{c^2}{d} \nabla \zeta_w = 0, \tag{1}$$

where ζ_w is the surface elevation, U_w the wave depth-averaged velocity vector $\mathbf{U}_w$ $=(U_w, V_w)$, d the depth, $\mathbf{Q}_w = \mathbf{U}_w h_w = (Q_w, P_w)$, h_w the total depth ($h_w = d + \zeta_w$), c the celerity and c_g the group velocity.

The above equations, derived by Copeland (1985a), are able to compute the combined effects of wave refraction, diffraction and total or partial reflection.

The model is extended to the surf zone in order to include the breaking effects which provide the equations with a suitable dissipation mechanism by introducing a dispersion term in the right-hand side of momentum eqn (1):

$$\nu_h = \nabla^2 \mathbf{U}_w, \tag{2}$$

where ν_h is an horizontal eddy viscosity coefficient estimated from Battjes (1975):

$$\nu_h = 2d\left(\frac{D}{\rho}\right)^{1/3}, \tag{3}$$

in which ρ is the water density and D is the energy dissipation given by Battjes & Janssen (1978):

$$D = D = \frac{1}{4} Q_b f \rho g H_m^2, \tag{4}$$

with f the mean frequency, H_m the maximum possible wave height and Q_b the probability that at a given point, the wave height is associated with the a breaking or broken wave (Battjes & Janssen, 1978).

3. Wave-induced circulation model

Taking the horizontal axes x_1 and x_2 on the still water surface, and the z axis upward from the surface, the definition of the radiation stress S_{ij} component is:

$$S_{ij} = < \int_{-d}^{\zeta_w} \left(p\delta_{ij} + \rho u_i u_j\right) dz > -0.5\,\rho g\left(d + < \zeta >\right)^2 \delta_{ij}, \tag{5}$$

where δ_{ij} is the Kroneker's delta, $u_i(z)$ is the wave horizontal velocity component in direction x_i , ζ is the mean sea level, p the pressure and < > denotes a time average quantity.

The total pressure p is obtained from the vertical momentum equation:

$$p = \rho g(\zeta - z) - \rho u_3^{\,2} + \frac{\partial}{\partial x_1}\int_z^0 \rho u_1 u_3 dz + \frac{\partial}{\partial x_2}\int_z^0 \rho u_2 u_3 dz + \frac{\partial}{\partial t}\int_z^{\zeta} \rho u_3 dz, \tag{6}$$

where u_3 is the z-velocity component.

Based on the above eqn (6) and after the substitution of u_i and p, from model results (eqn 1) using linear wave theory, Copeland (1985b) derived the expressions for S_{ij} without the assumption of progressive waves. Those expressions are used in the present model.

The radiation stresses are the driving forces of a 2D horizontal wave-induced current model:

$$\begin{aligned}
&\frac{\partial \zeta}{\partial t} + \frac{\partial (Uh)}{\partial x} + \frac{\partial (Vh)}{\partial y} = 0,\\
&\frac{\partial U}{\partial t} + U\frac{\partial U}{\partial x} + V\frac{\partial U}{\partial y} + g\frac{\partial \zeta}{\partial x} = -\frac{1}{\rho h}\left(\frac{\partial S_{xx}}{\partial x} + \frac{\partial S_{xy}}{\partial y}\right) + \frac{1}{h}\frac{\partial}{\partial x}\left(\nu_h h\frac{\partial U}{\partial x}\right)\\
&\qquad + \frac{1}{h}\frac{\partial}{\partial y}\left(\nu_h h\frac{\partial U}{\partial y}\right) - \frac{\tau_{bx}}{\rho h},\\
&\frac{\partial V}{\partial t} + U\frac{\partial V}{\partial x} + V\frac{\partial V}{\partial y} + g\frac{\partial \zeta}{\partial y} = -\frac{1}{\rho h}\left(\frac{\partial S_{xy}}{\partial x} + \frac{\partial S_{yy}}{\partial y}\right) + \frac{1}{h}\frac{\partial}{\partial x}\left(\nu_h h\frac{\partial V}{\partial x}\right)\\
&\qquad + \frac{1}{h}\frac{\partial}{\partial y}\left(\nu_h h\frac{\partial V}{\partial y}\right) - \frac{\tau_{by}}{\rho h}
\end{aligned} \tag{7}$$

where h is the total depth h=d+ζ, U, V are the current horizontal velocities and τ_{bx}, τ_{by} are the bottom shear stresses.

In the current model, the treatment of the bottom stress is critical (all longshore current models employing radiation stress solve for the mean current velocity through its role in the bottom friction term). The general expression for the time-average bottom shear stress in the current model is written:

$$\tau_{bx} = \rho C_f < (U + u_b)\sqrt{(U + u_b)^2 + (V + v_b)^2} >$$

$$\tau_{bx} = \rho C_f < (V + v_b)\sqrt{(U + u_b)^2 + (V + v_b)^2} >, \tag{8}$$

where C_f is the friction coefficient which depends on the bottom roughness and at the orbital amplitude at the bed (Karambas, 1998), and u_b and v_b are the wave velocities at the bottom.

Inside the surf zone the existence of the undertow current that is directed offshore at the bottom cannot be predicted by a depth-averaged model. However, representing the cross-shore flow is essential for a realistic description of the sediment transport processes. The present model calculates the near-bottom undertow velocity V_b using the analytical expression proposed by Putrevu and Svendsen (1993):

$$\frac{V_b}{\sqrt{gh}} = \left(\frac{V_m}{\sqrt{gh}} - \frac{A}{6} + \frac{\tau_{sb} h}{2\rho v_{tz} \sqrt{gh}} \right) (1 + R)^{-1}, \tag{9}$$

where V_m is the mean undertow velocity, τ_{sb} is the steady streaming term, v_{tz} is the eddy viscosity coefficient, and A, R are coefficients given by Putrevu and Svendsen (1993).

The above expression (9) is also valid outside the surf zone, where the undertow profile has a small seaward-directed velocity at the bed.

4. Sediment transport in the surf zone

The prediction of the sediment transport is based on the energetics approach, in which the submerged weight transport rates, i_{xt} in the x direction and i_{yt} in the y direction, are given by Karambas (1998):

$$i_{xt} = < \left[\frac{\varepsilon_b}{\tan\phi} \left(\frac{u_o}{u_{ot}} + \frac{d_x}{\tan\phi} \right) \omega_b + \varepsilon_s \frac{u_{ot}}{w} \left(\frac{u_o}{u_{ot}} + \varepsilon_s d_x \frac{u_{ot}}{w} \right) \omega_t \right] >$$

$$i_{yt} = \left\langle \left[\frac{\varepsilon_b}{\tan\varphi} \left(\frac{v_o}{u_{ot}} + \frac{d_y}{\tan\phi} \right) \omega_b + \varepsilon_s \frac{u_{ot}}{w} \left(\frac{v_o}{u_{ot}} + \varepsilon_s d_y \frac{u_{ot}}{w} \right) \omega_t \right] \right\rangle \qquad (10)$$

where w is the sediment fall velocity, ϕ is the angle of internal friction, ε_b and ε_s are the bed and suspended load-efficiency factors respectively (ε_b =0.13, ε_s=0.01), $u_{ot}=\sqrt{u_o^2+v_o^2}$ (u_o, v_o are the total flow velocities at the bottom), d_x and d_y are the bottom slopes and ω_t is the total rate of energy dissipation given by Leont'yev (1996) incorporating both the power expenditures due to bed friction and excess turbulence penetrating into the bottom layer from breaking waves.

Since the present model is a linear one, the total flow velocity at the bottom is considered to be the sum of the steady U, V, U_b and the oscillatory u_b, v_b components which include two harmonics:

$$u_o = U + u_{bm}\cos(\omega t) + u_{b2m}\cos(2\omega t),$$

$$v_o = V + V_b + v_{bm}\cos(\omega t) + v_{b2m}\cos(2\omega t + a), \qquad (11)$$

in which ω is the wave frequency, a is the phase shift and u_{bm}, u_{b2m} , v_{bm} and v_{b2m} are the velocity amplitudes given by Leont'yev (1996, 1997).

5. One-line model

The nearshore morphological changes can be calculated by solving the conservation of sediment transport equation:

$$\frac{\partial d}{\partial t} = \frac{\partial q_x}{\partial x} + \frac{\partial q_y}{\partial y}, \qquad (12)$$

where d is the still water depth and q_x, q_y are the volumetric sediment transport rates, related to the immersed weight sediment transport through:

$$q_{x,y} = \frac{i_{x,y}}{(\rho_s - \rho)gN} \qquad (13)$$

in which N is the volume concentration of solids of the sediment (N= 0.6) and ρ_s and ρ are the sediment and fluid densities.

Under certain assumptions, eqn (12) can be transformed into a 1D equation (one-line model). Let us define the total longshore sediment transport Q and the mean (cross-shore) water depth $\bar{d}$ by the equations:

$$Q = \int_0^{y_s} q_x dy \ , \quad \bar{d} = \frac{1}{y_s}\int_0^{y_s} d\,dy, \qquad (14)$$

where y_s is the width of the nearshore zone.

Using the Leibnitz relation, the integration of eqn (12) over the width of the nearshore zone from its outer boundary (y=0) to the shoreline ($y=y_s$) leads to the following equation:

$$\frac{\partial\left(y_s\bar{d}\right)}{\partial t}=\frac{\partial Q}{\partial x}-q_x(y_s)\frac{\partial y_s}{\partial x}+q_y(y_s) \tag{15}$$

where we have supposed that the following conditions are valid: d=0 at the shoreline ($y=y_s$) and the transport rates $q_x(0)=0$, $q_y(0)=0$ at the outer boundary (y=0) are zero.

Eqn (15) differs from a standard one-line model in the last two terms. The second term of the right hand side of the equation is related to the longshore transport rate near the shoreline while the last term incorporates the cross-shore related seasonal shoreline variation.

Karambas (1999) used for the cross-shore transport rate near the shoreline $q_y(y_s)$ the formula proposed by Sunamura (Yamamoto et al., 1996). The formula is written:

$$q_y(y_s)=K\,U_r^{0.2}\,\Phi\,(\,\Phi-0.13\,Ur)\,w\,d_{50}\,, \tag{16}$$

where U_r is the Ursell parameter $U_r=gHT^2/h^2$ (H is the wave height and h is the wave set-up at shoreline), $\Phi=H^2/shd_{50}$ (s is the specific gravity of sediment) and K is a coefficient of the sediment transport rate:

$$K=Ae^{-Bt/T} \tag{17}$$

where the coefficient A and B are given by (Yamamoto et al., 1996):

$$A=1.61\cdot10^{-10}\,(d_{50}/H_o)^{-1.31}$$

$$B=4.2\cdot\,10^{-3}\,(\tan\beta\,)^{1.57} \tag{18}$$

where H_o is the deep water wave height.

The coefficient K of eqn (17) is a function of time since the rate of cross-shore sediment transport decreases with time and the beach profile approaches equilibrium.

It can also be expected that the mean depth $\bar{d}$ is a relatively conservative characteristic in comparison with the local shoreline position y_s, and consequently, it can be considered as a constant in eqn (15).

6. Application to a submerged groin system

The present application aims to quantitatively describe the circulation and sediment transport patterns in the area of a submerged groins system normal to a sandy beach (Figure 1).

The effects of the submerged groins on the longshore current are shown in Fig. 2. The beach slope is 1/30 with mean grain size d_{50}=0.3 mm. The modeled wave condition is a moderate storm (incident wave height H=1.7m, period T=8 sec, angle of incidence 30°). The distance between groins is 25 m and the height of the groins d_g=1m. The restructuring of the known longshore current in the breaker zone is obvious. Circulation cells bifurcate from the mainstream of the longshore current.

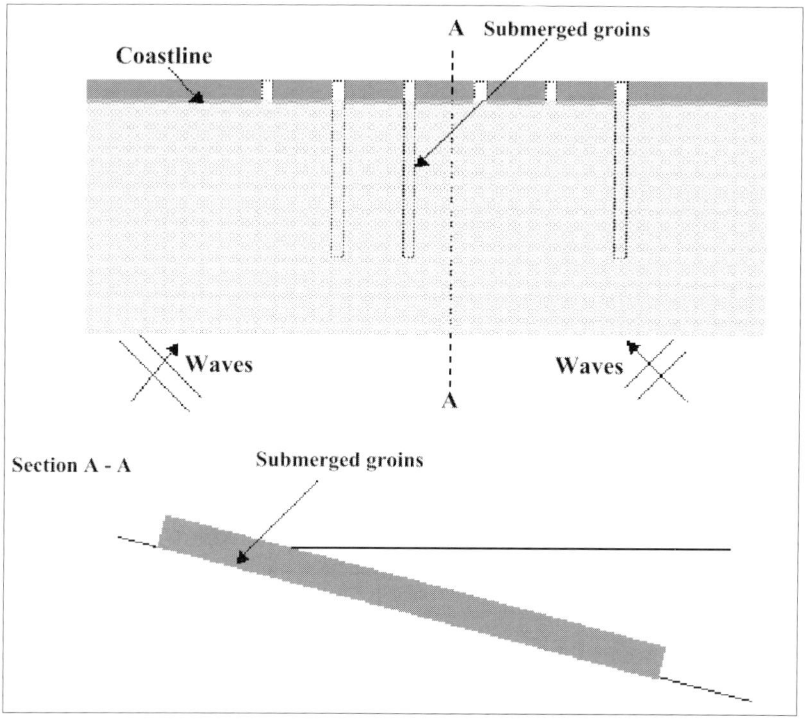

Figure 1. A submerged groin system

The sediment transport rates (q_x and q_y) are shown in Fig. 3. The circulation cells are capable of transporting sediment inshore and trapping a proportion of it inshore, contributing to shore accretion. The effects of undertow, which is an erosion mechanism since it transports the suspended sediment offshore, are reduced, while accretion is promoted. The shoreline evolution after 50 hr of wave action is shown in Fig. 4. The waves (although erosive ones) cause accretion at the space between successive groins. Model runs under different conditions indicate that the properties of the accretion process are controlled by structural characteristics (length, distance and height) as well as the sediment size and the wave characteristics (height and period).

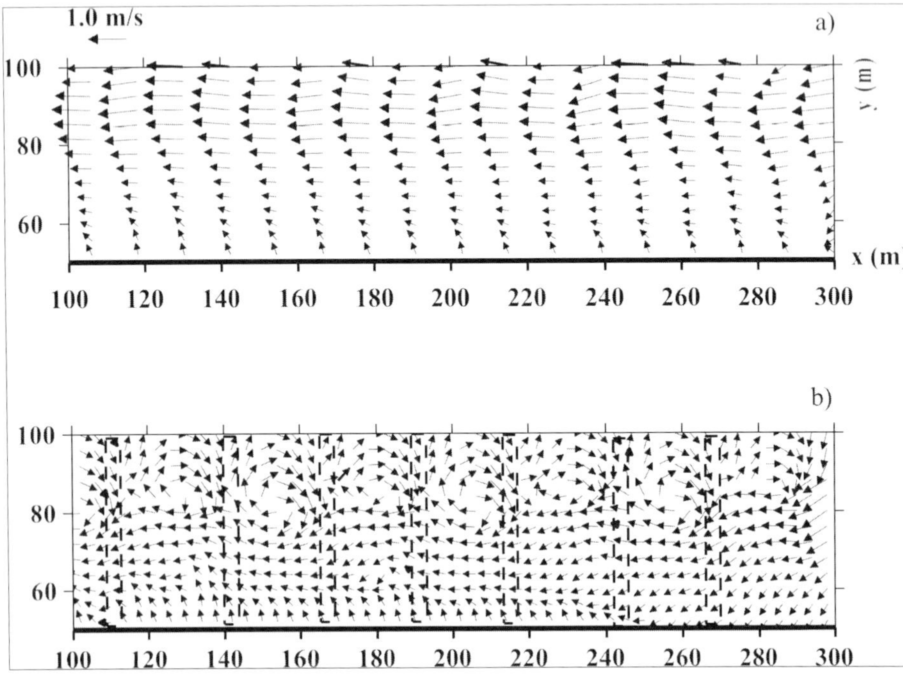

Figure 2. Effects of submerged groins on the longshore current:
a) without groins and b) with groins

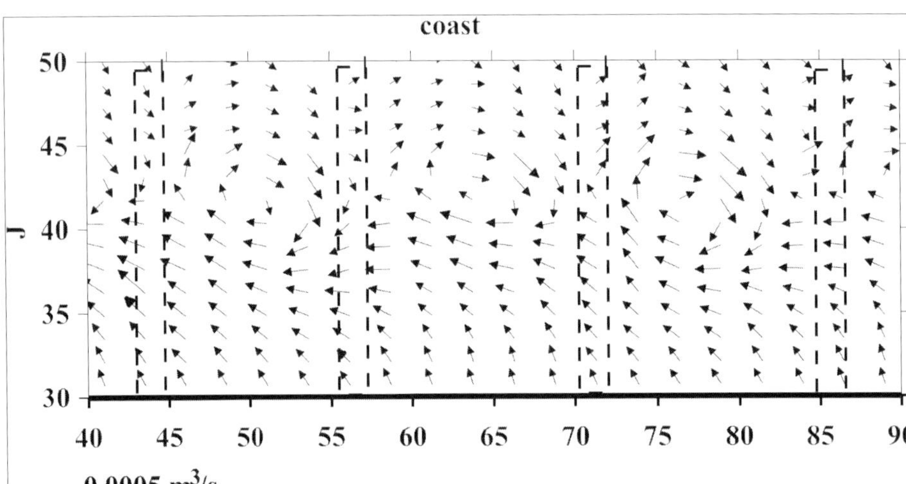

Figure 3. Effects of a groin system on the sediment transport rate

7. Conclusions

The theoretical investigation reveals that the wave-induced current field, due to the presence of a groins system, is capable of transporting sediment inshore and trapping a proportion of it inshore, contributing to shore accretion. In this way the erosion mechanisms are reduced and accretion mechanisms are enhanced.

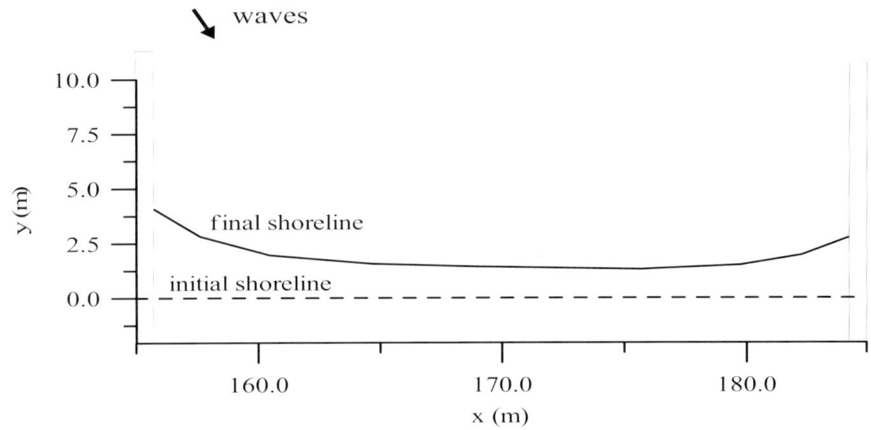

Figure 4. The shoreline changes between two groins after 50 hr of wave action

REFERENCES

Battjes, J. A., 'Modelling of turbulence in the surf zone', proceedings of the Symposium on *Modelling Techniques*, California, ASCE, pp. 1050–1061, 1975.

Battjes J. A. & Janssen, J. P. F. M., 'Energy loss and set-up due to breaking of random waves', International Conference *on Coastal Engineering '78, ASCE*, pp. 569–587, 1978.

Copeland G. J. M., 'A practical alternative to the "mild-slope" wave equation', *Coastal Engineering*, 9, pp. 125–149, 1985a.

Copeland G. J. M., 'Practical radiation stress calculations connected with equations of wave propagation', *Coastal Engineering* 9, pp. 195–219, 1985b.

Karambas Th. V., '2DH non-linear dispersive wave modelling and sediment transport in the nearshore zone', International Conference on *Coastal Engineering 1998*, ASCE, Copenhagen, 2940–2953, 1998.

Karambas Th. V., 'Numerical simulation of linear wave propagation, wave-induced circulation, sediment transport and beach evolution', in *Coastal Engineering and Marina Developments*, Wessex Institute of Technology Press, Eds C.A. Brebbia and P. Anagnostopoulos, pp. 253–274, 1999.

Leont'yev, I. O., 'Numerical modelling of beach erosion during storm event', *Coastal Engineering*, 29, pp. 187–200, 1996.

Leont'yev, I. O., 'Short-term changes due to cross-shore structures: a one-line numerical model', *Coastal Engineering*, 31, pp. 59–75, 1997.

Putrevu U., & Svendsen I.A., 'Vertical structure of the undertow outside the surf zone', *Journal of Geophysical Research*, vol. 98, no C12, pp. 22707-22716, 1993.

Yamamoto Y., Horikawa K. & Tanimoto K., 'Prediction of shoreline change considering cross-shore sediment transport', International Conference on *Coastal Engineering* 1996, ASCE, pp. 3405-3418.

DEVELOPMENT OF THE COASTAL EMBANKMENT SYSTEM IN BANGLADESH

MARKUS SAARI[1] AND SAEEDUR RAHMAN[2]

[1] *Chief Design Engineer, 2nd CERP, Jaakko Poyry Development Ltd.*
[2] *Project Director CERP, Bangladesh Water Development Board*

Abstract

The coastal lands of Bangladesh are in the process of being formed by the rivers which criss-cross them, bringing sediment from further inland, and by various marine, estuary and coastal processes which continuously shift, sort and modify the sediments. New alluvial land is continuously formed in the process. During recent decades, fast population growth has forced people to cultivate new agricultural land, and the coastal areas and islands have become densely populated. Embankment crest levels typically provide protection from the 5 to 11-year storm surge return period only (2% wave overtopping level). Erosion of the foreshore and the embankment system is a ongoing problem. Embankments face erosion caused by cyclone surges and accompanying waves, monsoon waves and rains, river currents and human and animal activities.

Many of the existing hard protection systems have also failed due to imperfect design or low quality of construction. The Coastal Embankment Rehabilitation Project (CERP) has improved planning and design methods and started development of a sustainable embankment system. The project has improved community participation in operation and maintenance schemes by introducing the Embankment Settler Scheme, where settlers are given a piece of the embankment for plantation and housing. The results are very good so far. GIS has been applied to form a comprehensive polder database and it has been used in planning, design and analysis of the embankment system. Modelling (MIKE 21) has been used to analyse the frequency of storm surge levels and to estimate flooding levels.

A lot of development and new investment is needed to improve the current situation. A new programme, the Coastal Zone Water Management Programme, is being planned. The proposed 10-year programme, which aims *inter alia* for better integration of the coastal activities, is proposed to be effective in 2003. The emphasis of the programme will be on improving protection for life and property by creating a sustainable embankment system and improving the cyclone shelter system, increasing agricultural production by creating better water management structures, and improving the environment by afforestation, erosion control and better water management.

1. Short description of the zone

The coastal lands of Bangladesh are in the process of being formed by the rivers which criss-cross them, bringing sediment from further inland, and by various marine, estuary and coastal processes which continuously shift, sort and modify the sediments. New alluvial land is continuously formed in the process.

C. L. Goudas et al. (eds.), Soft Shore Protection, 115-125.

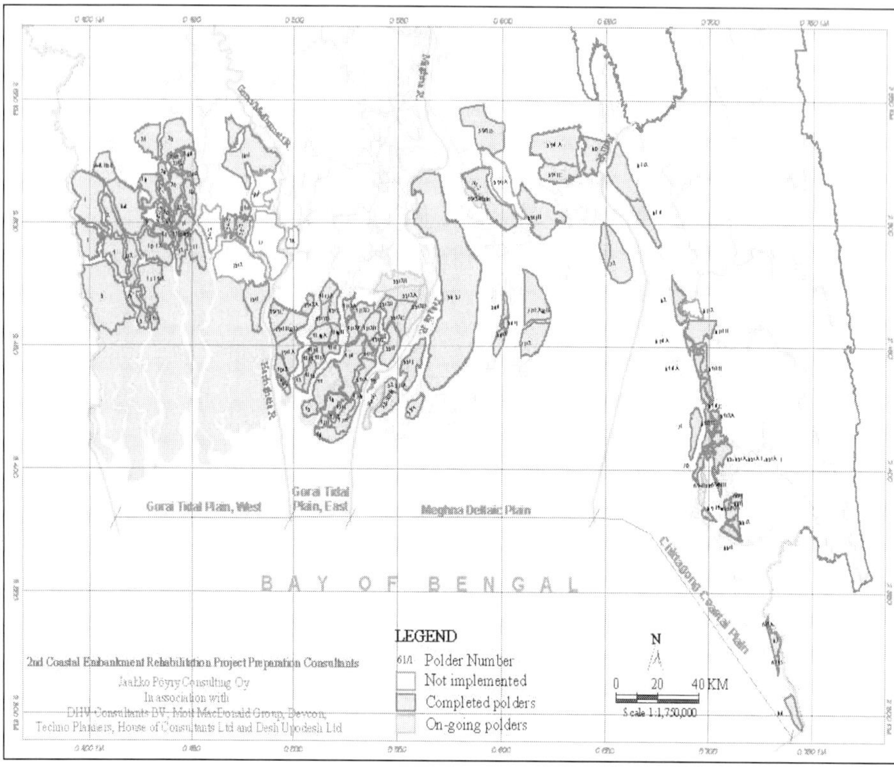

Fig. 1. Coastal polders on the Bay of Bengal

The area is subject to high tidal variation, monsoon flooding and tropical cyclones, the larger of which tend to damage the infrastructure and cause heavy loss of life. Three of them (1970, 1974 and 1991) had storm surge height 8 m or over. As a consequence about 500 000 people lost their lives.

During the 20th century, fast population growth forced people to cultivate new agricultural land, and the coastal areas and islands were densely populated. The polder development had started already by the end of 18th century, when farmers built their own small embankments in the zone in order to conquer land from the sea for their agricultural purposes, but the productivity of the land was low due to salty tidal water intrusion. The first publicly built embankments were built in 1960s to prevent salt-water intrusion and to further facilitate agricultural development. During the later half of the 20th century, the number and strength of the cyclones hitting the area increased. Consequently, during the last decades, cyclone protection has become one of the major functions of the embankments. Nowadays there are 125 completed polders with about 2000 km of embankments along the coastal belt.

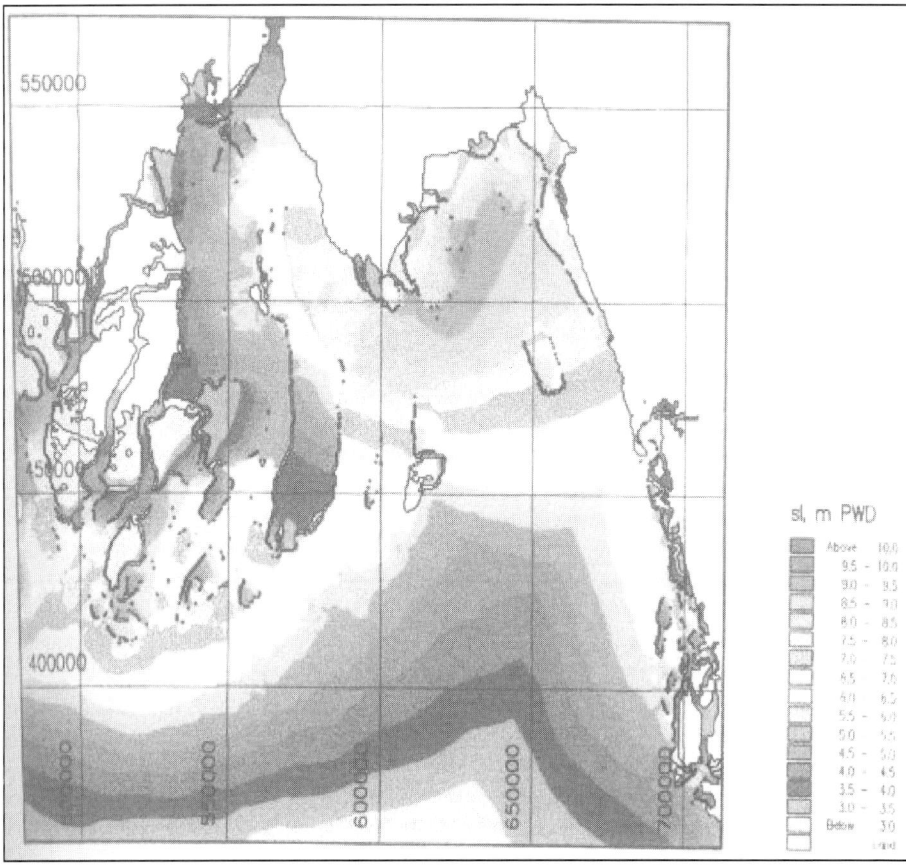

Fig. 2. Highest cyclone surge levels 1960 - 1996

The National Water Policy 1999 aims to protect coastal lands from tidal flooding and cyclone surges. National policy also has an emphasis on planning and implementing schemes for reclamation of land from the sea and rivers.

2. Problems

The main problems that the coastal polder system is facing are insufficient protection against cyclones, sub-optimal agricultural production due to missing or faulty water control structures and fast deterioration of the embankment system due to erosion and lack of maintenance. The institutional capacity of the organisations responsible for implementation, operation and maintenance is also low, resulting in sub-optimal use of available resources, poor maintenance and low quality of implementation.

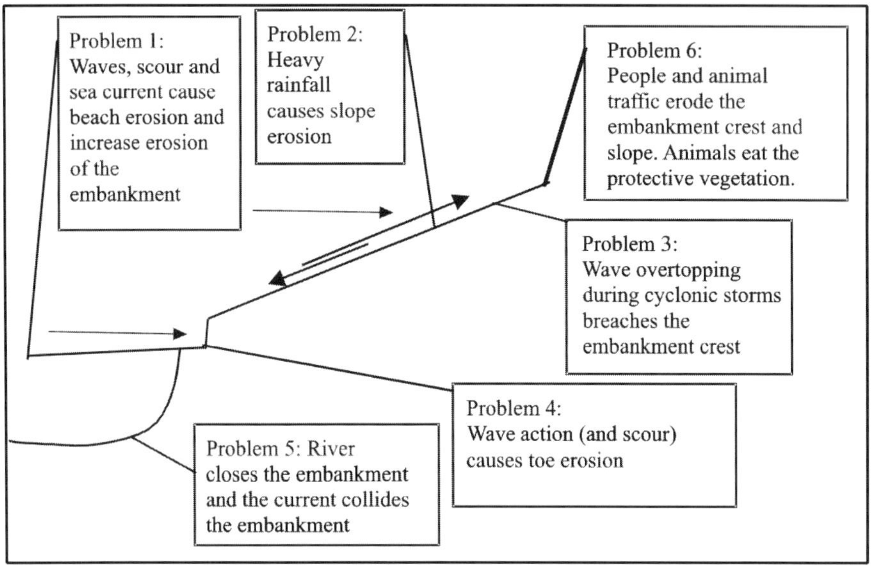

Fig. 3. Erosion of the embankments

Embankment crest levels typically provide protection from the 7 to 12-year storm surge return period only (2% wave overtopping level). Cuts in the embankment system may also let storm surge and tidal waves enter the polder. The funds available do not permit the construction of much higher embankments, nor an adequate number of cyclone shelters and effective communication and transportation systems in the polder area.

Agricultural production in many polders is low due to restricted drainage and salt-water intrusion. Faulty or missing water management structures force people to make their own cuts in the embankments, which create dangerous weak spots in them.

Erosion of the foreshore and the embankment system is an ongoing problem. The embankments face erosion caused by cyclone surges and accompanying waves, monsoon waves and rains, river currents and human and animal activities.

Destruction of the foreshore mangrove forests has also resulted in increasing beach erosion – and erosion of the embankments. As well as this, many embankments were constructed along the beach, which means that at normal high tide levels, the toe of the embankment was covered by water. There have not been sufficient funds to permit hard protection for the embankment. Many of the existing hard protection systems have also failed due to imperfect design or low quality of construction.

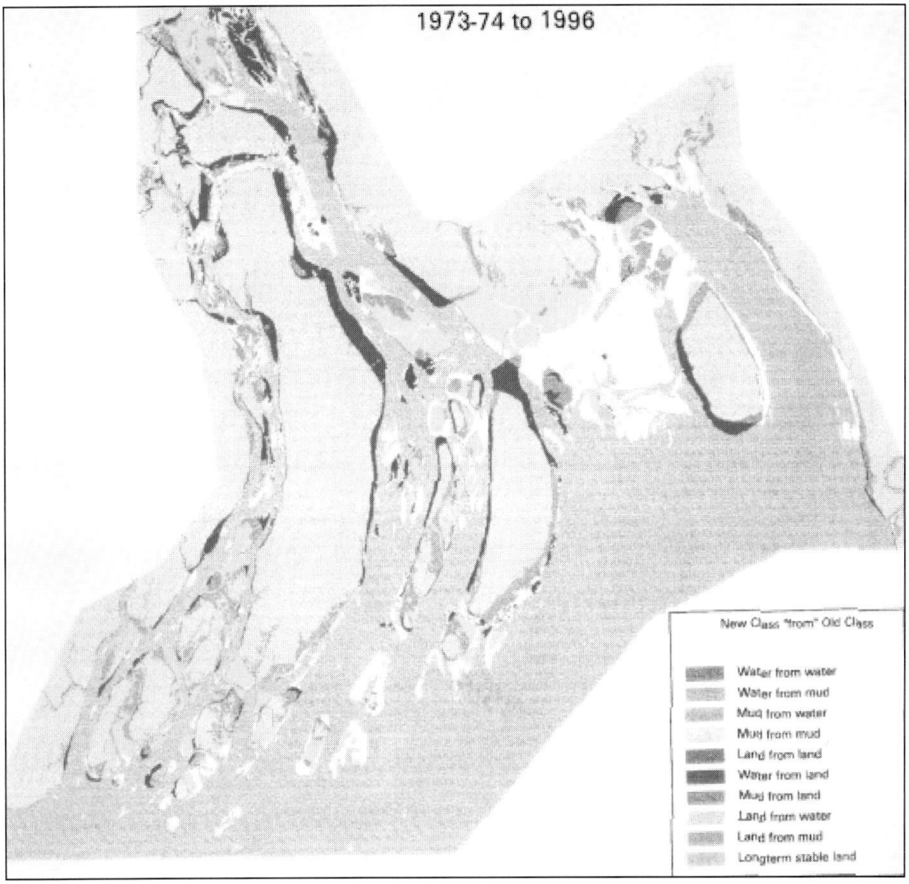

Fig. 4. Erosion and land accumulation on the Meghna river delta area

3. Coastal erosion protection in Bangladesh

There have been several projects to improve the embankment system, but it is still very vulnerable to erosion caused mainly by the monsoon season storms, currents and rains and to wave overtopping during the cyclone storms.

The ongoing Coastal Embankment Rehabilitation Project (CERP-II and 2nd CERP) has introduced new foreshore and embankment planting schemes to protect the embankment system against erosion. Community participation in operation and maintenance has been improved by introducing a community participation model called the Embankment Settler Scheme, where settlers are given a piece of the embankment for plantation and housing. The results are very promising so far.

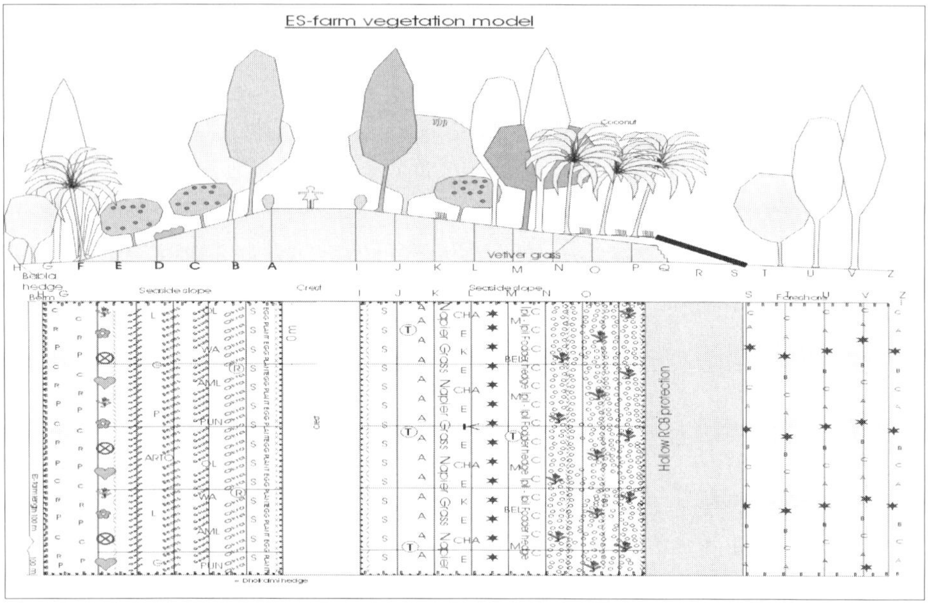

Fig. 5. Embankment settler farm planting model

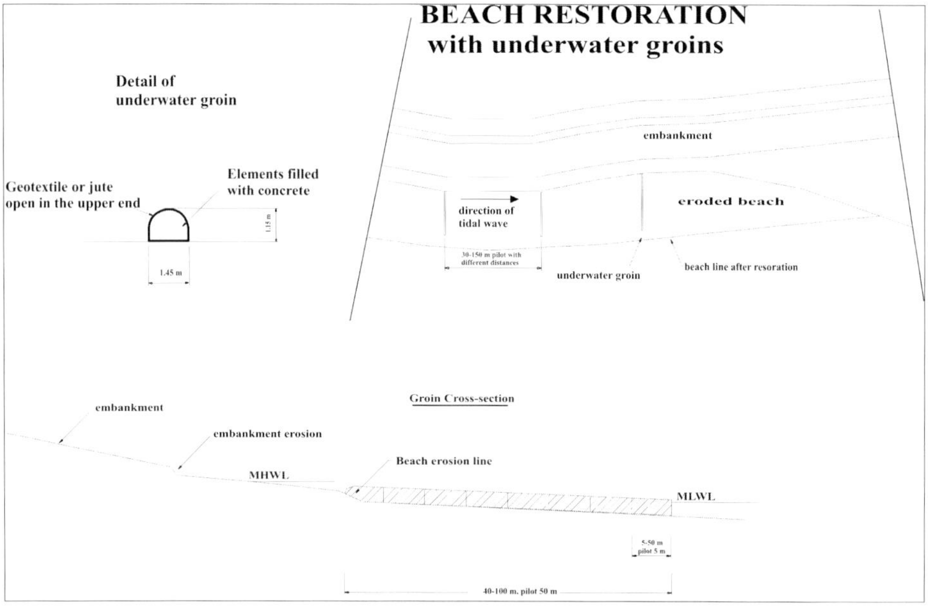

Fig. 6. Beach restoration using underwater groins

In the ES scheme, people are offered an opportunity to receive a part of the embankment (0.5 ha/settler) to set up a plantation for free. Land is given to landless people and people in the process of resettlement, which also reduces the level of poverty.

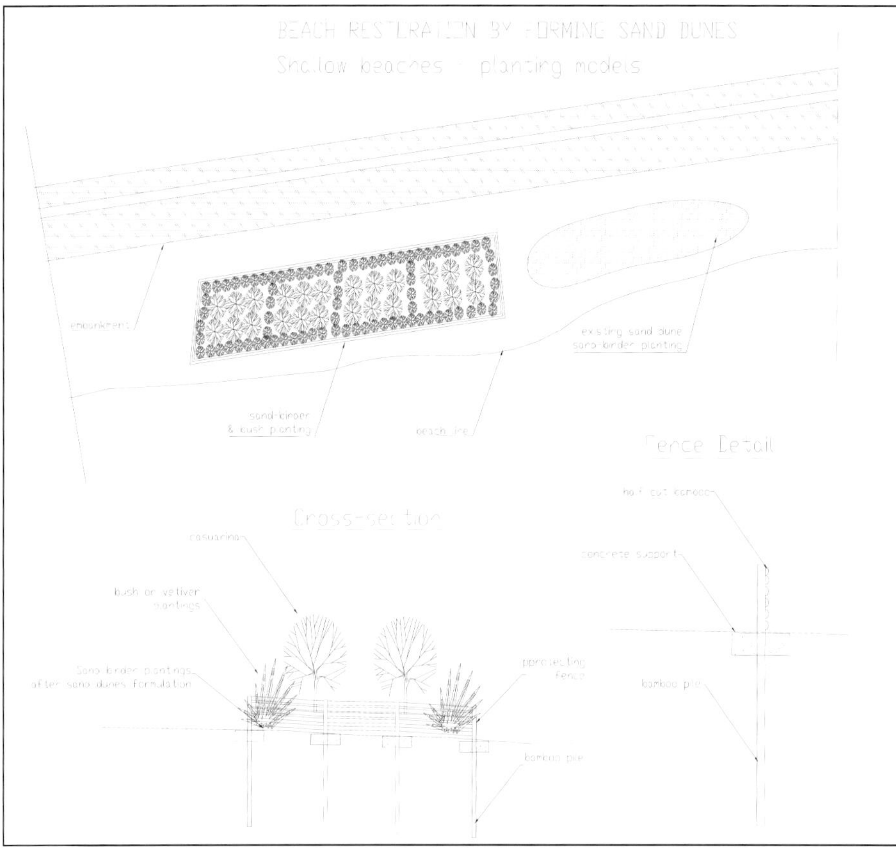

Fig. 7. Sand dune development

Settlers have duty to plant the seedlings given by the project and they have to maintain the embankment on their plantation area. They have right to built their houses behind the embankment or on a platform on the inside slope. In the initial stage they receive a subsistence allowance for the first 4 years (total ~ 600 USD/settler), and thereafter the plantation is expected to provide a full living for their families.

The project has developed methods which have moved away from hard and expensive embankment protection methods towards sustainable soft shore protection combining low-cost engineering and use of protective vegetation. This development includes the preliminary designs for several pilot projects.

3.1 PILOT 1.

Elevating the eroded beaches using underwater groins will be piloted on those beaches which are already lost or which are in the threatened zone. The concrete groins will be installed on the breaker zone between high and low tidal water levels. Mold for the groins will be constructed from strong jute fiber. Concrete casting will be done by manual labor.

3.2 PILOT 2.

New vegetation models on the foreshore area:

- Sand dune development on the sandy beaches using sand fence and sand catching vegetation
- Foreshore development by large mound afforestation.

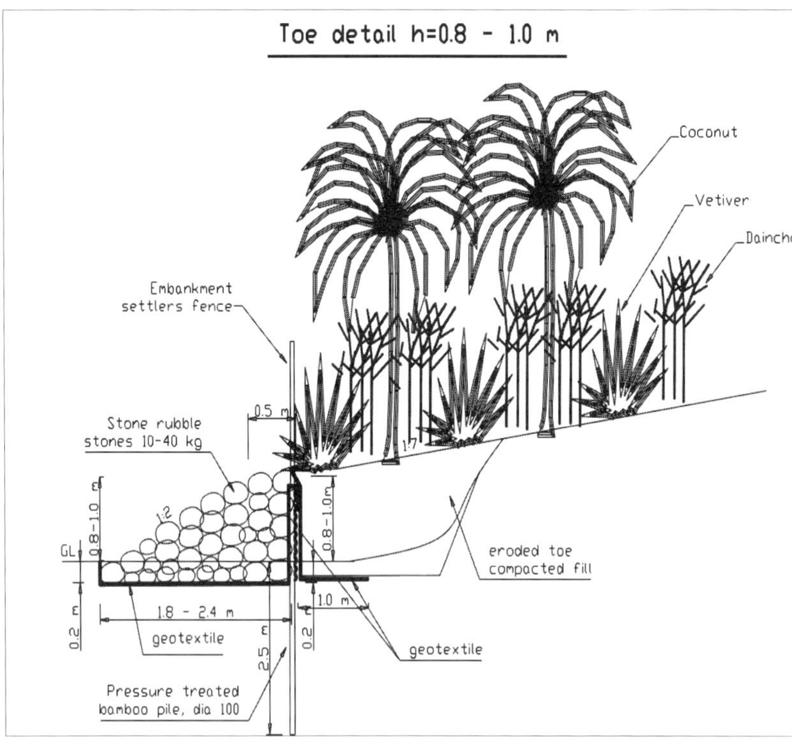

Fig. 8. Low-cost toe protection using stone polder and protective vegetation

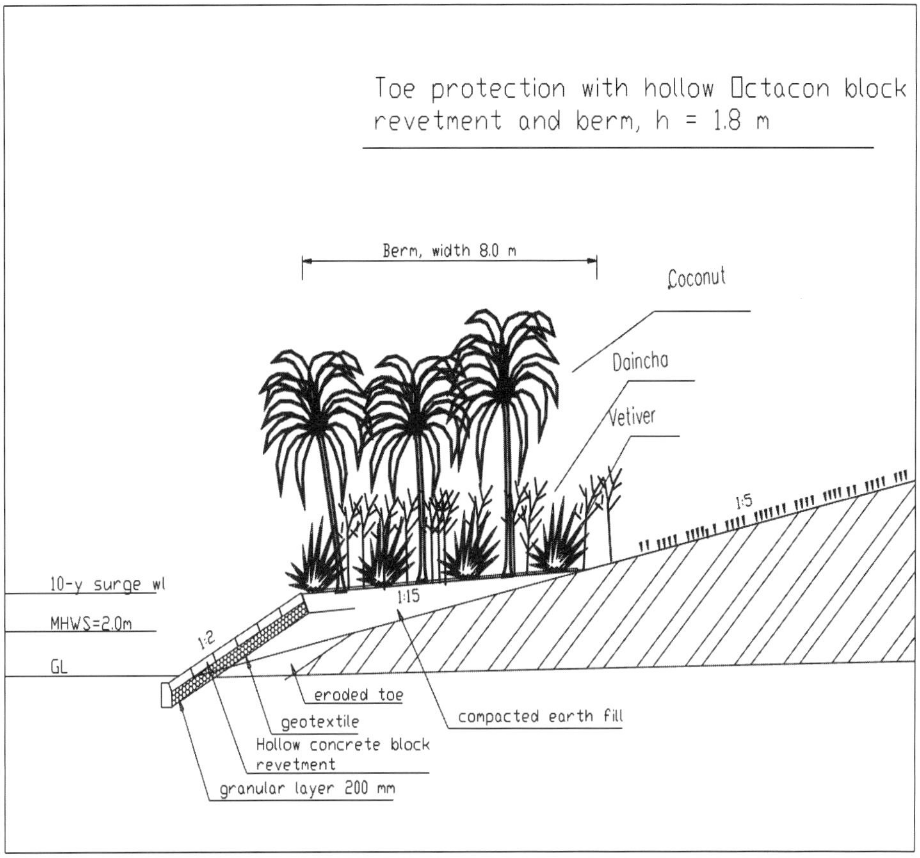

Fig. 9. Hollow concrete block revetment combined with berm and protective vegetation

3.3 PILOT 3.

New toe protection methods combining low cost engineering methods and protective plantings to protect the embankments against erosion and to reduce wave run-up.

3.4 PILOT 4.

Hollow concrete block revetment combined with protective vegetation.

The revetment is made of interlocked hollow octagon formed concrete blocks. The berm is situated at the 1/10-year storm surge level to be effective for cyclone surge and a related waves with a return period of 1/25-year or less. The vegetation on the berm is planned to give maximum resistant to erosion and wave run-up. The trees will also reduce the wind effect of the storms. The effective estimated reduction of the wave run-up is from 40% to 90%, depending on the water level and the wave height.

Installation will increase the estimated ability of the embankment to withstand breaching during the storms (return period for 2 % wave overtopping increased from 10 to 20 years).

Fig. 10. Embankment damage after a small cyclonic storm, October 2000

Other proposed pilots are:

- Slope protection using open stone asphalt for erosion protection of the sandy embankments
- Installing paved roads on the embankment crest to improve its strength and aid local transportation

The Second CERP project will end in December 2000. Implementation of the pilot schemes should take place during the proposed follow-up projects

4. Future of the coastal zone

A lot of development and new investments are needed to improve the current situation. A new programme, the Coastal Zone Water Management Programme, is in the planning stages. The proposed 10-year programme, which aims in *inter alia* at better integration of the coastal activities, is proposed to be effective by the end of 2003. The emphasis of the programme will be on improving protection for life and property by creating a sustainable embankment system and improving the cyclone shelter system, increasing agricultural production by creating better water management structures, improving the coastal environment by forestation, erosion control and better water management and improving operation and maintenance by taking a participatory approach.

Every monsoon season causes lot of erosion damage to the embankments and the cyclones in autumn and in spring increases the damage. The funds that the government of Bangladesh is able to make available for the coastal polder improvement will be very limited until the new World Bank-funded programme starts. According to the proposed timetable, the first investment projects on site could start in the end of 2003. If nothing is done in the meantime, many of the embankments in the danger zones will be completely lost, which will increase the likely cyclone and flood damage and result in an increased need of investment in the future.

The future of the coastal areas of Bangladesh is not rosy if the predicted rise in sea water level does occur. Extensive studies have shown that the sea water level is expected to rise about 50 – 100 cm during this century. Rougher seas with more frequent and devastating storms can be expected, and have already been experienced in Bangladesh. The embankments which now are above high tide levels would also start to suffer from wave erosion. People in Bangladesh cannot be blamed for this negative development, which is a result of global warming due to the green house effect which can be attributed to the industrial countries. Something needs to be done fast to stop the erosion and to improve sustainability and quality of the embankments, which are essential for safeguarding life and the future of the coastal zone of Bangladesh.

REFERENCE

Saari, M., and Rachman, S.: Technical Report, 2nd CERP, Jaakko Poyry Development Ltd., 2000.

CONFIGURATION DREDGING – AN ALTERNATIVE TO GROYNES & OFFSHORE BREAKWATERS

H.P. RIEDEL,[1] P. O'BRIEN,[2] AND R. SMITH[2]

[1] *Coastal Engineering Solutions Pty ltd, 561 Burwood Rd, Hawthorn, Victoria, Australia*
[2] *Coastal Engineering Solutions Pty Ltd, Level 9, 139 Leichhardt St, Brisbane, Qld, Australia*

Abstract

The impact of dredging in the near-shore zone on beach stability has all too often not been understood. The scenario typically involves dredging of a new channel for a port, where the commercial value of the port is high and the designers of the port are not aware that a localised deepening of the seabed can dramatically change the direction of waves passing over the dredged hole. Projects involving such 'errors' are rarely publicised. However, if dredging can change near-shore wave characteristics resulting in beach erosion, it must also be possible, by designing the shape of a dredged hole, to create changes to the near-shore waves which promote beach stability. This paper describes such an example and how 'configuration dredging' can influence longshore sediment transport in a manner similar to groynes and offshore breakwaters. The paper also outlines the general applicability of the concept.

1. Introduction

The senior author has first-hand knowledge of three ports in Australia where dredging of shipping channels or dredging for the winning of fill material has resulted in significant negative impacts on beach stability. Having been exposed to this concept, Riedel and Byrne (1982) utilised configuration dredging to optimise the access channel to a new supply base port. By dredging a scallop, somewhat like a turning basin, adjacent to the channel, waves that penetrated and focussed in the port with a parallel sided channel were now deflected out of the port. The configuration dredging took place in 1982 and has operated successfully. It deleted the need for a breakwater, which would have been needed if a parallel-sided channel had been dredged. Configuration dredging was considerably less expensive than the alternative breakwater concept. Figure 1 shows the channel with its dredged 'scallop'. Figure 1(a) shows typical wave orthogonals for a parallel-sided channel and Figure 1(b) shows how the waves are deflected away from the harbour as a result of configuration dredging.

In a similar manner, refraction and changing the direction of waves and their wave heights can be utilised to modify longshore transport rates along a beach. However, as with any form of change to the coastal system, the designer has to be aware that there may be down-drift impacts that need to be understood and managed so that the erosion problem is not transferred downstream.

This paper develops the concept of configuration dredging with reference to a beach which has in the past been de-stabilised by the construction of an impermeable jetty and other engineering works which interrupted the longshore sediment transport.

C. L. Goudas et al. (eds.), Soft Shore Protection, 127-136.

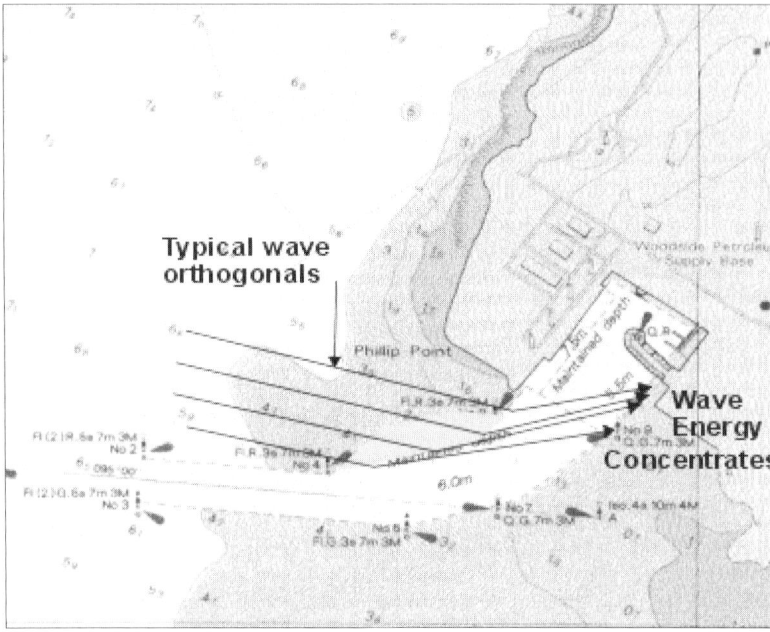

Fig. 1(a). Parallel sided channel

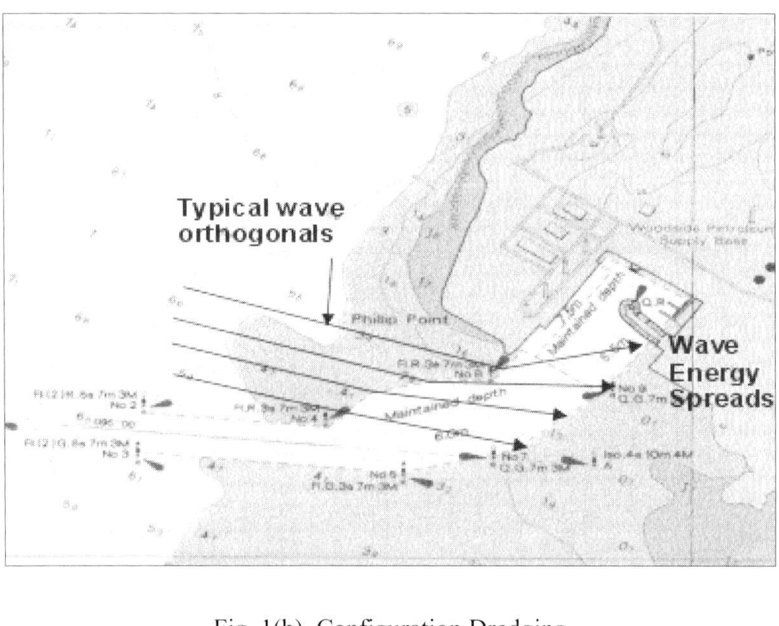

Fig. 1(b). Configuration Dredging

Configuration dredging is compared with the constructed remedial works which consisted of a new groyne with beach renourishment.

2. The Erosion Problem – Rye, Victoria, Australia

2.1 WAVE CLIMATE

Rye Front Beach faces north and is within a bay across which wave fetches are limited to less than 50 kilometres. Long-period swell waves do not reach the beach and the beach is protected from large waves by offshore banks. Figure 2 illustrates the setting.

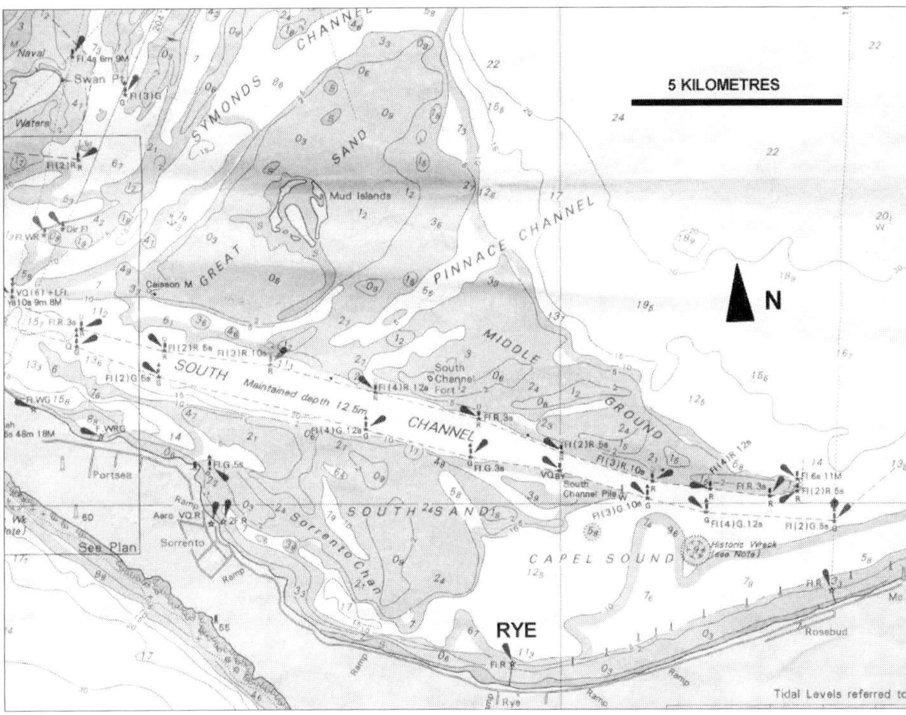

Fig. 2. Rye and offshore sand bars

As a result of the offshore banks, the wave climate approaching the shoreline is very mild and rarely will waves exceed one metre in height. Nevertheless, the beach has been subjected to ongoing erosion.

2.2 COASTAL PROCESSES

The gross capacity for longshore sediment transport is of the order of 5000 cubic metres per year with an average yearly imbalance of about 1000 cubic metres. That is, there is a net movement of sand from west to east. The net rate of sediment transport has been well established by surveys adjacent to an impermeable structure.

2.3 MAN-MADE CHANGES

The shoreline upon which Rye is located is a popular holiday destination. There were many structures built along the shoreline between 1850 and 1950, such as a coastal road, seawalls and groynes. Consequently the system was subjected to erosion in 1950 when a jetty was re-constructed using an impermeable abutment that extended 150 seaward from the natural waterline at high tide. Figure 3 shows the shoreline in 1959, and an accumulation of sand is evident to the west of the jetty.

Fig. 3. Rye front beach in 1959

Sand started to collect on the western side of the abutment and the erosion problem to the east was exacerbated. In the mid 1970s, the government reclaimed land on both sides of the jetty to provide facilities for boat launching, car parking and a parkland, and created an excellent beach. The volume of sand placed in the reclamation was about 200000 cubic metres. However, the net eastward drift of sand was not taken into account, and shortly afterwards the erosion problem re-surfaced. A rock seawall was constructed to prevent erosion. However, this seawall was constructed at such an angle that a beach could not be retained in front of it. The government was keen to re-establish a beach in this popular area. Figure 4, from a 1996 aerial photograph, shows these features.

3. Near Shore Seabed Characteristics

The seabed profile from the shoreline in a seaward direction is characterised by a wide shallow area – the beach grade below low water is about 1:300 from the 0 to the –1 metre contour. A series of low bars are superimposed over the near-shore area. The seabed slope then increases quite rapidly to water depths of greater than 10 metres. Another feature, due primarily to the mild wave climate and small tidal range (less than 1 metre), is that the position of bed forms has remained almost constant over the last 50 years (based on aerial photograph comparisons)

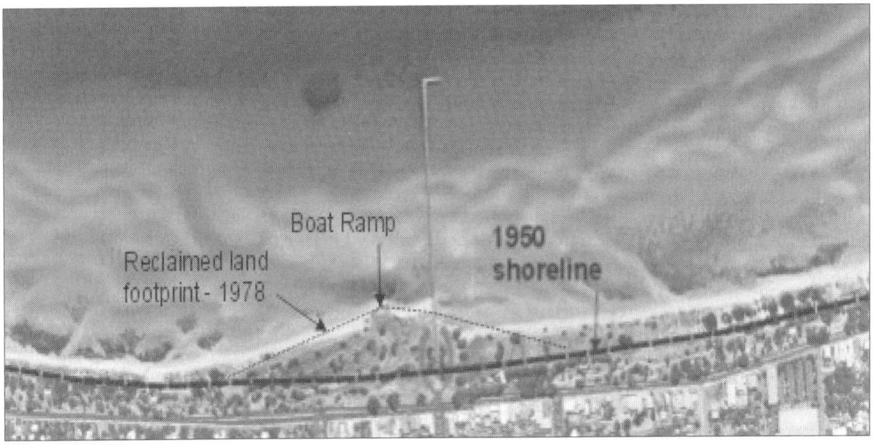

Fig. 4. 1996 aerial view of Rye beach

These features, together with the mild wave climate mean that sand-retaining structures such as rock groynes and offshore breakwaters are relatively inexpensive.

4. Beach Stabilisation Systems Investigated

4.1 ALTERNATIVES

As noted earlier, the beach is very popular over the summer months and there were numerous stake-holders with vested interests and strong opinions. Consequently, a very wide range of solutions based on the following types were investigated:

- Do nothing
- Remove all or some of the man-made structures
- Offshore breakwaters – submerged and surface piercing
- Groynes
- Beach renourishment
- Permutations and combinations of the above

Configuration dredging, the subject of this paper, was not considered because of the high capital cost resulting from the volume of sand that had to be dredged. There was no other nearby beach that required urgent renourishment, nor was there a budget available for dredging of about 200000 cubic metres of sand. Had configuration dredging been thought of in the mid 1970s when the reclamation was done at Rye, it would have been a very cost-effective solution.

4.2 OVERVIEW OF CONVENTIONAL BEACH STABILISATION TECHNIQUES

A comprehensive overview requires reference to a large number of aerial photo images which cannot be presented in the published paper. Therefore the paper focuses on the solutions that were seriously considered and for which detailed costings were undertaken. Further information will be given during the paper presentation.

The two serious contenders were a groyne or offshore breakwater, and in both cases the construction of these structures would be accompanied by beach renourishment and sand bypassing approximately every 3 years. Figures 5 and 6 show computer-generated images of the two schemes superimposed on recent aerial photography.

Solutions based on doing nothing or removing the entire infrastructure and allowing the shoreline to return to pre-development conditions would not solve the erosion problem. The cost of the infrastructure located on the reclaimed land was high, and if it was removed, all the facilities would need to be relocated elsewhere. In essence the problem would not be solved, and in addition a new section of coast would possibly be de-stabilised in order to relocate these facilities.

Fig. 5. Groyne plus nourishment

Beach renourishment on its own was not practical because of the shoreline orientation at the seawall. The seawall alignment is rotated about 30 degrees clockwise compared to an equilibrium beach alignment. The beach was required in front of the seawall. Consequently, whilst the net easterly drift is only about 1000 cubic metres per year, sand from in front of the seawall would be moved on at a rate of about 3000 cubic metres per year, and would in effect have to be placed there annually. This excess sand would become a nuisance downstream both with respect to excessive accretion and wind action. The natural sand on the beaches is fine-grained with a D50 of 0.25 mm. Finally, there was difficulty with an economic sand supply, together with reluctance on the part of the authorities to be constrained by an annual maintenance obligation. Sand was available from the western side of the jetty at an average annual rate of 1000 cubic metres. This sand was available as a by-product from maintenance dredging of the boat ramp channel and mooring area.

Fig. 6. Offshore breakwater plus nourishment

4.3 CONSTRUCTED WORKS

The groyne was chosen ahead of the offshore breakwater for reasons of cost and visual preference. 15 000 cubic metres of sand was placed on the beach from an onshore relic dune source and a fixed by-passing pipeline was located and buried at the back of the beach so that there was no disruption to the beach use when sand by-passing was undertaken. The total cost of the works was only $150 000.

5. Configuration Dredging Method

This is in essence a demonstration of how Rye foreshore development might have progressed if the knowledge available today had been accessible to the engineers undertaking the reclamation work in the 1970s.

5.1 WAVE REFRACTION

Two simple wave refraction model grids, using a 25 metre grid spacing, were set up to schematise the bathymetry approaching Rye front beach. The first was for existing conditions, and the second included a dredged, approximately triangular shaped hole. The reverse ray refraction technique for spectral waves was used and refraction origins were located at 50 metre intervals along the beach.

The Figure 7(B) curves show the near-shore wave directions at the centre of the eroding beach segment for the existing situation and the curves 7(A) show the directions after configuration dredging. The feature that stands out is the shift in the inshore wave angle after configuration dredging. For daily sea-breeze conditions (3 second waves) the inshore wave direction has rotated by some degrees clockwise and for longer period storm conditions (7 second waves) this rotation increases by a proportional number of degrees.

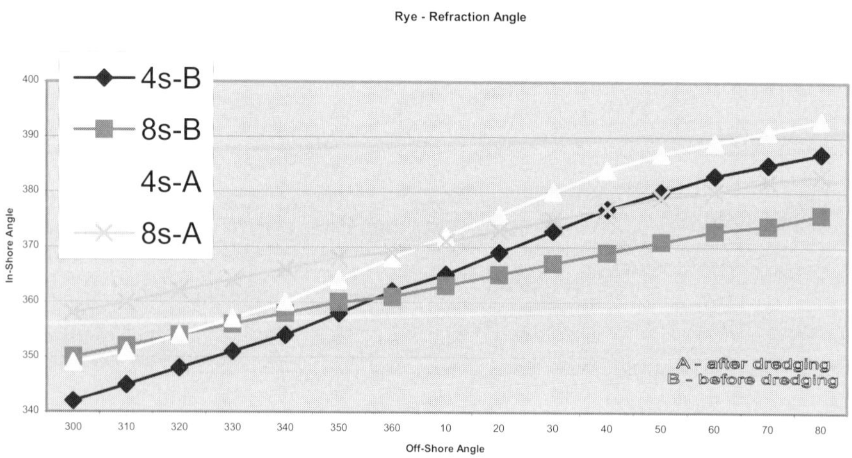

Fig. 7. Wave refraction – near-shore wave directions

A dynamic equilibrium beach alignment will tend to be perpendicular to the incoming waves. Utilising the output from the other wave refraction origins (50 metre spacing), a dynamically stable alignment for the beach with configuration dredging would be (very approximately) as shown in Figure 8.

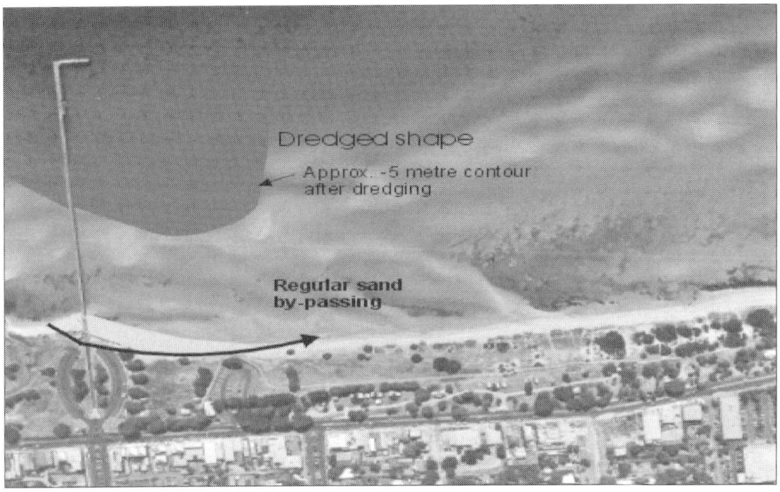

Fig. 8. Predicted beach shape with configuration dredging

6. Conclusions — Applicability of Configuration Dredging

At this stage, the range of problems that could be addressed by configuration dredging has not been investigated. It is clear that dredging in water depths whereby waves feel the bottom and are refracted will affect wave transformation to the shoreline and may impact on beach stability.

A key factor is that the dredged 'hole' needs to be sustainable, or at least cannot infill at a significant rate. Consequently, projects for which the technique should be considered include:

- Sites within bays or embayments with negligible swell wave penetration, where there is a considerable width of shallow water adjacent to the beach. The shallow water ensures that larger waves do not reach the beach and the mild short-period wave climate means that dredged area will infill very slowly.
- Sites within bays where there is a persistent low swell wave penetration superimposed on local sea conditions. A dredged hole in deep water will transform swell waves without affecting locally generated waves. The low swell would result in a slow infilling of the dredged hole; yet because of the long wave-period, transformation by refraction could still be significant.
- Open coasts with intermittent headland barriers, where there is already an established sand renourishment program. Usually sand is sourced from areas where there is no potential influence on refraction and coastal stability. It may be possible to source sand from closer to shore via configuration dredging and the net cost over the project may be lower.

REFERENCE

Riedel, H.P. & Byrne, A.P.: 'Dredging to minimize wave penetration into a harbour', Proceedings of the 18th International Conference on Coastal Engineering, Cape Town,1514–1521 ASCE, 1982.

NON-LINEAR WAVE AND SEDIMENT TRANSPORT MODEL FOR BEACH PROFILE EVOLUTION WITH APPLICATION TO SOFT SHORE PROTECTION METHODS

TH. V. KARAMBAS[1,2] AND C. KOUTITAS[2]

[1]*HYDROMARE, Ethn. Antistaseos 3A, Kalamaria, Thessaloniki, 551 34, GREECE*
[2]*ARISTOTLE UNIVERSITY OF THESSALONIKI, Dept. of Civil Engineering, Division of Hydraulics and Environmental Engineering, Thessaloniki, 540 06, GREECE*

Abstract

Non-linear wave transformation in the surf and swash zone is computed by a non-linear breaking wave model based on the numerical solution of the time-dependent wave-energy equation, which is incorporated into a Boussinesq model. The Dibajnia and Watanabe transport rate formula involving unsteady aspects of the sand transport phenomenon is adopted for estimating the sheet flow sediment transport rates as well as the bed load and suspended load over ripples. For calculation of the suspended load induced by wave breaking, the Bailard formula is used. Since the only dissipation mechanism is the wave breaking. The model is able to reproduce accretion and erosion without the use of a criterion to distinguish accretionary and erosional waves. The methodology is applied to simulate sediment transport and beach evolution in soft shore protection methods (beach nourishment, floating and submerged breakwaters).

1. Introduction

Long-shore and cross-shore sediment transport due to wave and wave-induced current action play an important role in various coastal engineering problems. In particular, the application of shore protection methods requires the evaluation of beach processes to be as accurate as possible.

There are two main approaches to estimating the sediment transport rate inside and outside the surf zone: the deterministic approach and the energetics approach. The deterministic models are based on the description of both the wave-induced mean flow and the concentration of suspended sediment, usually using quasi 3D models and linear wave theory (deVried & Stive 1987, Katopodi & Ribberink 1992, Briad & Kamphuis 1993). The energetics approach is based on the idea that the sediment transport is related to the rate of energy dissipation of the flow (Bailard 1981, Roelvink & Stive 1989, Karambas, 1998).

Karambas (1998) developed a 2DH non-linear breaking wave model based on the numerical solution of the time-dependent wave energy equation which is incorporated into a Boussinesq model. The model equations are capable of simulating non-linear wave refraction, shoaling, diffraction, reflection (presence of structures), breaking, dissipation after breaking, run-up, and breaking wave-induced quasi-3D current. In the present work, the above breaking-wave propagation model is used. Karambas (1998) also adopted Bailard's (1981) energetics approach for the evaluation of the sediment transport rate (i.e. bed load and suspended load). In the present work, the Dibajnia and

C. L. Goudas et al. (eds.), Soft Shore Protection, 137-149.

Watanabe (1992) transport rate formula involving unsteady aspects of the sand transport phenomenon is adopted for estimating the sheet flow sediment transport rates as well as the bed load and suspended load over ripples. For the suspended load induced by wave breaking, the Bailard formula is adopted since the only dissipation mechanism is wave breaking.

Since the model is a non-linear one, asymmetry of the near-bottom orbital velocity, as well as the presence of undertow, are automatically reproduced. The model is thus able to reproduce accretion and erosion without the use of a criterion to distinguish accretionary and erosional waves (transport onshore or offshore, respectively) as in some models (Larson, 1996, Leont'yev, 1996).

Most of the existing models predict the morphological changes of the submerged part of the sea beach slope. However bed deformations in slope sections located above the mean water level (swash zone) are as yet little studied. In this region, run-up flow is characterised by a relatively thin sheet of water moving back and forth along the beach profile. The high-energy dynamics of the swash zone provide very active sediment transport as well as rapid morphological changes. To calculate the sediment transport rate in the swash zone, an extrapolation technique is usually adopted (Larson, 1996, Rakha et al., 1997). Leont'yev (1996) proposed a different approach based on the Bagnold's bed load formula. In that approach, the swash zone contribution is parameterised in terms of run-up height and equilibrium beach slope gradient. In the present work, the wave model is extended to the swash zone so as to describe wave run-up and run-down. After the flow calculation, as far as the sediment transport is concerned, the swash zone is treated as though it were in the submerged part.

For three of the soft shore protection methods (beach nourishment, floating and submerged breakwaters) the methodology is applied to simulate:

- Breaking wave propagation over submerged breakwaters as well as the morphological effects of such structures on the cross-shore beach profile
- Breaking wave induced current velocity field and shoreline evolution behind a floating breakwater
- Coastline evolution in a beach nourishment scenario assuming a trapezoidal beach fill

2. Wave and hydrodynamic model

The wave model is based on the numerical solution of the time-dependent wave energy equation which is incorporated into a Boussinesq model (Karambas, 1996, Karambas, 1998). The governing equations are written in the form:

$$\frac{\partial \zeta}{\partial t} + \frac{\partial (Uh)}{\partial x} + \frac{\partial (Vh)}{\partial y} = 0,$$

$$\frac{\partial(Uh)}{\partial t}+\frac{\partial\left(\int_{-d}^{\zeta}u^2dz\right)}{\partial x}+\frac{\partial\left(\int_{-d}^{\zeta}uvdz\right)}{\partial y}+gh\frac{\partial\zeta}{\partial x}=\frac{d^2}{2}\frac{\partial^3(Ud)}{\partial x^2\partial t}-\frac{d^3}{6}\frac{\partial^3 U}{\partial x^2\partial t}+\frac{d^2}{2}\frac{\partial^3(Ud)}{\partial x^2\partial t}+$$
$$\frac{d^2}{2}\frac{\partial^3(Vd)}{\partial x\partial y\partial t}-\frac{d^3}{6}\frac{\partial^3 V}{\partial x\partial y\partial t}+\frac{\partial}{\partial x}\left(\nu_\tau h\frac{\partial U}{\partial x}\right)+\frac{\partial}{\partial y}\left(\nu_\tau h\frac{\partial U}{\partial y}\right)-\frac{\nu_{bx}}{\rho},\tag{1}$$

$$\frac{\partial(Vh)}{\partial t}+\frac{\partial\left(\int_{-d}^{\zeta}uvdz\right)}{\partial x}+\frac{\partial\left(\int_{-d}^{\zeta}v^2dz\right)}{\partial y}+gh\frac{\partial\zeta}{\partial y}=\frac{d^2}{2}\frac{\partial^3(Vd)}{\partial x^2\partial t}-\frac{d^3}{6}\frac{\partial^3 U)}{\partial y^2\partial t}+\frac{d^2}{2}\frac{\partial^3 V}{\partial y^2\partial t}+$$
$$\frac{d^2}{2}\frac{\partial^3(Ud)}{\partial x\partial y\partial t}-\frac{d^3}{6}\frac{\partial^3 U}{\partial x\partial y\partial t}+\frac{\partial}{\partial x}\left(\nu_\tau h\frac{\partial V}{\partial x}\right)+\frac{\partial}{\partial y}\left(\nu_\tau h\frac{\partial V}{\partial y}\right)-\frac{\nu_{by}}{\rho},$$

in which ζ is the surface elevation, d is the still water depth, h is the total depth, u and v are the horizontal velocity components in the x and y directions respectively, U and V are the depth-integrated velocities, ν_τ is the eddy viscosity coefficient and τ_{bx} , τ_{by} are the bottom shear stresses.

The equation of the conservation of the energy density E of the mean flow per unit horizontal area is written:

$$\frac{\partial E}{\partial t}+\frac{\partial E_{fx}}{\partial x}+\frac{\partial E_{fy}}{\partial y}=-D-u_o\tau_{bx}-v_o\tau_{by}+U\,BT+V\,BT\tag{2}$$

with

$$E_{fx}=\int_{-d}^{\zeta}\frac{1}{2}u^3\,dz+\int_{-d}^{\zeta}\frac{1}{2}uv^2\,dz+g\zeta\,Uh$$

$$E_{fy}=\int_{-d}^{\zeta}\frac{1}{2}v^3\,dz+\int_{-d}^{\zeta}\frac{1}{2}u^2v\,dz+g\text{æ}Vh$$

in which D is the dissipation of the mean energy (equal to minus the production of turbulent energy) and BT are the Boussinesq dispersion terms of the momentum equations (1). The dissipation of the wave energy D is given by (Karambas, 1996):

$$D=\Omega\;(c\;u_{ot})^3\tag{3}$$

in which c is the wave celerity, u_{ot} is the bottom velocity in the direction of the wave propagation and Ω a constant, Ω=0.03.

Equation (3) is solved simultaneously with equations (1) providing the values of the integrals in the two momentum equations. The model is described in detail in Karambas (1996) and Karambas (1998).

The near-bottom values of wave-induced current are obtained following the procedure proposed by Rakha et al. (1997). The time series of the bottom velocities calculated in the wave model are divided into a number of cycles N_c each consisting of N time steps. For each cycle the values of the bottom velocities u_o and v_o are time averaged over N time steps to obtain the mean currents U_c and V_c under the roller.

In the above calculations the presence of the mean undertow is automatically included. However, the model cannot predict the near-bed shoreward drift (steady streaming) generated by the phase shift in orbital motions due to bottom boundary-layer mechanisms (viscosity effects). That shoreward transport occurs more frequently in the offshore regions. Inside the surf zone, the phase shift mechanism is suppressed by both the developing turbulence and the undertow acting in the middle layer, and so the resulting near-bed mass transport should be directed offshore. The present model incorporates the above mechanisms by calculating the near-bottom undertow velocity V_b from the analytical expression proposed by Putrevu & Svendsen (1993) which is valid inside and outside the surf zone:

$$\frac{V_b}{\sqrt{gh}} = \left(\frac{V_m}{\sqrt{gh}} - \frac{A}{6} + \frac{\tau_{sb} h}{2\rho \nu_{tz} \sqrt{gh}} \right) (1+R)^{-1}, \tag{4}$$

where V_m is the mean undertow velocity, τ_{sb} is the steady streaming term, ν_{tz} is the eddy viscosity coefficient, and A, R are coefficients given by Putrevu and Svendsen (1993).

The wave model is extended to the swash zone by adopting a 'dry bed' boundary condition to simulate wave run-up and run down (Karambas, 1998). After the flow calculation, as far as the sediment transport is concerned, swash zone is treated as the submerged part.

3. Sediment transport and morphology model

A new transport rate formula involving unsteady aspects of the sand transport phenomenon has been presented by Dibajnia and Watanabe (1992). The formula estimates the sheet flow sand transport rates and has been generalised by them for the bed load as well as for the suspended load over ripples (Watanabe & Dibajnia, 1996). Dibajnia and Watanabe (1996) also extended the formula for mixed-size sands. The formula reads as follows:

$$\frac{q(1-\lambda)}{wd_{50}} = 0.001\,\text{sign}(\Gamma)|\Gamma|^{0.5}, \tag{5}$$

where q is the cross-shore net transport rate, w is the settling velocity, d_{50} is the grain diameter and λ is the porosity of the sediment. The quantity Γ is defined by:

$$\Gamma = \frac{u_c T_c\left(\Omega_c^3 + \Omega_t'^3\right) - u_t T_t\left(\Omega_t^3 + \Omega_c'^3\right)}{\left(u_c + u_t\right)T}, \tag{6}$$

where u_c and u_t are the equivalent root-mean-square amplitudes and T_c and T_t are the periods of the onshore and offshore velocity respectively and $T=T_c+T_t$ is the wave period. Values of the Ω_j are determined as follows:

$$\text{if } \omega_j \le \omega_{cr} \qquad \Omega_j = \omega_j T_j \sqrt{\frac{sg}{d_{50}}},\ \Omega_j' = 0$$

$$\text{if } \omega_j > \omega_{cr} \qquad \Omega_j = \omega_{cr} T_j \sqrt{\frac{sg}{d_{50}}},\ \Omega_j' = \left(\omega_j - \omega_{cr}\right) T_j \sqrt{\frac{sg}{d_{50}}}, \tag{7}$$

where the subscript j is to be replaced by either c or t, and

$$\omega_j = \frac{1}{2}\frac{u_j^2}{sgwT_j}, \qquad s = \frac{\rho_s - \rho}{\rho}$$

$$\omega_{cr} = 1 - 0.97\sqrt{\Lambda},\ \Lambda = \left\{1 - \left[\left(\Psi_{rms} - 0.2\right)/0.4\right]^2\right\} \min\left(1, 2\lambda_p / d_o\right), \tag{8}$$

in which ρ and ρ_s are the densities of the water and sediment, respectively, Ψ_{rms} is the Shields number estimated in terms of u_c and d_{50}, λ_p is the pitch length of ripples if present, and d_o is the near-bottom orbital velocity.

The above relations are derived to predict the net transport rate under sheet flow conditions. Experiments indicate that especially for highly asymmetric waves, the sand which had been entrained during the positive cycle was brought back into the negative direction by the successive negative cycle. This mechanism is in some cases strong enough to cause the net transport to be in the negative direction. The above formula takes into account the above mechanism. Finally, the formula also covers suspended load over ripples as well as bed load transport (Watanabe & Dibajnia, 1996).

The suspended load q_s induced by wave-breaking is simulated adopting the Bailard formula after the consideration that the only dissipation mechanism is the wave breaking. The formula is written:

$$q_s=0.02 <D> V_d/w, \tag{9}$$

where <..> denotes a time-average quantity.

The sediment transport rate in the swash zone is calculated as in the submerged part using only the Watanabe and Dibajnia formula, according to the assumption that the primary sediment transport mechanism in this zone is the sheet flow (Leont'yev, 1996).

The above formulas are applied in the cross-shore case. In the 2DH calculation, the energetics approach developed by Karambas (1998) is adopted.

Following the sediment transport calculations, the morphological changes of the sea bed are updated in the model according to the conservation equation of sediment mass.

Figures 1 and 2 show the beach profile change due to regular and irregular wave attack in comparison with experimental data. Two different experiments are reproduced: Dette et al. test B2 (Dette et al., 1998) and CRIEPI test 1-1 (Kajima et al., 1982). The first test corresponds to an erosional condition (TMA spectral waves with height H_{m0}=1.2m, period T_m=5.0 s and median diameter of sand d_{50}=0.3 mm) whereas the second one corresponds to a depositional case (regular waves with height H_o=0.46 m, period T=6.0 sec, initial slope 1/20 and median diameter of sand d_{50}=0.47 mm). Comparisons between model results and experimental data indicate the effectiveness of the model in reproducing erosion and bar formation as well as accretion and berm formation. The distinction between erosion and accretion was made automatically without the use of a criterion as proposed by many researchers (Larson, 1996, Leont'yev, 1996). In addition, the morphological changes in the swash zone are well described by the model despite the simple approach adopted.

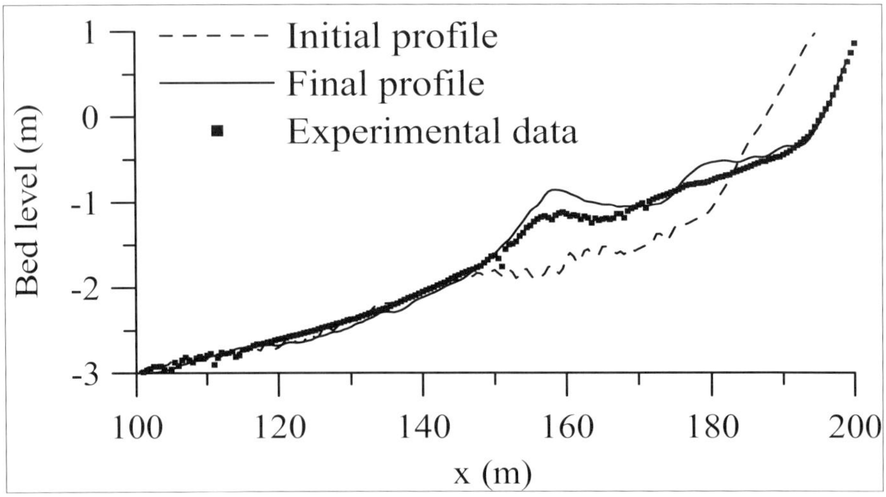

Fig. 1. Beach profile change due to random wave attack: erosional condition, Dette et al. test B2 (Dette et al., 1998)

4. Application to soft shore protection methods

In the following, the model is applied to predict the morphological effects of a submerged breakwater on the cross-shore beach profile, the breaking-wave-induced current velocity field and shoreline evolution behind a floating breakwater and the coastline evolution in a beach nourishment scenario assuming a trapezoidal beachfill.

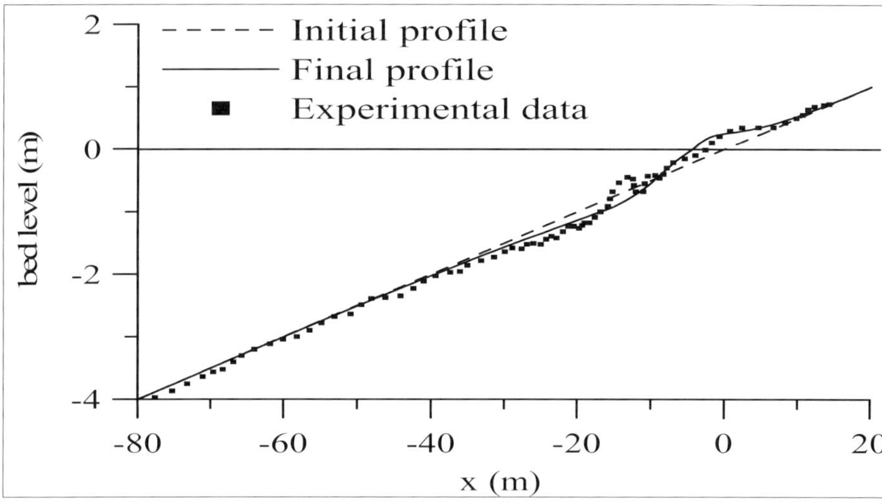

Fig. 2. Beach profile change due to random wave attack: depositional case, CRIEPI test 1-1 (Kajima et al., 1982)

4.1 SUBMERGED BREAKWATER

One of the most promising methods of coastal protection is the use of submerged breakwaters. Due to the presence of the structure, a part of the wave energy is reflected offshore, while a part of it is transmitted shoreward. A significant amount of wave energy is dissipated due to friction and breaking on the structure. Consequently, a small part of the energy is transmitted behind the breakwater and hence the structure can be used to protect the shoreward area from wave action. The present model is applied to predict the effects of a submerged breakwater on cross-shore evolution (Fig. 3). In the following, the TU Delft experimental data (Dekker, 1996) are reproduced. The slope was 1 in 15, the water depth in front of the structure was 0.4 m, the height of structure 0.3m and the freebroad 0.1 m. The generated wave spectrum was of JONSWAP type with significant wave height H_{sig}=0.098 m and peak period T_p=1.55 s (experiment B). The bed consisted of sand with D_{50}=95 μm and fall velocity w=0.01 m/s.

The model effectively describes the accumulation of the sediment behind the breakwater as well as its effects on the longshore bar formation.

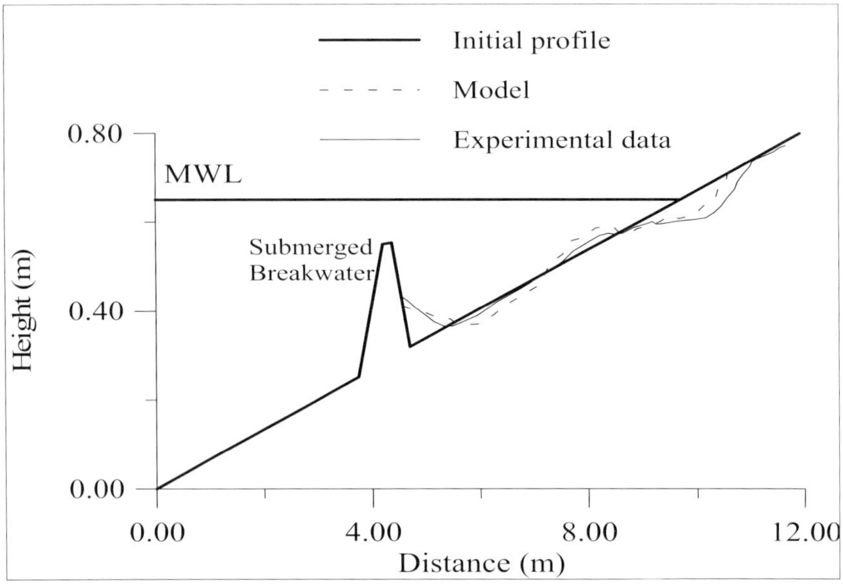

Fig. 3. Bed evolution behind a submerged breakwater: comparison between measurements and simulation

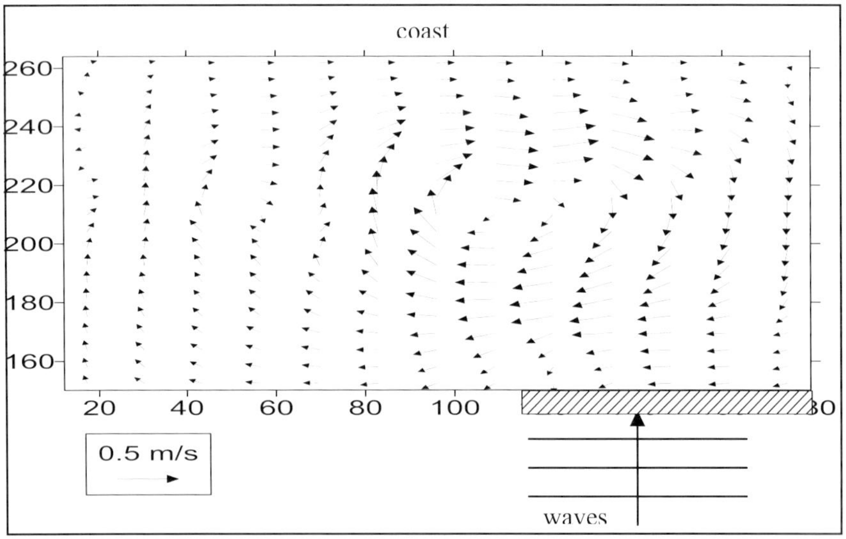

Figure 4. Breaking wave induced velocity filed behind a floating breakwater

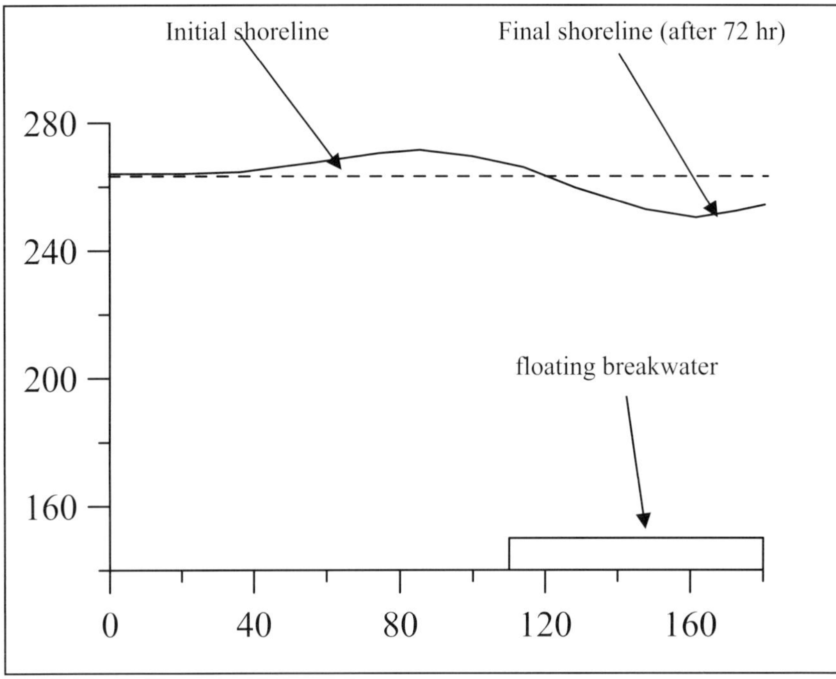

Figure 5. Shore-line evolution behind a floating breakwater

4.2 FLOATING BREAKWATER

Floating breakwaters can be used for coastal protection and restoration in regions with limited tidal mean water variations (Mediterranean sea). Their deployment and withdrawal in different periods of the year offer dynamic opportunities for full control of the hydro-morphodynamic conditions in the coastal zone. They can offer a satisfactory – not high – level of wave attenuation and protection from external wave disturbance.

The presence of a floating breakwater can be introduced in a Boussinesq model assuming that the water flux beneath the floating breakwater can be simulated in the same way as flux in a tube (for more details, see Kriezi et al., 1997). In the following numerical experiment, waves are diffracted and transmitted behind the floating breakwater in a coast with slope 1/25 and mean grain size d_{50}=0.2 mm. The breakwater length is 120 m. The incident waves (with period T=8 s and height H=1.6 m) are common on the breakwater and the beach. The breaking-wave-induced velocity field (Fig. 4) is very similar to that in which there is a wall instead of a floating breakwater. In Fig. 5, the shoreline evolution behind the floating is shown after 72 hr of wave action. As expected, the floating breakwater effects are similar to the detached breakwater ones.

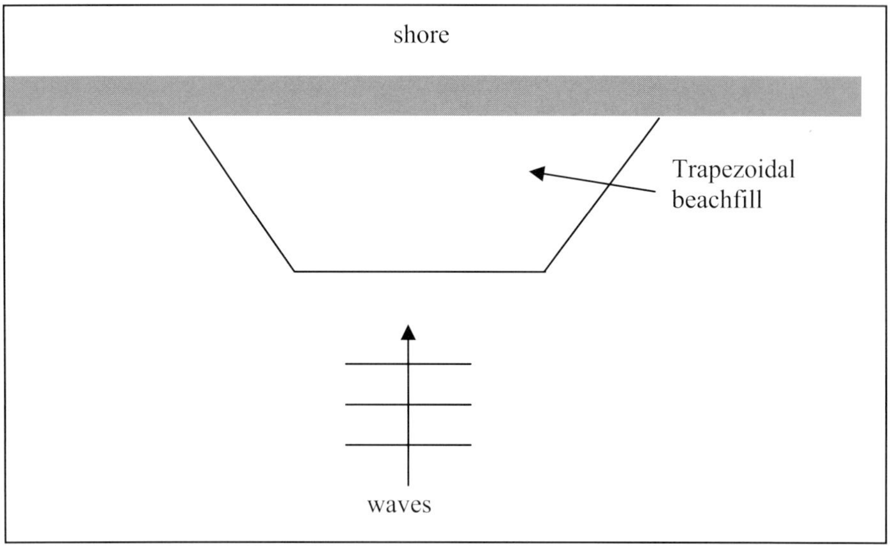

Fig. 6. Beach nourishment scenario (trapezoidal beachfill)

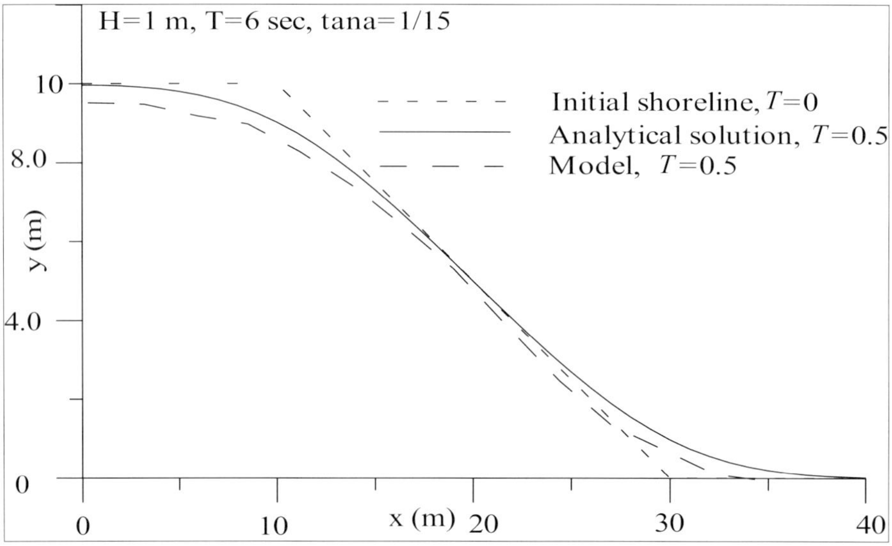

Fig. 7. Shoreline change of a trapezoidal beachfill: comparison between analytical solution (Work & Rogers, 1997) and the present numerical model

5. Beach nourishment scenario

The model is also applied to simulate the coastline evolution in a beach nourishment scenario assuming normal wave incidence on a trapezoidal beachfill (Fig. 6). In Figure 7, the predicted shoreline change after non-dimensional time T=0.5 is compared with an analytical solution of the Pelnard-Considere equation (one-line model, Work & Rogers, 1997).

The non-dimensional time T is defined as: $T=4(Gt)^{0.5}/l_1$, where G is ‘longshore diffusivity’ parameter (Work & Rogers, 1997) and l_1 is the longshore length (l_1=20m in the present case). The depth at the toe of the beachfill is h_t=1.5m, the slope 1:15, the incident wave height H=1m, the period T=6 sec and the fall velocity w=0.03 m/s. In Figure 7 the difference between the present model and the analytical solution is the shoreline displacement due to offshore transport. Under the applied conditions, the non-dimensional fall speed N=H/(wT)=5.55 (N is also known as the Dean number) is greater than the critical value N_c=3.2 and consequently the direction of the cross-shore sediment transport is expected to be offshore (erosion).

6. Conclusions

The 2DH energy and the Boussinesq equations are simultaneously solved for the simulation of breaking-wave propagation in the surf zone. Model results are used in an energetics sediment transport model based on the Watanabe and Dibajnia and the Bailard formulas.

Comparisons with experiments and analytical solution show that the model is capable of accurately predicting morphological changes in the nearshore area, including the swash zone.

The model can be also applied in soft-shore protection projects where the evaluation of beach processes has to be as accurate as possible.

REFERENCES

Bailard, J.A.: ‘An energetics total sediment transport model for a plane sloping beach’, *Journal of Geophysical Research* 86, No C11, 1981, pp. 1093—10954.

Briad, M-H.G. & Kamphuis, J.W.: ‘Sediment transport in the surf zone: a quasi 3-D numerical model’, *Coastal Engineering* 20, 1993, pp.135—156.

Dette H.H., Peters, K. & Newe, J.: ‘Large wave flume experiment ’96/97’ MAST III – SAFE Project, Report No. 825, 1998.

Dekker R.: ‘Submerged breakwater influence on beach hydrodynamics. A 2DV approach’, master thesis, 1996, UPC, Barcelona, TU Delft.

Dibajnia, M. & Watanabe, A.: 'A transport rate formula for mixed sands', *International Conference on Coastal Engineering 1996,* ASCE, 1996, pp. 3791–3804.

Dibajnia, M. & Watanabe, A.: 'Sheet flow under nonlinear waves and currents', *International Conference on Coastal Engineering 1992,* ASCE, 1992, pp. 2015–2028.

Kajima, R., Shimizu, T., Maruyama, K. & Saito, S.: 'Experiments on beach profile change with a large wave flume', *Int. Conf. on Coastal Engineering 1982, ASCE*, 1982, pp.1385–1404.

Karambas, Th.V., Southgate, H.N., & Koutitas, C.: 'A Boussinesq model for inshore zone sediment transport using an energetics approach', *Coastal Dynamics '95*, Gdansk, 1995, pp. 841–849.

Karambas, Th.V.: 'Nonlinear wave energy modelling in the surf zone', *Nonlinear Processes in Geophysics* 3, 1996, pp. 127–134.

Karambas, Th.V., Prinos, P., & Kriezi, K.K.: 'Modelling of hydrodynamic and morphological effects of submerged breakwaters on the nearshore region', *Coastal Dynamics '97*, ASCE, 1997, pp. 764–773.

Karambas, Th.V.: '2DH non-linear dispersive wave modelling and sediment transport in the nearshore zone', *International Conference on Coastal Engineering 1998, ASCE*, Copenhagen, 1998, pp.2940–2953.

Katopodi, I. & Ribberink, J.S.: 'Quasi-3D modelling of suspended sediment transport by currents and waves', *Coastal Engineering* 18, 1992, pp.83–110.

Kriezi, E.E., Karambas, Th.V., Koutitas, C., & Prinos, P.: 'Interaction of floating breakwaters with the 2D-H hydrodynamic processes in the coastal zone', *Coastal Dynamics '97*, ASCE, ed. E.B. Thorton, 1997, pp. 724–733.

Larson, M.: 'Model of beach profile change under random waves', *Journal of Waterway, Port, Coastal, and Ocean Engineering*, vol. 122, no 4, 1996, pp. 172–181.

Leont'yev, I.O.: 'Numerical modelling of beach erosion during storm event', *Coastal Engineering*, 29, 1996, pp. 187–200.

Putrevu, U. & Svendsen, I.A.: 'Vertical structure of the undertow outside the surf zone', *Journal of Geophysical Research*, vol. 98, no C12, 1993, pp. 22707–22716.

Roelvink, J.A. & Stive, M. J. F.: 'Bar-generating cross-shore flow mechanics on a beach', *Journal of Geophysical Research* 94, No C4, 1989, pp. 4785–4800.

de Vried, H.J. & Stive M.J.F.: 'Quasi-3D modelling of nearshore currents', *Coastal Engineering* 11, 1987, pp. 565–601.

Watanabe, A. and Dibajnia, M.: 'Mathematical models for waves and beach profiles in surf and swash zones.', *International Conference on Coastal Engineering 1996*, ASCE, 1996, pp. 3105–3114.

Work, P.A. and Rogers, W.E.: 'Wave transformations for beach nourishment projects', *Coastal Engineering*, 32, 1997, pp. 1–18.

'SOFT' SOLUTIONS FOR BEACH PROTECTION AND REVIVAL ALONG THE HERZLIYA ERODED BEACHES, CENTRAL ISRAEL MEDITERRANEAN COAST

YAACOV NIR[1] AND MORRIS PERPIGNAN[2]

[1] *Hydraulics Engineer, 15 Shimeoni St., Rehovot 76248, Israel*
[2] *Hydraulics Engineer, 50 Pinkas St., Tel Aviv, Israel*

Abstract

The Israeli coastline along the Mediterranean has one of the densest offshore systems of structures of all the countries bordering the Mediterranean , its 190 km long coast having some 50 sites with artificial structures such as harbours, marinas, cooling basins, detached breakwaters, groins, sea walls, and so on. This large disproportionately large number should be seen as the result of a basic misunderstanding of the coastal environment on the one hand, and the search for quick solutions on the other (as a result of short-sighted planning in relation to the natural sources).

Most of the structures mentioned have caused considerable damage to their immediate vicinity, as well as to the entire sand transport system.

However, the increasingly unsuccessful attempts to resolve the problem, as well as the beginning of what looks like a more logical approach on the part of the authorities, has led to new and more modern ways of tackling beach protection. The first site where sand replenishment is expected to be done on a relatively large scale is the Herzliya beach, located north and next to the Herzliya marina.

The decision to nourish with sand this shore is a crucial one and will be discussed in this paper.

1. Introduction

The Israeli coastline along the Mediterranean belongs to the so-called 'Nile Littoral Cell' that commences east of Alexandria at the Abu-Kir Bay, and extends along the Nile delta and North Sinai to the Acre promontory (Figure 1). Some 640 km long, this littoral cell is the longest in this eastern sector of the Levantine Basin. The Nile is the main sediment supply source of this cell. The nearby Sinai and Israeli mainland contributes to a minute degree to the sediment along their coasts. The1902 and 1964 damming at Aswan in Upper Egypt put a complete stop to the supply of sand to the delta's coasts. As a consequence, the Nile is practically sediment-free, and no sediment reaches the Mediterranean through the Rosetta and Damietta mouths. The average yearly amount of sand that has been supplied to the system by the Nile for thousands of years has been much larger than the longshore sand carrying and transport capacities. A huge amount of sand has thus been deposited along the delta coasts and in its shallow shelf. This reserve or a 'bank' of sand still plays a role as the main sediment supplier to the coastal system of the south-eastern corner of the Levantine Basin.

C. L. Goudas et al. (eds.), Soft Shore Protection, 151-156.

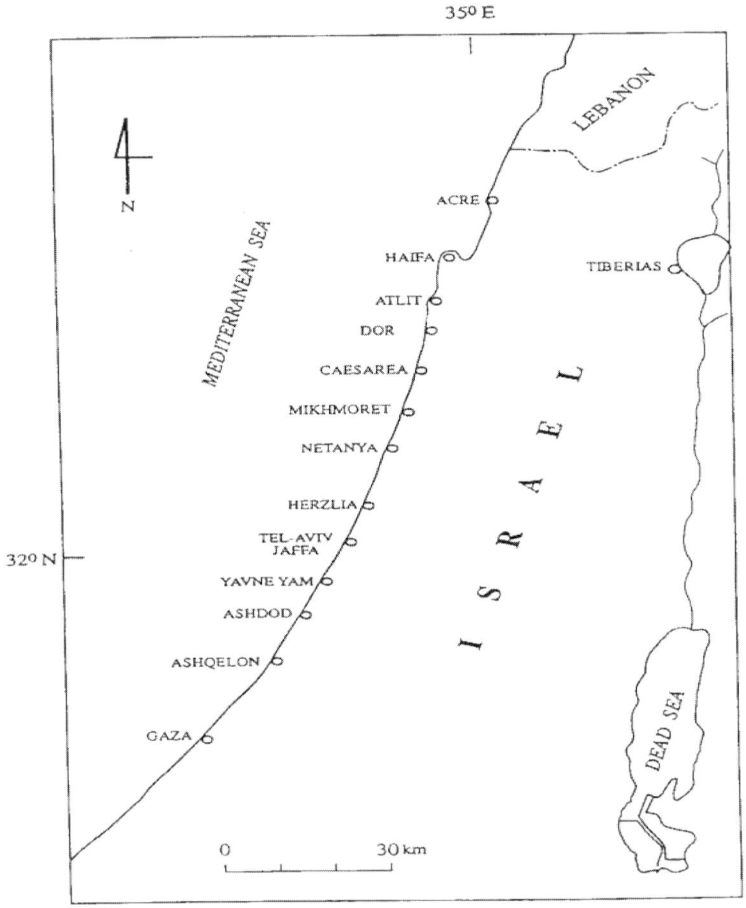

Fig. 1. Location map of the Herzliya and Ashqelon marinas

Until recently, the damming of the Aswan had not had much effect on the sedimentological conditions along the Israeli sector of this cell. Most erosion and coast-narrowing processes were caused by local offshore structures. The delta coasts, on the other hand, have suffered from erosion in many places and there has been severe damage to property: land, houses and roads have been covered by the advancing seawater.

The Mediterranean coastal region of Israel is densely populated, land is very expensive, and the rapid rise in the standard of living of the past few decades has resulted in, amongst other things, a considerable demand for berth places in marinas. This in turn has put considerable pressure on the planning authorities of the Ministry of the Interior, and a subsequent preliminary list of no less than 13 planned marinas. In the meantime, more than 50 offshore structures along its coasts have been erected: 3 big harbors, 4 marinas, 4 cooling basins of power plants, more than 20 detached breakwaters, groins, sea walls, and so on. This disproportionately large number is to some extent a result of a basic misunderstanding of the fragility of the coastal environment and its limited capacity. The quick solution to the so-called recurrent problems is to some extent the result of the wrong strategies and shortsighted planning on the part of the authorities in relation to the natural sources.

Fig. 2. Oblique aerial photograph of the Herzliya Marina, and the three detached breakwaters to the north of it. Photo taken immediately after the termination of construction (end of summer, 1993). Notice the very narrow beaches extending along the cliff at the backshore. (Photo by 'Ofeq')

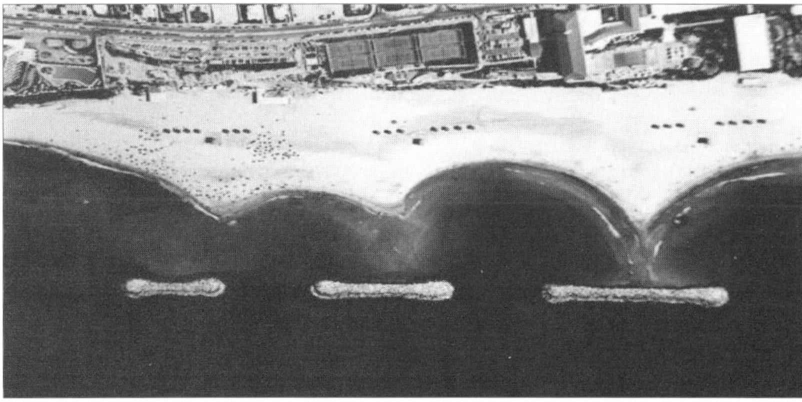

Fig. 3. Aerial photograph of the 3 detached breakwaters north of the Herzliya Marina. 6/11/93.(Photo by 'Ofeq')

Most of the structures mentioned have caused considerable damage to the nearby areas as well as to the local sand transport system. All of the large structures effectively became a dam wall, resulting in sand accumulation on the up-stream side, and large scale erosion down-stream. Until very recently, the only solution to the erosion downstream (which could have been expected) was building a series three detached breakwaters near the marinas (Figure 2). These breakwaters were constructed north of the marinas in order to protect the foci of the expected erosion as proposed by the physical model (Finkelstein, 1987). This model forecasted a minute influence of the combined

Fig. 4. Herzliya northen beaches: waves reaching the backshore during a very low storm. 19/1/2000 (photo by Y. Nir)

Fig. 5. Exposed kurkar rocks along the northern sector of Herzliya beaches during a stormy day. 19/1/2000 (photo by Y. Nir)

structure of the marina and its three breakwaters on their downstream beaches. In fact, the effects were much more severe, and shortly after construction, these breakwaters had accumulated a huge amount of sand (Figure 3), and were functioning as a new barrier to the natural flow of the sand, resulting in a 'domino' beach erosion effect (Figure 4). The downstream beaches lost a large amount of their sand, and were permanently narrowed. Baird (1998) estimated that some 400000 m^3 of sand has been deposited in various sites around the Herzliya marina and its three breakwaters. Although these figures have not been entirely proved, there is no doubt that the Herzliya beaches have lost a large amount of their permanent sand. Another example of downstream damage to an offshore structure is the southerly marina at Ashqelon (Figure 1), which caused most severe erosion to the beaches and their backshore kurkar cliffs ('kurkar' is a local term for carbonate cemented quartz sandstone dating back to the Pleistocene age).

2. Sand replenishment scheme for the Herzliya coast

The increasingly unsuccessful attempts to resolve the problem and a more rational approach on the part of the authorities has led to a new and more up-to-date beach protection philosophies. Despite this, however, many, including Baird Associates, are still advocating the 'hard' solution, that is the construction of another series of detached breakwaters (built in slow successive steps).

The first site where sand replenishment is expected to take place on a relatively large scale is Herzliya Beach north and adjacent to the Herzliya marina and its three breakwaters. Here, as a result of the above-mentioned marina constructed during the early nineties, large-scale erosion 'denuded' the 2.5-km long wide sandy beaches of the city (Ephrati & Madpis, 1994; Zvieli & Klein, 1998; Baird, 1996, 1998). The beaches lost more than 30% of their original width. Kurkar and beach rock that used to be exposed only during strong winter storms were exposed (Figure 5), sea and bathing conditions deteriorated and it became increasingly obvious that the coasts had to be reconstructed. As a result of this critical loss of sand on the one hand, and the greater risk to the constructed backshore cliff on the other, the authorities started to search for an environmentally friendly solution. In spite of the fact that all realize that some of the sand might be lost in the early stages and the whole plan is to some extent risky, the Herzliya municipality advocates sand replenishment.

This decision is quite an innovative one for some of the Israeli coastal experts and city councils since up to now they have advocated compact, 'sure', dogmatic and conservative solutions. On the other hand, all nature preservation bodies endorsed the measures long ago. *The new approach took set to usher in a new era in coastal protection in Israel.*

Sand replenishment in the 2.5 km-long Herzliya coast is currently in the planning stages. The following issues need to be tackled before the project can get underway:

1. Basic bathymetric, geophysical and sedimentological studies of the entire site and its neighbouring areas need to be done.
2. A basic biological study of the above area needs to be done.

3. Research into sand-supply sources needs to be done.
4. Methods of transportation of the sand to the site need to be found.
5. The right locations for the artificial sand deposition need to be identified.
6. The right quantities of sand to be fed need to be determined.
7. The correct follow-up and monitoring (bathymetry, aerial photography, etc.) need to be determined.

While the costs and benefits associated with these aspects of the project are currently still being examined, it can be assumed that the government agencies involved will endorse the planned more modern and much 'softer' and environmentally friendly coastal management.

REFERENCES

Baird, W. F.: Preliminary seminar on the physical characteristics of the shoreline of Israel with emphasis on sand transport. Interim report to the City of Herzliya, 1996, p.54..

Baird, F. W.: Design investigation for shore protection, wave agitation, and Sedimentation at Herzliya marina. Interim report to the City of Herzliya, 1998.

Ephrati, A., and Madpis, A.: Monitoring the Herzliya Marina. Unpublished internal report, 1994, p. 2 (in Hebrew).

Finkelstein, A.: Herzliya Marina – study of a sedimentological model. Coastal and Marine Engineering Marine Institute. Hebrew Technion, Haifa, 1987 p. 35 (in Hebrew).

Zvieli, D., and Klein, M.: The Herzliya marina influence on the neighbouring beaches – the predicted vs. the factual, Israel Oceanographic and Limnologic Research, Haifa; Annual Meeting, 1998, pp. 26-29 (in Hebrew).

FIELD EXPERIMENTS IN RELATION TO A GRAVITY–DRAINED SYSTEM AS A SOFT SHORE PROTECTION MEASURE

KAZUMASA KATOH[1], SHINN-ICHI YANAGISHIMA[1], IWAO HASEGAWA[2] AND AKIYOSHI KATANO[2]

[1] *Hydraulic Division, Port and Harbor Research Institute, Nagase 3-1-1, Yokosuka, Japan*
[2] *ECOH CORPORATION, Kita-Ueno 2-6-4, Taito-ku, Tokyo, Japan*

Abstract

Many years of observing the fluctuation in the beach width at Hasaki Beach, which overlooks the Pacific Ocean, have shown that rapid beach retreat occurs under storm conditions. Spectral analyses of incoming waves and beach profile surveys have revealled that infragravity waves in high wave trains play an important role in rapid beach retreat. Infiltration of swashing water due to infragravity waves raises the water table below the beach, and this gives rise to fluidisation of the beach surface.

Based on these observations, the authors have developed a gravity-drained system, which does not require any power supply from the outside the system. Field tests of this system confirm that the system keeps the water table from rising and effectively fulfils the function of beach stabilization and beach amenity preservation.

1. Introduction

Since World War Two, natural disasters have taken their toll on Japanese coasts and shore protection works have been extensively carried out to prevent coastal erosion. Although shore protection using various hard structures carried out since the early 1950s, has kept the damage to a minimum, people have been prevented from accessing stretches of coastline which should be available for everyone.

Seaside recreation has been one of the most popular marine pastimes in Japan. The bathing beaches, however, have eroded gradually since the 1960s, and many sandy beaches have disappeared. In addition, shore protection works of the hard type have disturbed the scenery of the coasts to a considerable extent. These considerations prompted us to develop a gravity-drained system providing both shore protection and beach amenities.

A series of field experiments were carried out to deepen our understanding of the physics of rapid beach erosion under rough seas and to examine the functions of a gravity-drained system installed on a sandy beach facing the open sea.

2. Study site

The field experiments were conducted at the Hazaki Oceanographic Research Station (HORS), which has a research pier 427 m long on gently sloping natural sandy beach facing the Pacific Ocean (Figure 1 and Photo 1).

C. L. Goudas et al. (eds.), Soft Shore Protection, 157-170.

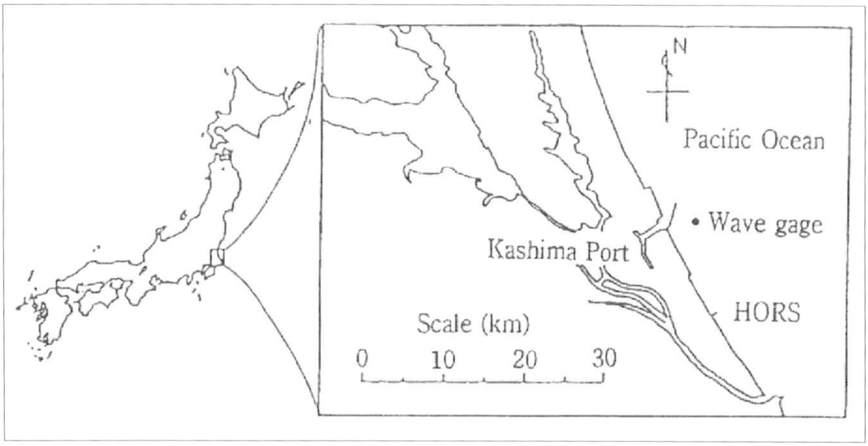

Fig. 1. Location of HORS

The median diameter of the sand near the shoreline is 0.18 mm. The foreshore averages 1/50, while the mean bottom slope in the surf zone is 1/60, slightly less than the foreshore. The tidal range is about 1.5 m. Aerial photographs taken from 1949 to 1984 indicate that the shoreline position has been stable since 1979.

Photo 1. Side view of HORS and the research pier

The beach profile of 500 m long from the tip of the pier to the backshore was measured once a day, at a cross-shore interval of 5 meters. The beach topography above the low water level was surveyed once a month. The sea bottom topography around HORS was surveyed also once a month.

Waves in the surf zone were measured every hour at 7 locations along the pier. The water table below the beach was measured using an array of pipes 12.5 cm in diameter, placed at regular intervals normally to the beachline. The water level inside the pipes corresponded to the local water level.

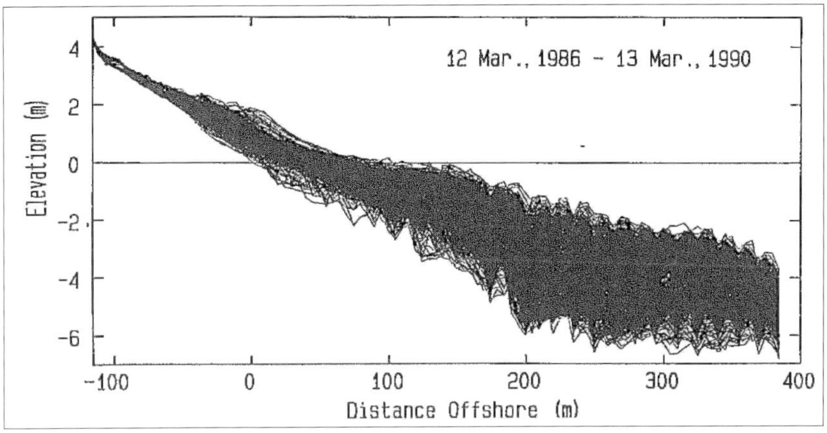

Fig. 2. Long-term variations in the beach profile

3. Evolutional features of the study beach

Figure 2 shows the variations in the beach profile along the research pier during a period of about four years, between 1986 and 1990. It will be seen that there was a considerable variation in the depth of the sea bottom (of about 4 m) and a fluctuation of about 50 m in the shoreline position at the high water level, 1.4 m above the datum level.

Figure 3 depicts the daily variations in the shoreline position. We observed a tendency for the shoreline to retreat in the high wave energy months of March, April, and August, and a gradual advance at a average rate of 0.68 m/day during the low wave energy period from May to July.

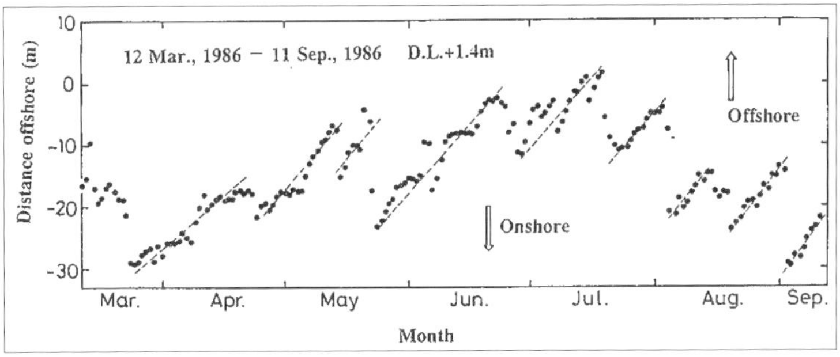

Fig. 3. Daily variations in the shoreline position between March and September in 1986

In addition, the shoreline constantly advanced and retreated in line with the daily variation in incident wave conditions. These beach profile changes are typical of sandy beaches facing the open sea.

In particular, we paid attention to the rapid shoreline retreat of about 20 m over a period of one or two days due to high waves, because understanding the physics of such a rapid retreat was crucial to determining how, where, and when the beach could be stabilized.

4. The physics of rapid erosion

Based on observations, Grant (1948) pointed out that the water table within the beach has a significant influence on the deposition and erosion of the foreshore and backshore. Since then, many researchers have investigated the interaction between the beach water table and the beach profile in field, laboratory, and numerical experiments (e.g. Packwood, 1983; Yanagishima et al, 1991; Oh & Dean, 1994; Baird et al., 1996). Here, we will also discuss the influence of ground water level behaviour on sediment transport in the swash zone.

Engineers used to think that violent wind waves play a leading role in rapid foreshore erosion. Wind waves, however, lose most of their energy as they propagate toward the shore. Accordingly, wind waves alone are unlikely to cause rapid erosion of the foreshore.

Spectral analyses of incoming waves and beach profile surveys were made simultaneously to elucidate which component of the incident waves is predominant in foreshore and backshore erosion. The incident wave energy was computed from wave data obtained at nine wave-gage locations. Of these, three wave gages were temporarily set on the extension line of the pier at locations of 3.2 km (24 m deep), 2.1 km (14 m deep), and 1.3 km (9 m deep) offshore from the shoreline.

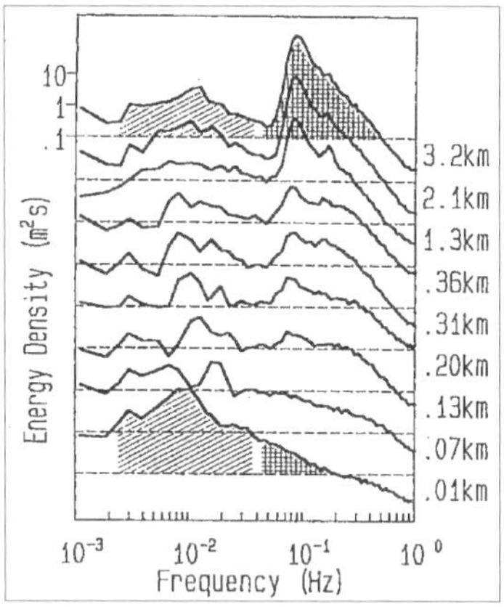

Fig. 4. Cross-shore variations in spectral energy density

Figure 4 shows the cross-shore variations in spectral energy densities for two hours; the offshore significant wave height was 3.7 m. The spectral energy density in the offshore has two clear peaks. The largest peak has a frequency of 0.1 Hz, corresponding to the frequency of wind waves, and the other peak at 0.01 Hz corresponds to that of infragravity waves.

Figure 4 demostrates that the wind wave energy decreases greatly as the waves approach the shoreline due to wave-breaking, while that of the infragravity waves increases because they do not break in the surf zone and become largest around the shoreline as a result of wave shoaling. It therefore seemed likely that the infragravity waves bring about severe beach erosion, and contribute greatly to the beach evolution, especially during rough sea conditions. The infragravity waves maintain their high energy and run up higher beyond the foreshore crest and flow over the backshore. The uprushing water piles up for a while on the backshore (Photo. 2).

Photo 2. Uprush of the infragravity waves

Figure 5 is a wave record during Typhoon No. 13 in 1987, which passed near the study site. The upper figure shows variations of significant offshore wave height H0 and period T. The offshore wave height started to increase on 11 September and ran to about 3 m on the 14th. In the following two days, the wave height decreased slightly. During these days, there was a predominant wave-field swell of wave periods from 12 to 13 sec. After the 14th, wind waves prevailed and the largest wave height attained was about 6 m on 17th. In the latter stage of the field experiment, the significant period of the wind waves was 10 sec.

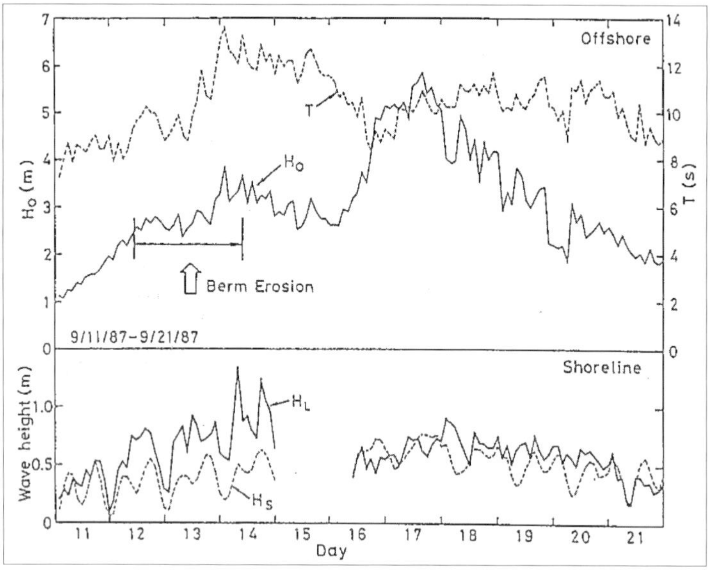

Fig. 5. Wave records during Typhoon No. 13 in 1987

As shown in the lower figure, the height H_S of the wind waves near the shoreline, changes periodically independently of the offshore wind waves, since the water depth varies with the tide, and it controls the wave height in the surf zone. On the other hand, the height H_L of infragravity waves near the shoreline, was about 1 m on the 13th and exceeded 1 m on the 14th. When the offshore wind waves reached their maximum height on the 17th, the infragravity wave height was no more than 0.7 m.

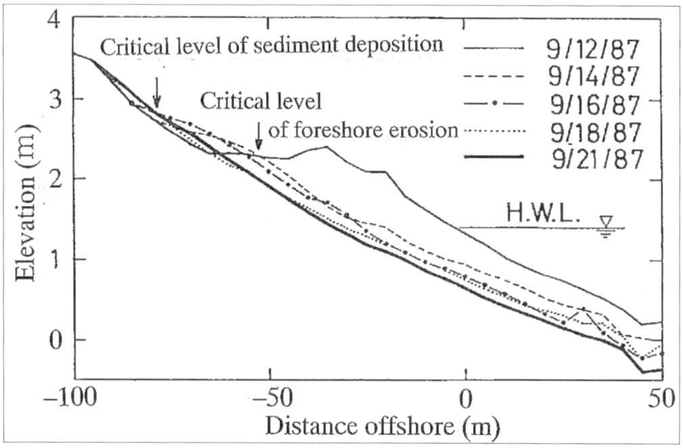

Fig. 6. Successive comparisons of the foreshore and backshore profiles during Typhoon No. 13 in 1987

Figure 6 compares the successive foreshore and backshore profiles during Typhoon No. 13. The fully developed foreshore (12th Sept.) was formed during a period of calm wave conditions before the typhoon, but it was washed away within only 24 hours, from the 12th through the 14th September, denoted by a horizontal bar in Figure 6. If Figures 5 and 6 are compared, we notice that the rapid erosion of foreshore and backshore occurred when the infragravity waves predominated.

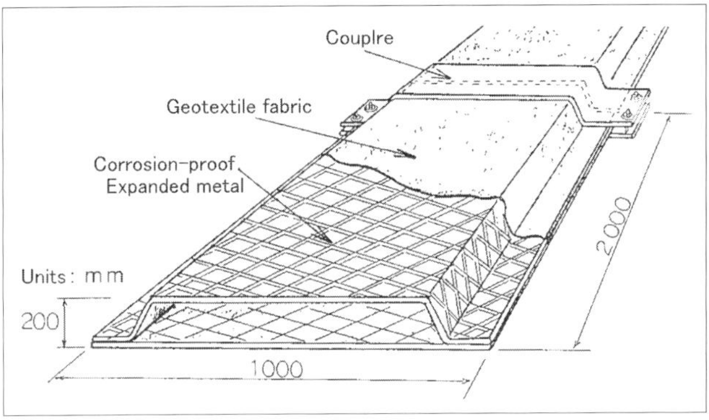

Fig. 7. Permeable layer unit of the gravity-drained system

During the rapid erosion, the wind wave height near the shoreline was less than 0.5 m, almost the same height as that under the calm wave conditions before the typhoon. Consequently, it would seem that the wind waves do not contribute to a great extent to rapid erosion. On the other hand, the infragravity wave height near the shoreline increased and attained a maximum height of 1.3 m in morning of 14th, when the rapid erosion occurred.

Moreover, Figure 6 provides very interesting evidence that sediment transported up from the foreshore zone is deposited around the upper limit of uprush, even when the foreshore erosion is in progress.

If the water table is far below the backshore surface, the swashing water percolates into the beach and thus reduces the swash velocity. This results in sediment deposition around the upper limit of wave run-up. Infiltration of the swashing water raises the water table, which is sometimes contiguous with the foreshore surface. As a result, the seaward edge of the water table slopes seaward, so that the effluent from the foreshore surface not only causes fluidization of the sediment at the beach surface but also enhances the backwash. The enhanced backwash removes the sediment on foreshore seaward (Grant, 1948). Figure 10, Section 6, shows this clearly.

It would therefore seem that infragravity waves in storm conditions significantly contribute to the rapid erosion of foreshore and backshore and high wind waves play a lesser role in rapid erosion. The field experiment indicates that one of the most essential and effective measures for preventing rapid erosion of foreshore and backshore is to let down the water table within the beach (Kawata & Tsuchiya, 1986; David et al., 1992).

After feasibility studies, a gravity-drained system which did not need any power supply from outside of the system was proposed.

5. Gravity-drained system

A gravity-drained system is composed of a hollow square tube unit 1 m wide, 2 m long, and 0.2 m thick (Figure 7). The unit is made of corrosion-proof expanded metal, and enclosed in a fine geotextile fabric 3 mm thick. A permeable layer is formed by connecting the units with a steel coupler. A drainpipe 0.4 m in diameter is connected to the permeable layer. The ground water drained into the permeable layer is discharged into the surf zone from the outlet pipes in the middle of the surf zone, 144 m offshore from the shoreline (Figure 8).

Photo 3. Gravity-drained system at HORS

In August 1994 the gravity-drained system was installed at a depth of 3 m below the beach surface within the bounds of the foreshore and the backshore (Photo. 3). The area covered was 88 m by 7.8 m.

After installation of the gravity-drained system, it was covered with beach sand, and the beach profile over the system was resorted to resemble a natural beach.

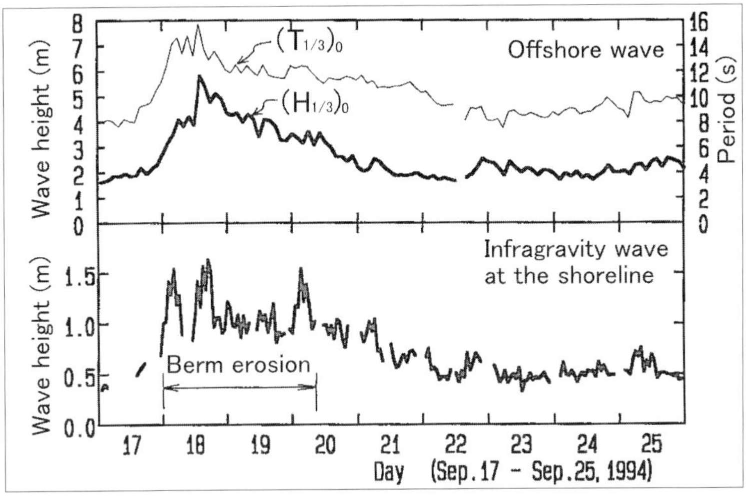

Fig. 8. Cross-section of the gravity-drained system

6. Field experiment using the gravity-drained system

About one month after setting up the system, Typhoon No. 24 attacked the study site. Figure 9 is a wave record of the typhoon. The upper figure shows the temporal variations in the offshore wave height and period, and the lower that of the infragravity wave height near the shoreline.

Severe berm erosion occurred from 18 to 20 September, denoted by a horizontal bar. Within that period, the maximum height of significant offshore waves was 5.8 m. The height of fully developed infragravity waves near the shoreline was about 1.5 m.

Photo 4 shows the uprush of infragravity waves on the beach below which the gravity-drained system (drained beach). The line of poles indicates the centerline of the drained beach.

Photo 5 shows the backwash of the infragravity waves. While the drained beach is dry, down-rushing water still remains on the natural beach on the right-hand side of the picture. The line of poles indicates the side boundary of the drained beach.

Figure 10 compares the beach profile and the ground water table between the natural and drained beaches. The thick arrow in the figure indicates the upper limit of wave run-up. The water table within the natural beach, denoted by triangles, ascends and is contiguous with the backshore surface, while that within the drained beach, denoted by circles, is 0.6 m lower than the beach surface and its gradient is almost level.

Photo 4. Uprush of infragravity waves on the drained beach

Photo 5. Backwash of infragravity waves

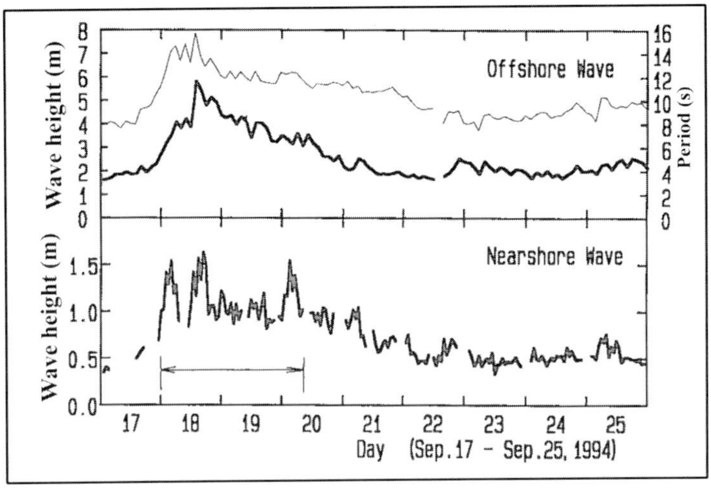

Fig. 9. Wave record during Typhoon No. 24, 1994

Figure 11 shows the temporal variations in the tidal level at the outlet pipe, the water table within the drained beach, at a location of 25 m onshore from the high water shoreline, and the mean outlet discharge in the drainpipe for 20 minutes. The drained discharge was very small before the typhoon attack on 17 September, while it increased greatly during the 18th and 19th, when the typhoon's effects were still being felt. The mean drained discharge was found to be roughly proportional to the elevation difference of the water table within drained beach and the mean sea level in surf zone.

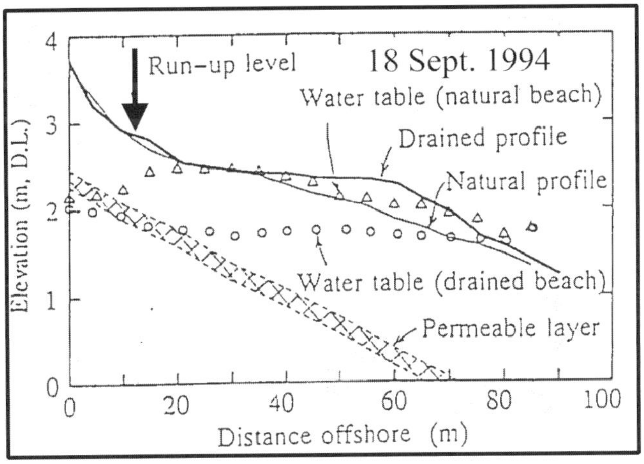

Fig. 10. Comparisons of the beach profile and the water table between natural and drained beaches

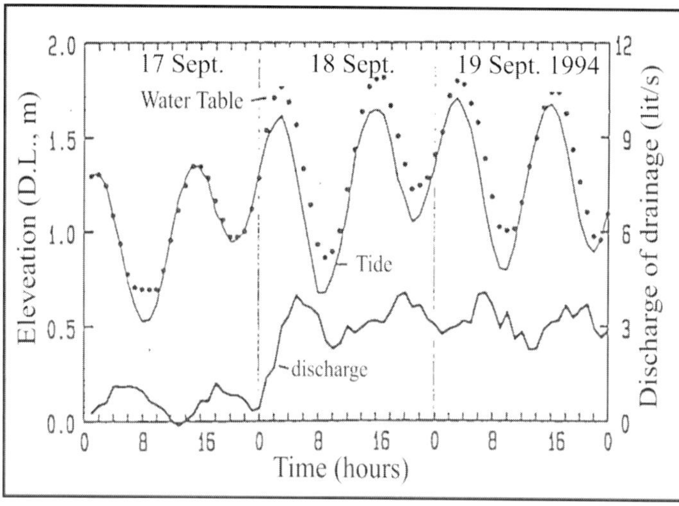

Fig. 11. Temporal variations in the water table tidal level, and mean drained discharge

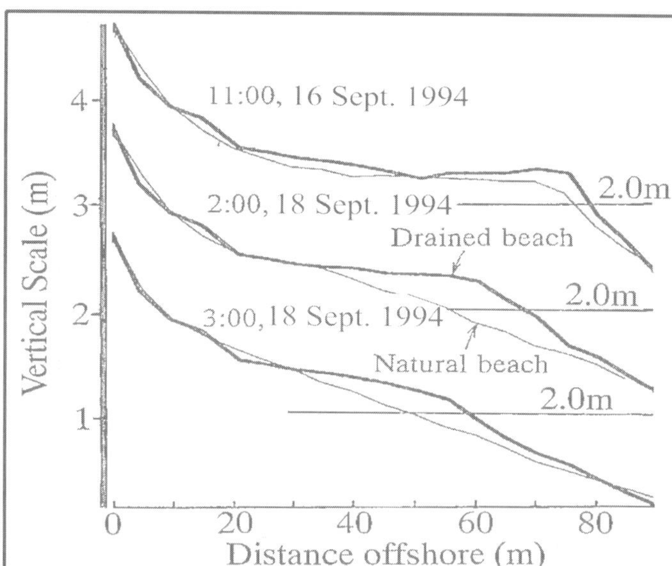

Fig. 12. Beach profile comparisons between the natural and drained beaches

Figure 12 compares the successive foreshore and backshore profiles between the natural and drained beaches before and under the typhoon. Two days before the typhoon, there was fully developed foreshore on both the beaches. However, the typhoon attack caused both the foreshore and backshore of the natural beach to be completely washed away. On the other hand, the foreshore crest on the drained beach still remained and there was no erosion of the backshore.

The field experiments confirm that a gravity-drained system is effective in stabilizing sandy beaches, even under high wave conditions.

7. Summary

The main findings of this study are as follows;

(1) Infragravity waves in storm conditions contribute significantly to the rapid erosion of foreshore and backshore. High wind waves play a lesser role in rapid erosion.

(2) Infragravity waves bring about a rise in the water table of the beach. Infiltration of swashing water raises the water table, which is sometimes contiguous with the foreshore surface. The foreshore surface effluent not only causes fluidisation of the sediment on the beach surface but also enhances backwash. The enhanced backwash removes the sediment on the seaward foreshore.

(3) Rapid erosion of the foreshore and backshore can be prevented by lowering the water table below the beach.

After feasibility studies, the gravity-drained system shown in Figure7, composed of a hollow square tube unit and not requiring any power supply from outside the system, was installed.

Field experiments confirmed that the system prevented the water table from rising and effectively stabilised the beach.

REFERENCES

Baird, A.J., Mason, T.E. & Horn P.H.: Mechanisms of beach ground water and swash interaction, *Proceedings of the 25th International Conference on Coastal Engineering*, 1996, pp. 4120–4133.

Davis, G.A., Hanslow, D.J., Hibbert, K., & Nielsen, P.: Gravity drainage: A new method of beach stabilization through drainage of the water table, *Proceedings of the 23rd International Conference on Coastal Engineering*, 1992, pp. 1129–1141.

Grant, U.S.: Influence of the water table on beach aggradation and degradation, *Journal of Marine Research*, 7(3), 1948, pp.655–660.

Kawata, Y. and Tsuchiya, Y.: Application of sub-sand system to beach erosion control, *Proceedings of the 20th International Conference on Coastal Engineering*, 1986, pp.1255–1267.

Oh, T.M. and Dean, R.G.: Effects of controlled water table on beach profile dynamics, *Proceedings of the 24th International Conference on Coastal Engineering*, 1994, pp.2449–2459

Packwood, A.R.: The influence of beach porosity on wave uprush and backwash, *Coastal Engineering*, 1983, 7, 29–40

Yanagishima, S., Katoh, K., Katayama, T., Isogami, T., & Murakami H.: Effects of depressing water table on changes of foreshore profile, *Proceedings of the 38th International Conference on Coastal Engineering*, 1991, pp.266–270 (in Japanese)

SHORT-TERM FIELD EXPERIMENTS ON BEACH TRANSFORMATION UNDER THE OPERATION OF A COASTAL DRAIN SYSTEM

MICHIO SATO, RYUICHIRO NISHI, KAZUO NAKAMURA AND TAKAO SASAKI

Department of Ocean Civil Engineering, Kagoshima University, 1-21-40 Korimoto, Kagoshima, 890-0065,JAPAN

Abstract

The objective of this study is to add prototype data for estimating the performance of a coastal drain system. Small-scale model experiments aimed at understanding the mechanisms that make coastal drain systems effective were carried out. The model experiments gave us qualitative information on the relevant mechanisms, as well as encougaring results. However, our experiments were done under conditions in which the role of the drain in sediment movement was magnified compared to the effect of wave motion. Some caution is thus advisable when the experimental results are extended to estimate the performance of prototype-scale coastal drain systems. In order to be able to evaluate the performance of a coastal drain system and establish the designing procedures, prototype scale-detailed data that includes information on waves, sediments, beach profiles, drainage and so forth, together with the development of numerical models, is necessary. This paper, firstly briefly discusses where to install drainpipes, then discusses the experimental results.

The results obtained are as follows: 1) Comparisons of the profile changes did not show any definite differences between the dewatering and the non-dewatering part of the beach. 2) Movement of the tracer reflected the flow induced by the drain. 3) Compaction may be a way of preventing erosion. An important issue is why our results and those of Vesterby (1991) differ.

1. Introduction

We used small-scale laboratory experiments to examine the mechanisms that make a coastal drain system effective in stabilizing beaches following wave action. The results of the laboratory experiments have helped us obtain a general view of the relevant mechanisms in a qualitative sense as well as encouraging results. However, our experiments were done under conditions in which the role of the drain of sediment movement was magnified compared to the effect of wave motion. If the results are extended to estimating the performance of prototype scale coastal drain systems, some caution will be necessary since detailed information on waves, sediments, beach profiles, drainage and so forth, are required together with the development of numerical models. Such data will also be needed to verify laboratory experiments and to check the numerical model.

The objective of this study is to add prototype data on the performance of a coastal drain system obtained by means of a short-term field experiment.

C. L. Goudas et al. (eds.), Soft Shore Protection, 171-182.

2. Some preliminary considerations relating to where to install

Where to place pipes is one of the main problems to be resolved from a practical point of view, especially at sites where the tidal range is not small. If tidal range is the determining factor, there are three pipe placement options, as shown in Figure 1. Judging from the results obtained with our small-scale experiments, all three of these positions may be effective in preventing beach erosion, although the dominant mechanism is different for each position.

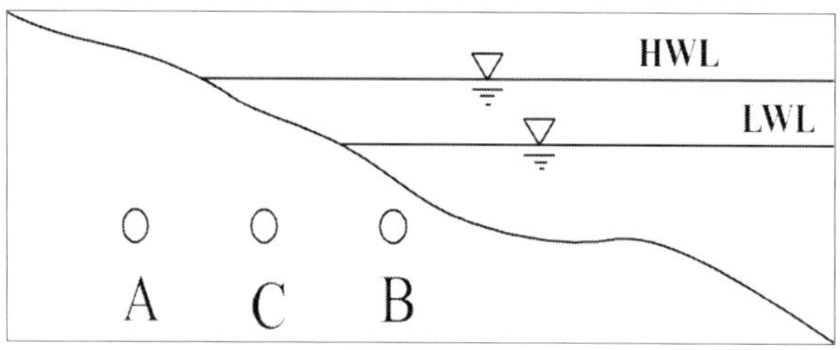

Fig. 1. Possible pipe positions

Position A seems to be preferred in conventional methods. In such systems, (BMS, GDI), dewatering lowers groundwater adjacent to the pipes and increases the unsaturated area of the backshore region. This enhances the infiltration of up-rushing and down-washing water into the beach face and quickens sediment deposition. Discharge is expected to increase proportionately to the depth a pipe is buried based on an existing formula as shown in Figure 2. The drained water contains not only infiltrated water along the swash zone but also that from the landward region of the pipe.

However, in the new method that Parks (1990) refers to in his leaflet on the 'Beach-Advancer System', the pipe is placed at B position. The method would seem more effective in that it collects groundwater beneath the littoral zone effectively. In this case, the discharge is proportionally larger the thinner the layer of sand above the pipe (Figure 3). The laboratory observations revealed in this case that, unlike in position A, the influence of the drain on the process in the swash zone was weak, and modifications of the flow field in the surf zone and in the bottom boundary layer induced by dewatering became important in preventing erosion.

One of the present authors has reported an experimental result that showed a fairly large amount of deposition of sediments around adjacent area of the pipe in a short time, and also gave two examples from our laboratory experiments which showed that in the same erosive wave conditions, different locations of the pipe resulted in different beach profile evolution (Sato et al, 1994). The latter experimental results gave us the impression that in the case of position B the drain system protects the area landward of the pipe. They seemed to endorse the 'New Method' and position B. Moreover, the coastal drain results were encouraging. However, they were obtained with small-scale physical experiments in which the effects of seepage were overestimated in comparison with the role of wave action.

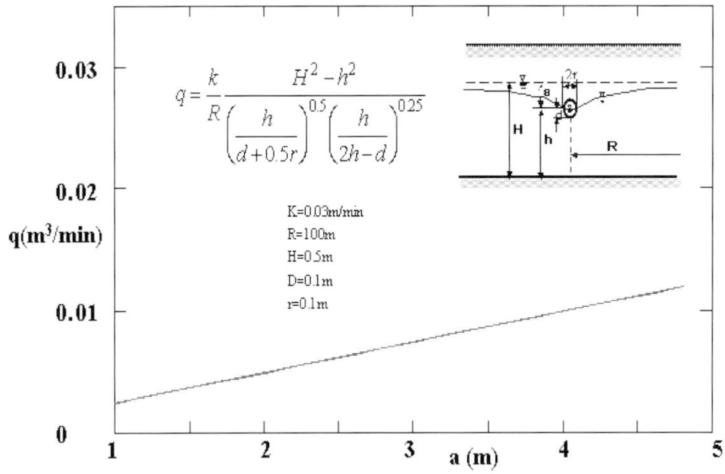

Fig. 2. Depth of the pipe and discharge (position A)

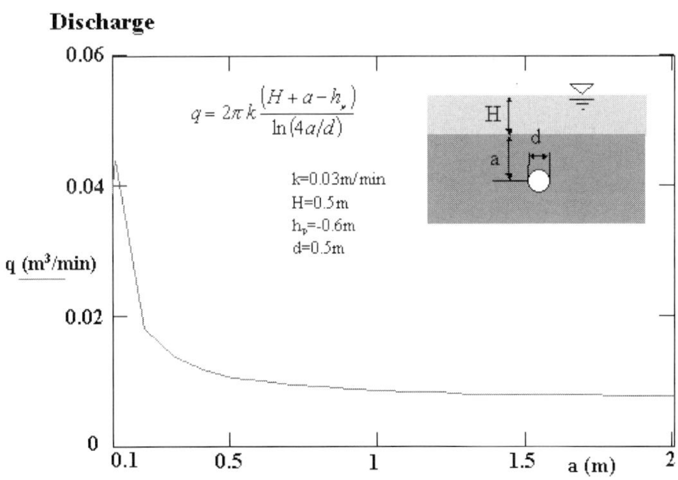

Fig. 3. Depth of the pipe and discharge (position B)

How to drain is another issue. With the conventional method, ground water is drained into a sump by gravity, and the water is then pumped out of the sump intermittently when the water volume attains a certain level. Consequently, the pump is not always driven at the indicated pump capacity. In our experiments, the pumps were joined to pipes and drained directly with continuous suction. The direct pumping drain method that we adopted would not, however, seem to be commercially viable as it has the drawback that long-term operation causes clogging up of the filter. However, we adopted the method as it was expected to enable the drain discharge to become larger and was acceptable for a short-term experiment.

3. Outline of the field experiments

The field experiments were conducted during the periods 24 November to 4 December, 1998, from 8 October to 25 October, 1999 and from 12 September to 11 October, 2000 including installation and withdrawal. These limited periods were mainly due to annual budget factors. However, Vesterby reported several meters of shoreline advance after two weeks of installation. This encouraged us to plan the field experiments.

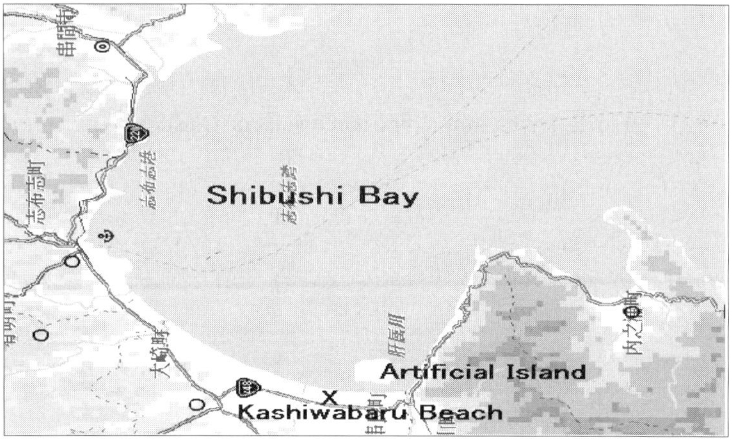

Fig. 4. Chart of the site of the experiments (Kashiwabaru Beach), at the southern end of Shibushi Bay in Kagoshima, Japan

The site of the experiments of 1998 and 1999 was Kashiwabaru Beach, which is the southern end of the innermost coast of Shibushi Bay in Kagoshima, Japan (Figure 4). The beach faces the Pacific Ocean. In 2000, the experiment was conducted at Iso Beach, which is located on inner margins of Kagoshima Bay, where the wave conditions are not as severe as at Kashiwabaru Beach.

In this paper, the main description relates to the 1999 experiment. The experimental site is a sandy beach. The beach had slope of about 1/30, and the beach material was well-sorted sand with median diameter of 0.5mm. During the field experiment, tides were semidiurnal with a maximum range of 1.8m (Figure 5). Figure 6 shows the wave height and period of significant waves measured near Biro Island by the Ministry of

Transportation. Although steeper waves were recorded from 11 October through to 12 October, on the other days, there was a consistent significant wave height of about 0.4 m with a period of 7-8s .

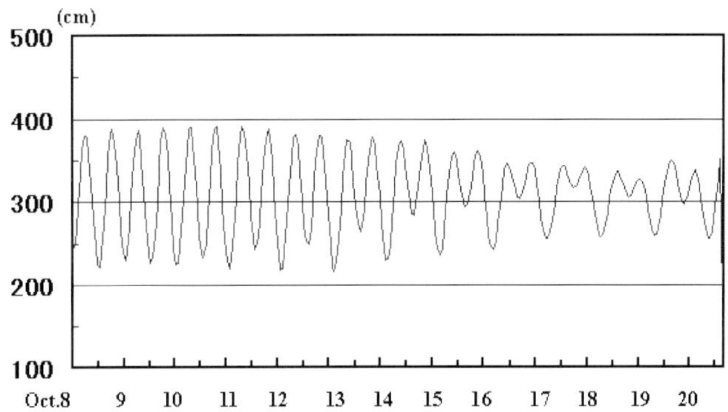

Fig. 5. Tide during the experiment

Four units of perforated pipes wound with a filter 1cm thick were buried parallel to the shoreline under the foreshore using a backhoe. The position of pipes was between the low tide shoreline and the high tide shoreline (Position C in Figure 1). This position was selected for ease of installation. A ditch 25m long, 3m wide and 1m deep was excavated with the backhoe. Four pipes, each of them attached to a steel frame, were then placed in the ditch, then covered with sand using the backhoe. The steel frames were anchored in order to cope with stormy wave conditions. Each unit consisted of a 20m long pipe with a pump at the both ends (Figure 7).

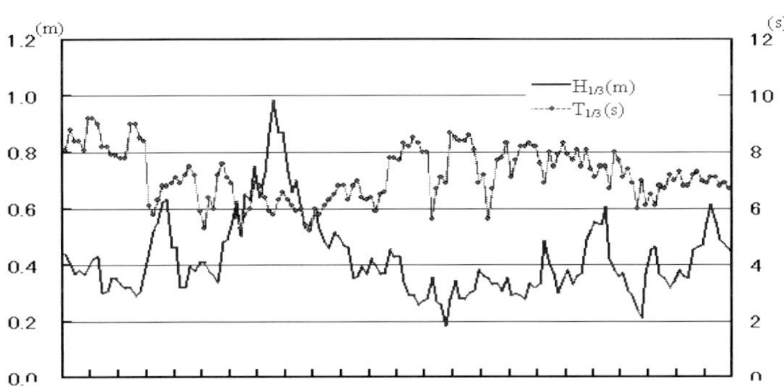

Fig. 6. Wave conditions during the experiment

The discharge was measured in terms of the time needed to fill a bucket of 100-liter capacity. We also tried to use an electromagnetic flow meter, but it did not work well, perhaps because of air bubbles. The average discharge of the each pump was $27m^3/h$, and eight pumps were operated. The discharge per unit beach width was thus

$11 m^3/h/m$. Turner and Leatherman described the pump capacity and the longshore extent of the prototype dewatering system as it was in 1995. According to them, discharges per unit beach width were estimated to be 0.5 to 4.4 $m^3/h/m$ at the most.

In order to measure the change in beach face elevation, 154 iron rods 2m in length, were driven into the beach face at an interval of 5m and parallel to the shoreline, and at 2m intervals and perpendicular to the shoreline. This made a grid system of 40m by 26m.

The repeated levelling of the beach face was examined every low tide by measuring the length of the rods exposed. A profile survey of the non-dewatering area, which was 20m from the south edge of the grid system, was also done to compare and contrast it with the dewatering area. Measuring lines parallel to the shoreline were numbered from 1 to 14. Those perpendicular to the shoreline were marked alphabetically from A to I. Two lines in the non-dewatering part were marked Y and Z.

In order to track sediment movement, the tracer technique was applied using dyed sand with 3 fluorescent colors as the tracer. Compaction of the surveyed area was also measured using a penetration-type device.

As the experiment neared its end, the total discharge decreased gradually. Inspection showed two of the pipes had filled with sand through a filter torn during the backhoe operation. The discharge was eventually reduced to half the initial discharge. This was probably partly due to the clogging up of the filter.

Fig. 7. View of the site of the experiment during installation

4. Experiment results

4.1 BEACH PROFILE CHANGES

The time histories of the averaged beach profiles of the dewatered and the non-dewatered parts during the experiment are summarized in Figures 8 and 9. At first sight there are no recognizable differences between either part.

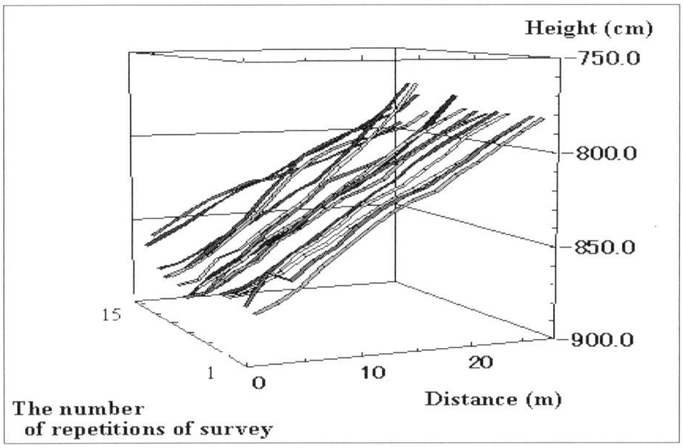

Fig. 8. Temporal beach profile change during the experiment (dewatered)

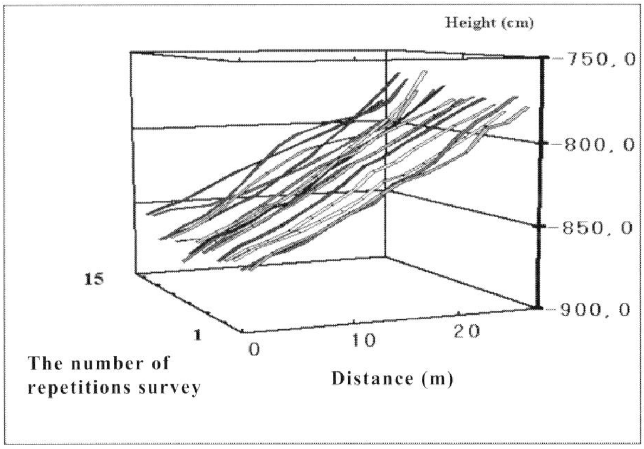

Fig. 9. Temporal beach profile change during the experiment (non-dewatered)

Figure 10 shows the difference between the averaged final profile and the initial one. Positive variation indicates deposit. The horizontal axis is the landward distance measured from the most seaward measuring line. The pipes were positioned 16 to 18m from the axis. The changes to both parts are roughly similar. However, when we examine the profile change in detail, sand deposition on the seaward side of the pipes in the dewatered part is larger than in the non-dewatered part, and the loss of sand on the landward side of the pipes in the dewatered part is slightly smaller than in the non-dewatered part. The difference in the variation of each side can be considered as being the amount of sand gained by the drain system during the experiment. It is estimated to be roughly 5m^3 for 20m of beach width.

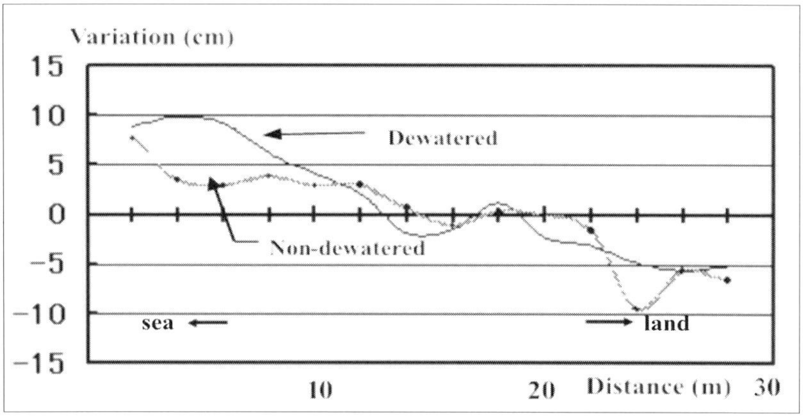

Fig. 10. The difference between the averaged final profile and the initial one

Figures 11 and 12 show the time-averaged mean profiles and mean profile changes expressed by rms values in relation to the variation in the dewatered part and the non-dewatered part, respectively.

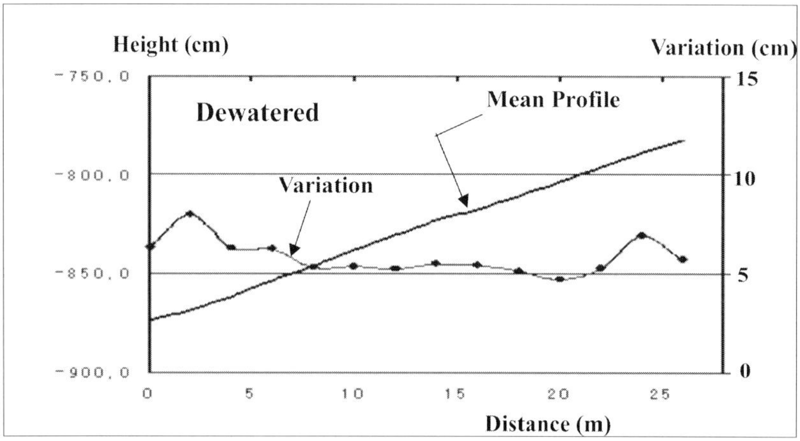

Fig. 11. Mean profile and variation during the experiment (dewatered)

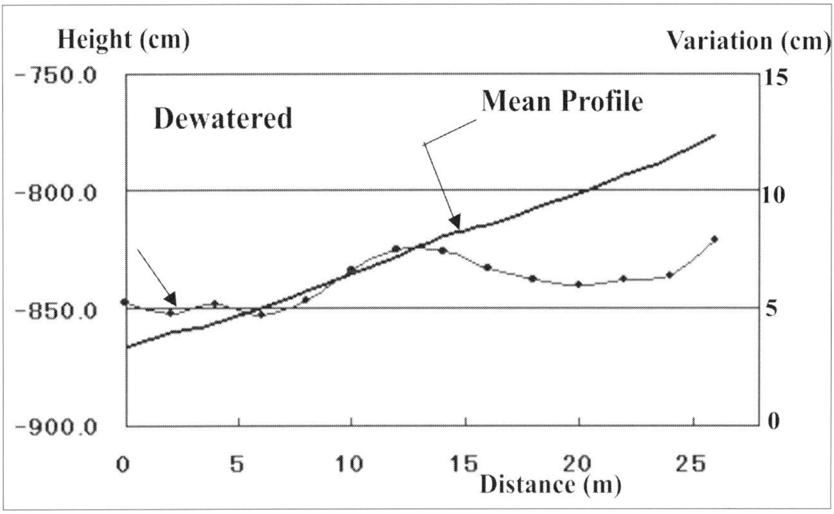

Fig. 12. Mean profile and variation during the experiment (non-dewartered)

The average sections are almost the same. However, the profile change in the seaward area of the pipes from 18m to 8m is smaller on the dewatered part.

The temporal changes in the averaged beach face height of the measuring lines parallel to the shoreline are shown in Figures 13 and 14.

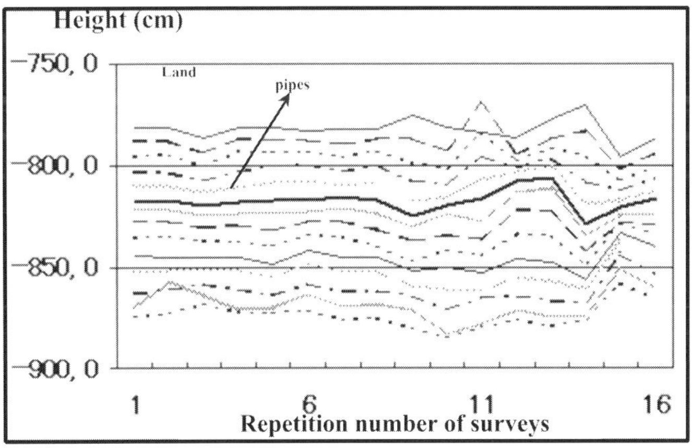

Fig. 13 Time history of the measuring lines parallel to the shoreline (dewatered)

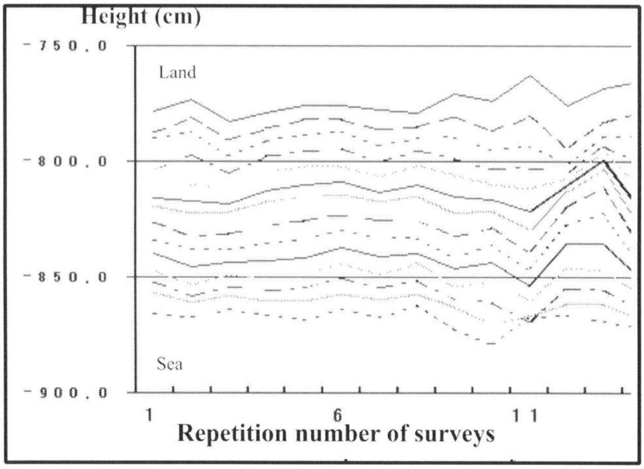

Fig. 14. Time history of the height of measuring lines parallel to the shoreline

4.2 COMPACTION OF THE BEACH FACE

During the laboratory experiments, it was sometimes observed that when the drain system started to work, the beach face of the area adjacent to the pipe became compacted. This seemed to indicate that the compaction of the beach face might play a role in stabilizing the beach under the operation of the coastal drain system.

In order to quantify the compaction, we attempted to measure the penetration length of a spear using a spear gun. Figure 15 shows the results. They show that at just seaward of the pipes, the compaction was considerable, but further seaward of the pipes where sediments deposition was large, there was less compaction. The beach face of the area landward of the pipes was dry when the measurement was done. Measurements should have been done during a higher tide. The compaction is considered to be due to our way of forced drainage with suction.

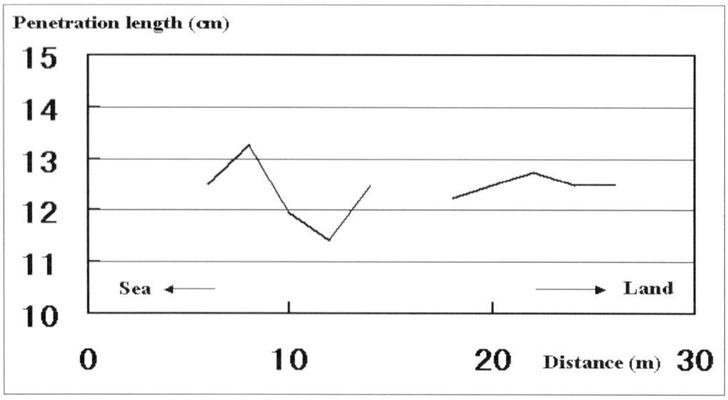

Fig. 15. Compaction of beach surface.

4.3 SEDIMENT MOVEMENT BASED ON TRACER TECHNIQUE

Figure 16 shows the movement at the centres of gravity of 3 tracers. Tracers were placed on the beach face just above the pipes on the dewatered side and the same position of the beach profile on the non-dewatered side. This measurement was done at the final stage of the experiment. As was mentioned earlier, two pipes were already not working, and consequently, the intensity of infiltration was not uniform along the pipes. This result reflects the non-uniform character of the infiltration on the dewatered side. On the other side, sediments moved landward on the non-dewatered side.

In our experiments, the flow induced due to the suction of the pipes seemed to have a significant influence on the sediment movement near the pipes.

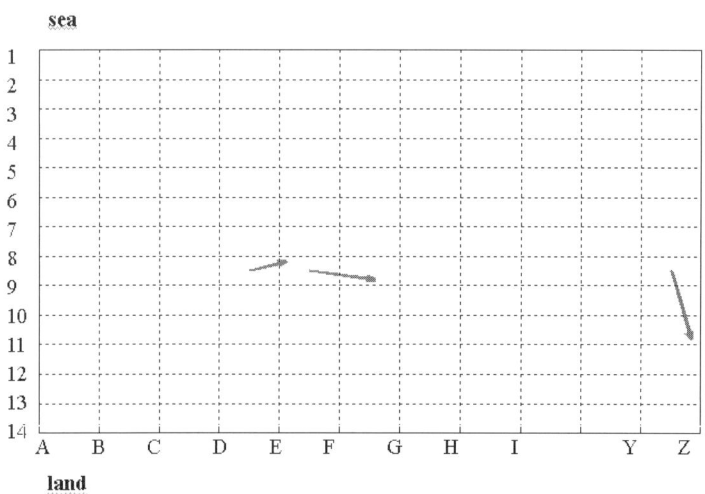

Fig. 16. Sediment movement

5. Conclusions

The following became clear from the field experiment.

1) Comparisons of the profile changes did not show any definite differences between the dewatering part and non-dewatering part of the beach, although a small difference was found in the seaward area of the pipes.

2) Based on the small difference, the amount of sand gained by the drain system was estimated as being roughly 5m^3 for 20m of beach width.

3) Movement of the tracer reflected the flow induced by the forced drainage.

4) Compaction was larger at just seaward of the pipes.

Acknowledgement

The authors would like to express their gratitude to the students of the Department of Ocean Civil Engineering, Kagoshima University, for their cooperation during the field experiments. This study was supported in part by a grant from the Ministry of Education, Science and Culture.

REFERENCES

Parks, J.: "Beach-Advancer System", Leaflet of DYNEQS Ltd., 1990, Tampa, FL., USA.

Sato M., Hata S. and Fukushima M.: An experimental study on beach transformation due to waves under the operation of coastal drain system, *Proceedings of the 24th international conference on coastal engineering*, 1994, pp.2571–2582.

Sato M., Fukushima M., Nishi R. and Fukunaga M.: On the change of velocity field in nearshore zone due to coastal drain and the consequent beach transformation, *Proceedings of the 25th international conference on coastal engineering*, 1996, pp.2666–2676.

Sato M., Fukunaga M., Nishi R. and Arima T.: *The effect of coastal drain on beach process, assessment & monitoring of marine systems*, Faculty of Science & Technology, University College Terengganu, 1999, pp.302–311.

Turner I.L. and Leatherman S.P.: Beach dewatering as a 'soft' engineering solution to coastal erosion – a history and critical review. *Journal of Coastal Research*, 13, 4, 1997, pp.1050–1063.

Vesterby, H.: Coastal Drain System: A new approach to coastal restoration. *Proceedings of GEO-Coast '91*, 1991, pp. 651–654.

ALTERNATIVE TO TRADITIONAL WAYS OF TREATING SHORELINE EROSION

DICK HOLMBERG

President, Holmberg Technologies Inc., 1800 Second St. Suite 714, Sarasota, Florida, 34236 USA

Abstract

In the United States national attention by major media has focused on America's 'threatened coastlines'. Misdirected engineering methods and efforts to control erosion of beaches have been proven to be wrong and counterproductive. 'Armoring' the shore with structures such as sea walls or riprap is unavailing and unsightly. Expensive and temporary beach dredged nourishment is coming under increased attack for causing and increasing erosional damage. Corrective action requires a sound analysis of the ecology of natural beach formation. Man's alteration of shorelines has created unnatural water currents, often remotely situated, which now divert the inbound littoral supply of sediment away from shore. Normally this sand would ensure a positive balance among factors preserving natural beach configurations.

Manmade artificial structures such as jetties also create disruptive currents. Proven environmentally and ecologically harmonious methods for controlling shoreline and bluff erosion and for restoring the natural environment are now available. Successful restoration of beaches and dune lands has resulted from patented low profile undercurrent stabilizer filtration systems in installations designed for specific sites. All countries with shoreline erosion need to review their present policies and make administrative changes to encourage such innovation via large-scale planning and objectively monitored demonstration projects. Those who dictate policies of retreating from our coasts and express such views as letting nature take it's course are wrong. Nature is not the enemy and abandoning the seashore should never be considered as an option. In the United States and some other countries, streamlining permit procedures and a review of jurisdictional overlap is advocated to ensure a timely response to solving the current crisis.

1. Reversing Beach Erosion Is The Alternative to Traditional Ways of Treating Shoreline Erosion

Major news media have recognized the scope and national significance of 'America's threatened coastline' (see e.g. *Editorial Research Reports*, 1984, or *Newsweek*, 24 September, 1984). Losses in property values and in the recreational and aesthetic qualities of fine sand dunes and beaches have reached crisis proportions. Tragically and all too commonly, beaches that once extended hundreds of feet in width and elevation have receded and disappeared into lakes and oceans. Unless countervailing efforts succeed, this trend will lead to the destruction of our wondrous but finite heritage.

Paradoxically, our contemporary crisis of shore land eradication is not simply a natural process. The term 'erosion' can mislead as it implies a slow and steady process of displacement of sedimentary materials. Indeed, at least one dictionary defines erosion

C. L. Goudas et al. (eds.), Soft Shore Protection, 183-198.

as a 'natural' process, evoking an image of a relentless and ineluctable force, one which in examples such as the Grand Canyon is measured in geologic time (*The American Heritage Dictionary of the English Language*, 1969). But relatively recent records (up to around 1855) reveal a steady progression of expanded rather than contracted beaches (Dana, 1864, Door, 1971).

Excessive interest has been placed on a rising global oceanic high water table and its alleged effect on our shorelines. This is misleading, as evidenced by historical records indicating expanding and elevating coastlines created as a result of ice age melt (Dana, Door, ibid.).

Our thesis is that beach, dune and bluff erosion have resulted primarily from fairly recent human interventions rather than from natural ecological and environmental factors. The import of this is twofold. Firstly, from a scientific and engineering point of view, it is critical that the natural ecology of beach formation be understood so that remedial efforts can lead to positive results. Secondly, from a public policy point of view, it is important that those victimized by counterproductive interventions in shoreline ecology not be castigated as culprits by those responsible for their protection.

The public interest in the reclamation and preservation of the coastal environment comes first but includes private property. It also goes beyond public recreational areas, important as these are. Beaches function as purification instruments safeguarding against pollution of water as well as waterfront. The chain of organic interdependence, ranging from microorganisms to more complex fauna and flora, is also crucially benefited by the filtration and percolation functions of our sandy beaches. A common interest must be recognized in the need for a concerted effort and awareness if a vital national resource is to be restored and preserved.

2. The Healthy Natural Beach and Its Growth

Although it seems the process of shoreline beach formation is generally well known, I will emphasize a neglected aspect. Wave action in its visible impact on cliff, dune and beach is commonly emphasized. Less visible and too little appreciated is the formation of currents accompanying distant underwater and near-shore wave actions. Actually, waves generated by Southern Hemisphere storms have been recorded as travelling more than 5 000 miles before arriving on the California coast (Inman, 1982). Less dramatic subsurface currents and submerged wave pressures have equal significance for remote deposits or dislocations of sedimentary materials. This will bear upon our subsequent consideration of the influences of even distantly sited jetties, deepened channels, offshore dredging, and so on on shoreline erosion problems.

Beach areas are formed from sand, shell, coral and pebble deposits that are transported to them from their original upland or seabed source off by water, or hydro effects. The transport system primarily entails currents and waves generated by wind, gravity and other geophysical forces.

Rivers and streams venting into the ocean carry a vast quantity of material that contributes to perpetual nourishment of beaches. These streams flow through the land, picking up sediments that have been created by natural erosive processes. These sediments eventually reach the vents or mouths of the rivers and streams at the ocean front. As the river currents meet the littoral transport systems of the coasts, their power is lessened and the material they carry is deposited in the form of deltas. This process continually contributes to the shoaling in these areas. In many cases, deepwater channels have been dredged for ships to navigate the area. The deltas and shoals that are formed become the eventual nourishment supply, or feeder, for the coastal beach areas. Those sand deposits and sediments transported into the deeper water form the life-supporting sandy bottoms and have now become the main natural source for protection and future nourishment for beach creation.

Prevailing winds influence the directions of littoral or nearshore transport currents. Wind fetch, or the cumulative distance the wind has for building up without interference, dictates the formula for the wind's power feed. Another factor is 'constant direction measured in time'. The longer the wind blows in one direction, the bigger the wave energy it creates to drive toward shore. When larger waves approach the shore, they energize the currents, moving them parallel to the shore. This increased power allows the sediment transport system to move even larger volumes of materials and water along the paths that these currents follow.

Waves are unrelenting as they approach, increasing the amount and height of the water table at the shore. The increased height, weight, power and current that are assembled in the nearshore areas are managed in a positive way by a healthy beach.

A natural beach typically consists of a plentiful sand supply source, deltas, sandbars that are not dredged, shallow offshore sandbars that are unbroken, and, most important of all, a wide, slowly tapering beach with healthy vegetation growing above the high tide and wave wash-up zone.

A healthy, well-balanced beach area can withstand large storms with minimal damage. Indeed, turbulent waters actually contribute to beach maintenance by mobilizing the wet sea bed's unconsolidated sediments and moving the sand from deep water bottoms into the near-shore beach and dune system. The process entails three basic stages.

2.1 THE FIRST ENCOUNTER: SHALLOWING OFFSHORE PROFILE

a. The gravitational difference of an ever-shallowing offshore profile absorbs of some of the wave's energy by forcing it to climb an elevating bottom contour. This decreases the base support of the wave.

b. Frictional drag occurs when water meets a rough, rasp-like, bottom.

c. Water percolation and fluidization occurs as the wave adds its weight downward on the sandy bottom. Energy is absorbed through the percolation process. Soil becomes buoyant and unstable due to this percolation. Sediment becomes fluid and is now capable of being transported in the wave hydraulics toward shore. Obviously, larger waves are capable of creating even more dramatic activity due to their higher elevation and greater weight.

Fig. 1. Before-After photographs of an extremely eroded beach (Photograph A taken in 1983) at New Buffalo, Michigan, and its impressive recovery (Photograph B taken in 1999)

2.2 THE SECOND ENCOUNTER: THE SHALLOWS AND SHORELINE BEACH AREA.

a. Large amounts of material carried by these forces are suspended in flow. There are several factors impairing their conveying ability. Firstly, the wave's breaking and surging upon shallow bottoms slows this sideways movement. Secondly, the energy of currents and waves are sapped by gravity as they climb and flow onto higher elevations. These factors combine to cause the material to drop out of suspension, and the bottom profile becomes elevated, resulting in additional energy loss with more deposits and more frictional drag.

b. The shoreline beach area is always in a state of dynamic flux. Ebb tide and flowing river waters meet with ocean forces and either join the currents or are slowed, thereby causing backwater flows to slow and rise. Slowing sediment-laden currents causes the transported material to drop from suspension-creating accretion or an increase in bottom buildup or beach elevation. Accrued material creates new land formations of beaches, dunes or deltas, shoals, and wetlands around or near the mouth of the contributories where the flowing energies merge and interact.

2.3 THE FINALE ENCOUNTER: THE BACK SHORE

a. After storms subside, note the many new shells left on the beaches, and the flotsam and occasional artifacts from a distant offshore wreck. Organic residues or deposits become compost, habitat, nourishment for other seashore life forms such as benthic and plants.

b. Plants with their root systems are another asset for a healthy beach system. They bind soil, stabilize and slow down wind-driven material. Eventually dunes are created by wind-driven sand, and as the ecological process matures, dune grass, shrubs and trees further protect the beach and dunes by forcing the wind to break and lose its potential to erode. Its waste and surplus continue to develop top soils.

c. Another overlooked and little understood benefit of a natural self-building and maintained beach ecosystem is its water filtration and toxic-processing abilities. As undesirable residues are cast upon the beach they are exposed to sunlight's UV rays, which are known to destroy over two hundred toxic elements. Many harmful and threatening substances are neutralized or isolated by the new deposits of sand and the processes of its life forms. Thus many tangible as well as intangible and hidden factors affect the ever-shifting equilibrium between water and shore. A multiplicity of elements combines to produce a natural but vulnerable balance. Whether deposition or erosion will be predominant in any particular place depends upon a number of interrelated factors: the amount of available beach sand and the location of its sources; configuration of the coastline and of the adjoining ocean floor; and the effects of wave, current, wind and tidal action. The establishment and permanence of natural sand beaches are often the result of a delicate balance among a number of these factors, and any changes, natural or human-made, tend to upset this equilibrium (Inman, 1982).

Fig. 2. The spectacular transformation from condition A (before) to condition B (after) induced by the installed stabilizer system at the beach of Harold Dekker on Lake Michigan. The positive impact of this soft intervention upon the environment is from the extensive vegetation it induced

3. Causes of Shoreline Erosion: Natural and Manmade

A healthy beach, although temporarily vulnerable to the destructive power of storms, has a homeostatic rebuilding capability to restore its natural equilibrium. The increased supply of sedimentary materials disturbed and relocated from remote bottoms and made part of the littoral transport system nourishes the beach and provides protection from permanent storm damage. Natural factors such as storms are not the main source of lasting beach erosion, although all too dramatically manifested in the short run and at particular sites.

Manmade disturbances of the natural equilibrium of beach and sea are the prime source of our erosion problems, including the inability of altered beach profiles to handle storm turbulence. Recall the process whereby a natural and gradually elevated beach absorbs wave and current-driven energy. This is vital to our understanding of the unintended but deleterious consequence of artificial structures and dredged channels, Unless these are installed or dug after adequate analyses or countermeasures which take into account their impacts upon alongshore processes, sand supplies, current and other factors, they will upset the fragile balance of a natural beach system. Unfortunately, such care is rarely taken, since shoreline beach and dune erosion is considered by many to be a natural condition or ongoing process.

Comparison of beach profiles before and after storms suggests erosion of the beach (above MSL [mean sea level]) can amount to 5 to 24 cubic meters per kilometer (10000 to 50000 cubic yards per mile) of shoreline during storms having a recurrence interval of about once a year (DeWall, Pritchett, & Galvin, 1971; Shuyskiy, 1970). While impressive in aggregate, such sediment transport is minor compared to longshore transport of sediment. Longshore transport rates may be greater than 765000 cubic meter (1 million cubic yards) per year (U.S. Corps of Engineers, Department of the Army, 1984).

Deep channels were dug so ships could navigate safely into bays or rivers to discharge their cargoes, while jetties and breakwaters were installed to protect the vessels from angry seas. The jetties were attached to land and reached into the sea across beach areas, including sandbars, to create, in many cases, a scouring action to help in keeping the channels inside them deep and limiting the dredging of the shoals that normally would swallow the inlets. Since such areas are maintained deeper than the adjoining shores, loose sand from adjoining beaches are carried by gravity-created currents that began to flow downhill parallel to the coasts. Corps studies show sand starts to erode from beaches located more than forty miles from such channels and harbors, beaches that had been in a state of accretion for over a hundred years. (U.S. Army Corp of Engineers, 1990, Coastal Engineering Consultants Inc., 1991, Burch et al, 1992).

Fig. 3. The reverse process, namely sedimentation, activated by the stabilizer system installed in the eroded beach (photograph A) created the friendly and natural beach presented in photograph B. The visitor cannot even see evidence of the intervention tools. The results provide clear evidence of the fact that this arrangement is 'accepted' by nature and also that 'soft' does not mean weak, but powerful.

Since the landward sides of the deepwater jetties are anchored to the shore, the alongshore currents that cannot climb over the dikes in their paths are diverted seaward. As the current meets the structure, its direction is changed, reducing the ability of the transport currents to deposit sand, and some of its energy charge is absorbed. Some of the transported sand is deposited on the side of the jetty meeting the current. The balance of the material still in flow is diverted seaward. This riptide, or diverted current, blends with the wind-driven, powerfully energized waves, and increases their velocity at this stage to compound the scouring and eddying effects at their meeting point. Accurate measurement of such discharges of materials into deep water has yet to be fully documented. But the immense scope of this diverted sedimentary material plays a large role in the loss of bottom profile along our shrinking shores. This phenomenon best explains our eroding and disappearing beaches.

Currents that create 'scouring' cause a deepening and channeling effect, which further increases the erosion from adjacent channel edges and shorelines. Simply put, upland soils will tend to restore elevation differentials created by scouring by moving downhill toward the channel bottom. As an illustration, consider an alongshore current flowing in a continuous direction for a period of time with a certain velocity. The current is flowing in a north to south direction, passing a pair of relatively close discharging channels, both having jetties. Even at an ordinary rate of flow, the material picked up in transport from the northerly channel will quickly arrive at the southerly channel. This means that for the remaining flow time, scouring occurs beyond and seaward of the southerly channel. The influence of such jetties and similar structures has an authoritative basis (Office of the Chief of Engineers, 1969). The amount of displaced material is the only variable. A misleading and unproven view is that once such material is discharged into 18 feet of water or more, it becomes a non-returnable item (Bascom, 1980, Komar, 1976).

Some persist in downplaying the negative consequence of such structures. We often are told that a jetty maintains a total blockage of material and that the beach material simply will be held temporarily from reaching the other side of the channel. Then, they claim, as soon as the wind changes, it will proceed in the opposite direction. This is correct, but too commonly left out of the analysis is the tremendous amount of sand lost to the offshore areas having been diverted away from the beaches prior to wind redirection. The gradual loss of huge beach areas over a period of time has become a serious concern to the people living along these shores. Bottom profile and beach loss happen so slowly in the beginning that many people remain unaware of the major causes of the loss cycle. Understandably, they are unaware of seemingly remote structures as causal influences on their beach profile. The currents are invisible to the untrained eye, whereas the beach visibly deteriorates under the force of now unimpeded wave action.

When the closing sea endangers houses, roads and other constructions, owners are driven to action. The first thought is to build structures to withstand the onslaught of the waves. Typical efforts include diking the shorelines with revetments or seawalls or with rocks piled high enough so that even a 'hundred year' or major storm cannot go over them. Others build groins to try to hold the sand from being swept away by the currents running along the shore.

It is almost a primeval instinct for man to try to defeat the sea by building higher, bigger, stronger barriers. But these structures actually increase the current flows by becoming hardened parallel shoreline. Gone is the absorption effect of a gradually rising beach, which drains energy from the waves by percolation and friction. Instead, unimpeded waves strike upon a hardened surface which has no absorption power. Unabsorbed wave energy is thereby reflected and converted to scouring power and increased velocity of the current running alongshore. Increased velocity means more material in transport and an increased speed of the alongshore current between the two points of the shoreline in our example of jetty influences. The greayer the area hardened between these two points, the greater the velocity of the current between the points. This supercharged current flows parallel to them and scours a channel in a sandy bottom. Such a deep channel, forming at the front face of such structures, causes sand on the adjoining beaches to be pulled or to run downhill into the trench scoured out by the increased velocity. That is why, cuts into the shore are to be found at the end of sea walls and similar structures. As the bottom profile in front of such structures becomes deeper, reverse reflection occurs. Incoming waves that feed the alongshore current are no longer losing energy to an elevated beach profile, so they strike with more power and are all the more strongly reflected back to sea. This power that sweeps seaward tends to accentuate a flattening of any near-shore bar areas and causes breaks in their formations. This, in turn, reduces the ability of the bottom profile to dampen the power of the incoming waves. Again, the powers of reflection and alongshore currents are further increased.

Fig. 4. The author standing on a stabilizer installed and almost buried by sand in Lake Michigan

Groins were also installed along the shorelines with the thought of slowing and trapping sand that is flowing in the alongshore currents. Most groins were solid like a wall placed perpendicular to the shore with a fixed elevation above the sea. They, too, cross the currents with results similar to jetties. We find that the design of these structures also directs currents seaward and, where the structures are not properly tied into the shore, we find currents being diverted around the structures at both ends. Most times the seaward route is dominant. Again, as with the jetties, we find eddying, compressing, seaward diversion of materials, and a trench or scouring channel forming. This further increases the movement of sand away from the shores and a deepening of the profiles. Another type of groin, the permeable type, is intended to slow rather than stop the currents. However, these structures alter the currents so that, in many cases, they eddy and scour alongside the groins, eventually creating trenches running seaward and sometimes settling or toppling them. This type of structure often plays havoc with down-drift properties because of design problems. A tradition of fixed height has caused most of the difficulties.

It was recently disclosed that in the manuals of the US Army Corp of Engineers recommended structures were improperly designed because of inaccurate laboratory tests (US Corps of Engineers, 1994). The report stated that structures such as rock revetments and groins tend to accelerate down-cutting and erosion, in fact, directing materials offshore to a greater degree. Is there a structural solution? Moreover, can such structures harmonize with the natural environment of dune and beach? My affirmative answer derives both from theory and practical experience with successful applications.

4. Control and Restoration: Beach Development through Patented Undercurrent Stabilizers

Geo-fabric filter media for the purpose of soil stabilization has been developed and improved over the years. Before the use of rough, stable fabrics made from petrochemicals, many forms of filters were tried, including treated papers, vegetable and animal fiber, fiber glass, and even woven metallic fibers. With the advent of the petrochemical filter fabrics in the 1960s, vast improvements have been made, especially in durability.

Geo-filters were originally used in the Netherlands to assist in holding back ocean waters from farmlands wrenched from the sea. The use of geo-textile filters has since been adapted for other uses around the world. During the last 35 years, a new use of geo-fabrics has been under research and development by our company: for control of beach erosion. The systems have been tested and improved in applications along the Great Lakes and oceans with astounding success. An equilibrium has been restored at numerous locations, reversing critical deterioration of shorelines with natural beach nourishment systems.

The use of patented undercurrent stabilizer systems is being employed to work with nature by utilizing her wave, current and tidal energy carrying sand to the beach area from offshore deposits. This natural process actually accomplished the same result as artificial beach nourishment but without disturbing the natural ecology of the ocean bottom. An additional benefit is that it does not require continual replenishment, cause an increase in the erosion rate or have the harmful effects of offshore dredging and strip mining to maintain the beach (US Corps of Engineers, 1989).

Each area requires its own special design system. Basically, all effective systems involve one form or another of low profile structures adjusted to water and storm levels. Construction materials can vary from site to site, depending on conditions and the desired approach and final result. The most common materials employed are the geo-filter fabrics using concrete filler.

Sand accretion is obtained by creating a gentle gravity resistance to the onshore and littoral currents. This accretion or elevated profile induces wave energy and currents to deposit more materials, further increasing the near-shore supply. This is similar to the underwater part of a delta formation which absorbs some wave energy rather than redirects all of it. Sand is usually abundant in these wave-driven and longshore currents flowing along the shores. Only a certain amount will be removed from these streams whatever the method chosen. The balance flows onward with the currents until directions are reversed by natural means. Moreover, it is not true that the accretion of sand in one area permanently robs another. If filter systems were installed along an entire shoreline, the vast quantity of flowing sand would be minimally reduced, if at all. Undercurrent stabilizer systems will encourage volumetric sand accretion, but they will not cause accelerated erosion. The reverse actually occurs with remarkable speed. A beach that has been elevated and widened by a filter system will in time accelerate accretion downstream. Like natural deltas, it will slow down the currents and waves allowing sedimentary materials in transport to be deposited. The original sand deposits are unbalanced until enough sand has accumulated to uniformly straighten beach formations and build up the back shore and dunes.

The damaging effects of major storms cannot be prevented by armoring our shorelines. We can, however, minimize their powers by widening our beaches and restoring our dunes. This can best be accomplished by the installation of a system of filters laid on the near shore coastline bottoms in low profile and following the natural grade of the land. Undercurrent stabilizer system technology requires competent analysis and sophisticated design. An attenuated installation, for the sake of economy, will not succeed. The filter system program in most cases involves several steps. Each step generally includes a network of filters, the most visible of which are the fabric-core, tube-like filters consisting of heavy duty polyester media filled with cement. A less visible but equally vital part of a system includes sheet filter media placed horizontally under the cement-filled core, extending outward for a preditermined distance and provided with perimeter anchorage. All materials in the configuration blend harmoniously with the natural environment.

Step 1 involves a series of cement-filled tube forms laid perpendicular to an approximate, proposed beach line. The length of the stabilizers is established by design computation, with the landward set being tied into a toe protection system along an existing bluff line to protect against wave action cutting behind the system.

Step 2 begins with the development of dunes from dried sands being accreted during the initial stages of buildup. Special mixtures of organic and geo-textiles are utilized to enhance a growing medium. Vegetation should also be planted as soon as possible to develop a network of roots to hold the dunes against further wind movement.

The final result of a properly designed program is a wide, flat, graded beach gradually succeeded by vegetated dunes having an appearance in every respect akin to that which nature created before the advent of man's damaging structural systems. In a recent engineering summary of a four-year field evaluation of undercurrent stabilizers, they were described as being effective in protecting and even elevating the beaches they were intended to protect. Furthemore, they perform this function at no detriment, and have actually shown benefits, to adjacent downdrift beaches.

The low-profile cross-section of these structures and their placement along the bottoms mimics the desired natural slope of the beach and generates a fraction of the turbulence and scour that accompany other shore protection structures. They remain in place when facing liquefaction forces, preventing the landward migration of scour troughs during significant storm events.

The findings clearly suggest that these stabilizers need to be seen as a unique type of shore protection structure and as not having the negative effects of other shore protection structures. They should be recommended for shore protection needs over traditional methods and requests for permits to install them given due consideration. They should also be considered for use in counteracting or mitigating the negative effects of larger or more damaging marine structures (David Schultz, 2000).

Fig. 5. The before-after condition of the beach of Jim Westgate at Lake Michigan as shown in two photographs taken 23 months apart. Photograph A was taken immediately after installation of the stabilizer system (September, 1997). Photograph B, taken in August 1999, justifies the choice of method and the implementation

5. Implications For Changes in Public Policy

It has been argued that erosion is just as natural as the beach itself. Indeed, it is human opinion which categorizes beach formation as desirable and erosion of beaches as undesirable. But this value-neutral aspect of natural processes has led to a perverse distortion of its implications for public policy. Thus some claim that erosion is a natural phenomenon and that persons with a property, or environmental-aesthetic interest in beach conservation, must accept 'nature's' destruction of their property or of public recreational shorelands. Indeed, an underlying attitude among some personnel responsible for protecting this highly valued and threatened public and private resource is that shoreline property owners and recreational beach mangers are to blame for their erosion problems.

The primary source of erosion problems on the scale of the past century is projects, such as jetties and channels, permitted or constructed by public agencies. While there may have been reasons for the slowness of our coming to understand the non-natural, manmade causes of coastal erosion, it is now understood that structures, dredging and channels change natural processes or interfere with currents, diverting sand from normal littoral transport and beach deposition. Public policy and administrative regulations need to be adapted to our new understandings.

Beaches can and must be restored to their original configurations. Regulatory agencies need to change their permit procedures accordingly. Firstly, we need to recognize the futility, not to mention the wasted expenditure, of barrier constructions such as seawalls, groins, revetments and riprap. They are not only ineffective but they remove rather than preserve beaches. Secondly, they are aesthetically incompatible with the natural qualities which give value to beaches and dunelands. It is discouraging to find that recommended counter-erosion measures currently circulated by responsible public agencies include the use of materials such as automobile tires. The conversion of beaches into junkyards is scarcely in the public interest. We have witnessed iron rods protruding from concrete slabs called rip-rap used to line once beautiful beach areas. Such unsightly, dangerous, and counterproductive materials and methods should not be permitted.

We must not only prohibit harmful and outmoded methods, but need to encourage and facilitate the now available technology in a systematic program of beach restoration. This is not special pleading: there is a general public interest at stake. The scope of the problem has gone beyond the possibility of a lasting piecemeal solution, and the problem itself derives from a set of multiple and interrelated phenomena. As noted, individual sites are influenced by remote installations which frequently have been structures built by or under the aegis of public authorities. New techniques are best combined with a multi-pronged design for large-scale reclamation. The piers, jetties, and channels, as well as the immediate targets of planned beach restoration, need to be viewed in terms of their reciprocal relationships. This entails supportive worldwide public policy and innovative leadership.

REFERENCES

The American Heritage Dictionary of the English Language, Houghton-Mifflin, 1969.

Bascom, W.: *Waves and Beaches*, New York: Anchor Press/Doubleday, 1980.

Dana, J.D.: *Textbook of Geology*, Theodore Bliss & Co., London Cribner and Co., 1864.

Door, J.A. Jr., & Heitman, D.F.: *Geology of Michigan*. University of Michigan Press, 1971.

Editorial Research Reports. 'America's Threatened Coastlines', November 2, 1984

Inman, D.L.: 'Nearshore Sedimentary Processes', McGraw-Hill Encyclopedia of Science of Technology. VI.9, p.46. 1982.

Komar, P.D.: *Beach Processes and Sedimentation*, Prentice Hall, Inc., 1976.

Newsweek, '*The Vanishing Coasts*', 24 September, 1984, pp. 74-—76.

US Corps of Engineers, Department of the Army, *Shore Protection Manual,* Vol. 1, 1984.

Office of the Chief of Engineers, *Water Research Policies and Authorities: Prevention and Mitigation of Shore Damage Caused by Existing Federal Navigational Works,* Regulation 1165-2-309. Department of the Army, Office of the Chief of Engineers, Washington, D.C. June 23, 1969.

US Army Corp of Engineers, Jacksonville District South Atlantic Division, *Navigation Study for Canaveral Harbor, Florida, Final Feasibility Report and Environmental Impact Statement-81240*, August, 1990.

Coastal Engineering Consultants Inc., *Stump Pass Inlet Management Plan*, CEC File No. 89.170 December, 1991.

Burch, T.L., & Sherwood, C.R.: *Historical Bathymetric Changes Near the Entrance to Grays Harbor Washington*, PNL-8414/UC-000, prepared for US Army Corps of Engineers, Seattle District by Pacific Northwest Laboratory, Richland, Washington, 1992.

Schultz, D.: *Analysis of Lake Michigan Monitoring Surveys, Norton Shores, Michigan*, PE August 30, 2000.

US Corps of Engineers, *Lakebed Downcutting and Its Effect On Shore Protection Structures*, No 02124 Abstract form for all GSA Meetings, Charles N. Johnson, U.S. Army Corps of Engineers, North Central Division, Chicago, II, 1994.

US Corps of Engineers, *Beach Restoration Hearing*, Environment, Energy, and Natural Resources Subcommittee of the Committee on Government Operations, House of Representatives, April 28, 1989.

EVALUATING THE EFFECTIVENESS OF A SUBMERGED GROIN AS SOFT SHORE PROTECTION

PIERLUIGI AMINTI,[1] CHIARA CAMMELLI,[2] LUIGI E. CIPRIANI[3] AND ENZO PRANZINI[2]

[1] *Dipartimento di Ingegneria Civile - Università degli Studi di Firenze. Via di S. Marta, 3 - 50137 Firenze, Italy.*

[2] *Dipartimento di Scienze della Terra - Università degli Studi di Firenze. Via Iacopo Nardi, 2 - 50132 Firenze, Italy.*

[3] *Regione Toscana - Dipartimento delle Politiche Territoriali e Ambientali. Via di Novoli, 26 - 50127 Firenze, Italy.*

Abstract

The Marina di Carrara harbour (Tuscany, Italy), was built in the early 1920s and since then has been trapping longshore moving sediments coming from the north, the main cause of beach erosion along downdrift beaches. Since the 1930s, seawalls, groins and detached breakwaters have been built along 6.7 km of the coast south of the harbour, so that now each kilometre of beach is protected by 1.4 km of hard structures (protection ratio). Since these types of defence have induced some degree of shoreline progradation, the local tourist operators have asked that shore protection be extended southward, where the increasing erosion rates are due to the more northern structures.

In 1998, the administrative body of the region of Tuscany financed a project to test soft shore protection in the area. If the results are positive, these soft shore protection measures could be applied to the southern beaches, and eventually gradually replace the hard northern structures. In spring 1999, an experimental underwater groin was built approximately 1 km downdrift from the southern hard structure, along a stretch of beach where erosion rate over the last 20 years has been approximately 4 m/yr. The groin was built with 2.5 x 1.8 x 0.7 m polypropylene bags filled with sand. The groin is 180 m long; it is rooted inside the backshore and reaches the –3 m isobath with a mean depth of approximately 1 m.

Morphological and sedimentological beach evolution was monitored and four surveys performed during which cross- and long-shore bathymetric profiles were acquired as well as sediment samples. Preliminary results from this monitoring show that the submerged groin effectively interacts with nearshore dynamics, boosting natural processes and increasing seasonal bar cross-shore shifting. Due to the limited length of the structure, during severe storms, the bar system moves offshore of the groin head, and rip currents are stabilised by the structure itself. Under these conditions, beach erosion near the groin increases, but the end of the storm, enhances the efficiency of the structure to draw in sediments. Under fair weather, the protected beach is wider on the backshore and less steep on the foreshore. The erosion rate fell to 1 m/yr, and although a longer period is required to draw definite conclusions, the tourist operators are satisfied. In contrast to events resulting from emerged rock rubble/boulder groins, beach evolution is symmetrical on both sides, and no significant changes have been observed along downdrift beaches. Grain-size variations proved to be limited, but for a small increment in coarse material at the groin root.

C. L. Goudas et al. (eds.), Soft Shore Protection, 199-210.

Results of the first year of monitoring suggest that this type of soft defence should be longer and fully cross the bar system in order to prevent scouring at the groin head, where strong currents occur during storms. These currents are responsible for sediment loss in the protected area.

1. Introduction

Marina dei Ronchi is located in northern Tuscany (Figure 1). In this area, potential net longshore sediment transport has been estimated to be moving southwards at approximately 150000 m^3/yr (Aminti et al., 1999). The sediment source for the Marina dei Ronchi beach is the Magra River, outflowing at the northern margin of the physiographic unit and feeding the beaches to the south as far as Forte dei Marmi, as demonstrated by beach sediment petrography (Gandolfi & Paganelli, 1975).

The construction of the industrial harbour at Marina di Carrara in the early 1920s caused interception of the southward longshore drift, inducing rapid progradation in the updrift beach and erosion downdrift. The Marina di Carrara beach experienced shoreline progradation of approximately 300 m following construction of the harbour, although in recent years (1985–1998) the trend has reversed (Cipriani & Pranzini, 1999) and the shoreline has retreated as a consequence of a strong fall in the Magra River sediment load (Pranzini, 1995). Marina di Massa, which is located downdrift, has experienced severe erosion since the early 1930s (Albani, 1940), even though the harbour updrift jetty was then only 400 metres long, in contrast to its present length of 900 metres. In 1930, the first seawall was constructed in order to protect the coastal highway and in 1957 a series of breakwaters were added, even though a 2 kilometre stretch of the beach south of the harbour had already vanished (Berriolo & Sirito, 1977).

In the meantime, shoreline retreat was gradually shifting southward, and a series of hard structures such as seawalls, breakwaters, groins and submerged breakwaters were built along the coast (Figure 1). Today, a 6.7 kilometre-long stretch of coast south of the harbour is protected by 9.3 kilometres of hard structures (1.4 km of hard structures per km of coastline).

In 1970, a bypass system was projected in order to transfer approximately 200 000 m^3/yr sand from the northern side of the harbour to the south. After several interruptions, the experiment was finally abandoned in 1974 on account of the expensive maintenance costs and the structural instability of the harbour's northern jetty due to sand dredging at its foot.

During the late 1980s and early 1990s, artificial beach nourishment projects using sand dredged from inside and outside the Marina di Carrara harbour were performed along the Marina di Massa beaches, but the results were limited due to the small volume and fine texture of the sediments (Iannotta, 1997). This led to the local tourist operators requesting further hard structures in order to stabilise the shoreline. Indeed, between the Lavello and Frigido rivers (Figure 1), groins and submerged breakwaters were effective in stabilising the beach that was retreating at a rate of approximately 1 m/yr. during the period 1938–1978 (Cipriani & Pranzini, 1999).

However, beach erosion continued to increase southward. Figure 2 shows the evolution of the shoreline between 1938 and 1998 in the study area. Sectors 9 – 13 show an expansion of the dry beach after the construction of coastal defences, while sectors 14 – 17 show an increase in beach erosion as a consequence of the construction of the same structures.

2. The project

The study beach is located at Marina dei Ronchi, some 7 km south of Marina di Carrara harbour, and extends for less than 2 km between Magliano and Poveromo Creeks (Figure 1). In order to stabilise the shoreline retreat without increasing sediment starvation to downdrift beaches, the administrative body of the region of Tuscany financed a research project on coastal morphodynamics for the entire physiographic unit. Research included studies of the wave climate and sediment transport models, morphological and sedimentological studies, investigation of offshore sand reservoirs, testing experimental defence and evaluation of the environmental impact of the project. To test the experimental defence, it was decided to build a submerged groin along a stretch of coastline experiencing erosion and free of hard structures. The object was evaluation of the effectiveness of a submerged groin as soft shore protection downdrift to a highly protected coast (Figure 1). An accurate monitoring program was developed in order to gain information on coastal morphodynamics in the vicinity of the submerged groin and its effects on adjacent beaches.

Prior to the execution of the submerged groin, a study using a numerical model was undertaken, using the morphological and sedimentological data and wave climate gained from studies on the physiographic unit. The model tested several angles of wave approach, as well as groin length and angle to the shoreline, in order to measure the interaction of the groin with longshore currents and sediment transport (Aminti & Cappietti, 2000). After several options, it was decided to test a submerged groin 180 m long, perpendicular to the shoreline with a berm elevation ranging from 0 at the shoreline to 1.5 m at the groin offshore tip along the −3 m isobath. Numerical model simulations forecast aggradation in the foreshore and only limited areas of erosion. In particular, the model detected no beach erosion along downdrift beaches (Aminti, 1998).

The submerged groin was built between February and May 1999, approximately 1 km south of the last rock rubble/boulder construction, the jetties at the Magliano Creek mouth. The stretch of coastline where the submerged groin was built is characterised by a shoreline retreat of approximately 4 m/yr. over the last 20 years. The position of the groin on the beach offers a good opportunity to test its effectiveness as a soft shore protection.

The groin is 180 m long, running from the body of the backshore to the −3 m isobath, and thus it is completely covered by sand and water. Its environmental impact is therefore not relevant either as far as the view is concerned, or in relation to the circulation of the seawater. The project forecast an angle of 90 degrees to the shoreline, although the construction was oriented slightly to the south. The groin is made of 3 x 1.8 m polypropylene bags containing a volume of approximately 1.5 m^3 of sand for a mean weight of 3.7 tons. The weight of the sand bags should guarantee its stability under extreme storm conditions.

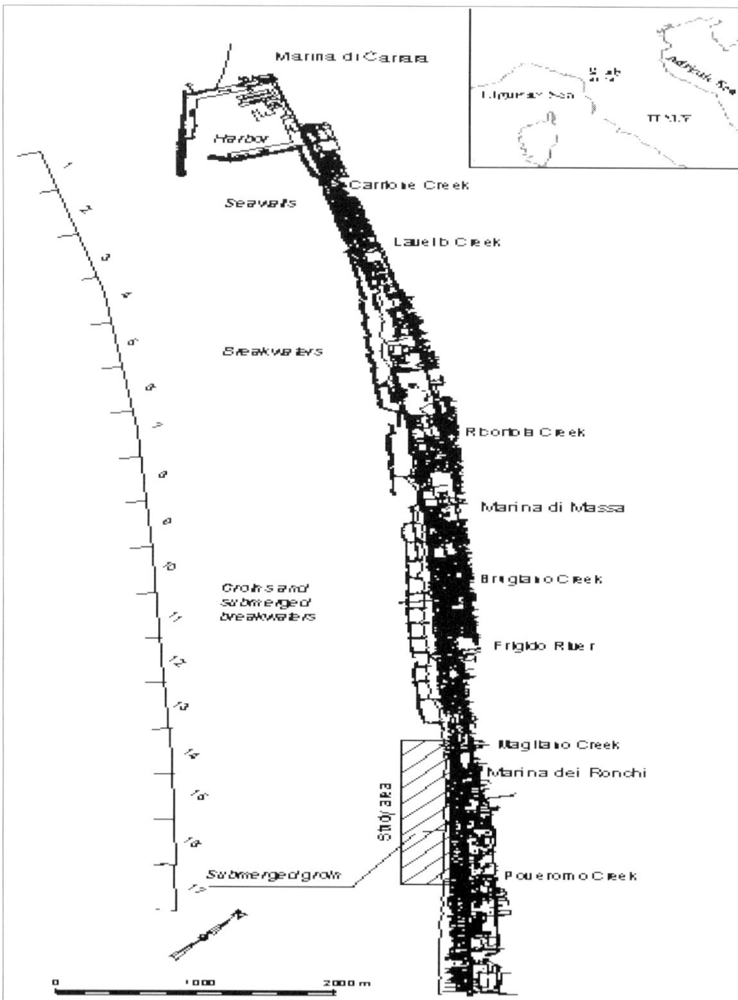

Fig. 1. Location of the study area and identification of the sectors in which the coast has been divided to analyse shoreline evolution

Land machinery was employed throughout the construction of the groin (Figure 3). The sand bags were deposited following the plan in Figure 4, so forming a 2, 3 or 4 layer structure, depending on the water depth. The help of divers assured the correct positioning of the sand bags in the foreshore. On the backshore, the sand bags were set inside a trench and covered with sand. Work ended in May 1999 and by the beginning of the summer season, the groin was already covered with sand even in the foreshore, so there was no visible trace of the works.

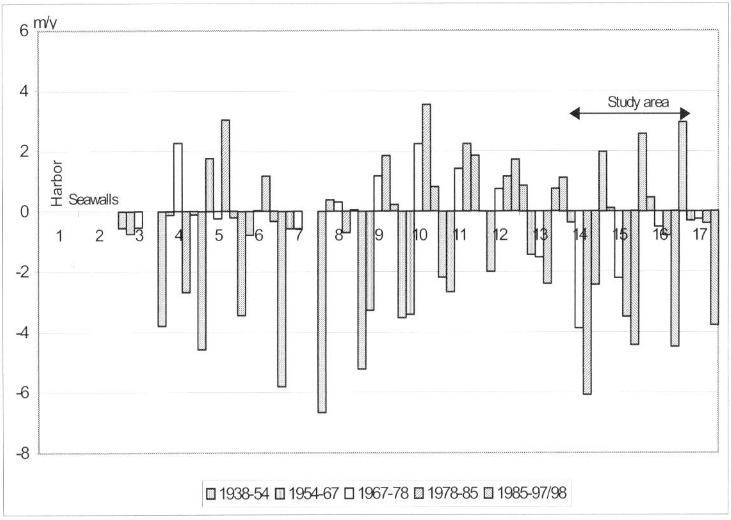

Fig. 2. Mean shoreline evolution in the different sectors (see Fig. 1) over the last 60 years

Fig. 3. Submerged groin construction

Fig. 4. Groin root on the backshore

3. Materials and methods

To gather information on beach response after the construction of the submerged groin, a monitoring program was carried out between February 1999 and April 2000. It consisted of four bathymetric surveys (Feb. 1999, Oct. 1999, Feb. 2000 and Apr. 2000) and two sedimentological surveys, taken during the bathymetric surveys in October 1999 and February 2000. Pre-work sedimentology was known from a survey performed in 1997. The shoreline position was also surveyed. The study area was investigated through 20 transverse sections from the backshore to the nearshore at the −8 m isobath and additional parallel sections in order to acquire further information on the nearshore morphology in the vicinity of the groin (Figure 5).

Using the SURFER software package, it was then possible to calculate changes in beach elevation by comparing beach profiles. Backshore surface variations were computed from comparisons of the shoreline positions.

During the October 1999 and February 2000 surveys, approximately 6 sediment samples were collected along 12 transverse sections, for a total of 159 samples. Each sample was dry-sieved at 1/2 phi intervals and Folk & Ward (1957) textural parameters were obtained. Grain size parameters were then mapped for each survey and the correlation between beach morphology and sediment texture was analysed.

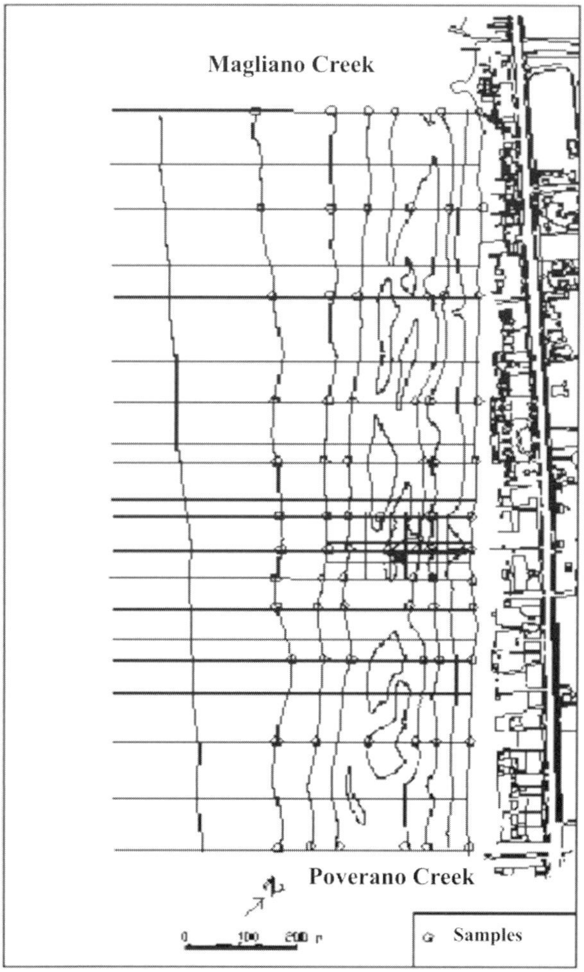

Fig. 5. Surveyed sections location and beach sediment samples position

4. Beach morphology evolution

Analysis of beach morphology evolution consisted of comparison of different bathymetric surveys in order to evaluate beach response following groin construction. The study revealed that the nearshore was characterised by a bar-trough system that moved within the −5 m isobath, which itself remains fairly stable (Figure 6).

Nearshore morphology after construction of the submerged groin (Oct. 1999) showed aggradation (sediment deposition inducing vertical elevation of the sea bottom) in the foreshore. In particular, the –2 m and –3 m isobaths shifted offshore and tended to form a cusp around the submerged groin (Figure 6b). On the other hand, nearshore morphology, surveyed in February 2000, was similar to the first survey (Feb. 1999), as a result of the similar seasonal wave climate (Figures 6a and 6c). In February 2000, the nearshore morphology survey was characterised by rhythmic features probably due to rip currents (Figure 6c). On the other hand, the October 1999 survey showed a crescentic nearshore as defined by Greenwood and Davidson-Arnott (1979) (Figure 6b).

Fig. 6. Contours maps acquired from the different surveys carried out during the monitoring period

The bathymetric data for each Digital Elevation Model (DEM) was subtracted from the DEM of a different survey. This revealed areas of accretion and erosion. As expected from the profile analysis, this resulted in the volume of sediment being redistributed inshore of the –5 m isobath and in particular between the –2 m and –3 m isobaths. This was probably due to the adjustment of a typical storm profile (dissipative) to a fair weather profile (reflective) between February and October 1999, rather than a systematic effect induced by the submerged groin (Figure 7A). On the contrary, as shown in Figure 7, maximum changes in nearshore morphology appeared near the groin. Comparisons of bathymetric surveys in February 1999 and October 1999 revealed maximum accretion values on both sides of the structure (Figure 7A). This indicates the effectiveness of the submerged groin to trap material without inducing sediment starvation to down-drift beaches. In addition, comparison of October 1999 and February 2000 surveys showed erosion in the nearshore along the −2 m and −3 m isobaths, with a localised erosion area around the submerged groin (Figure 7B).

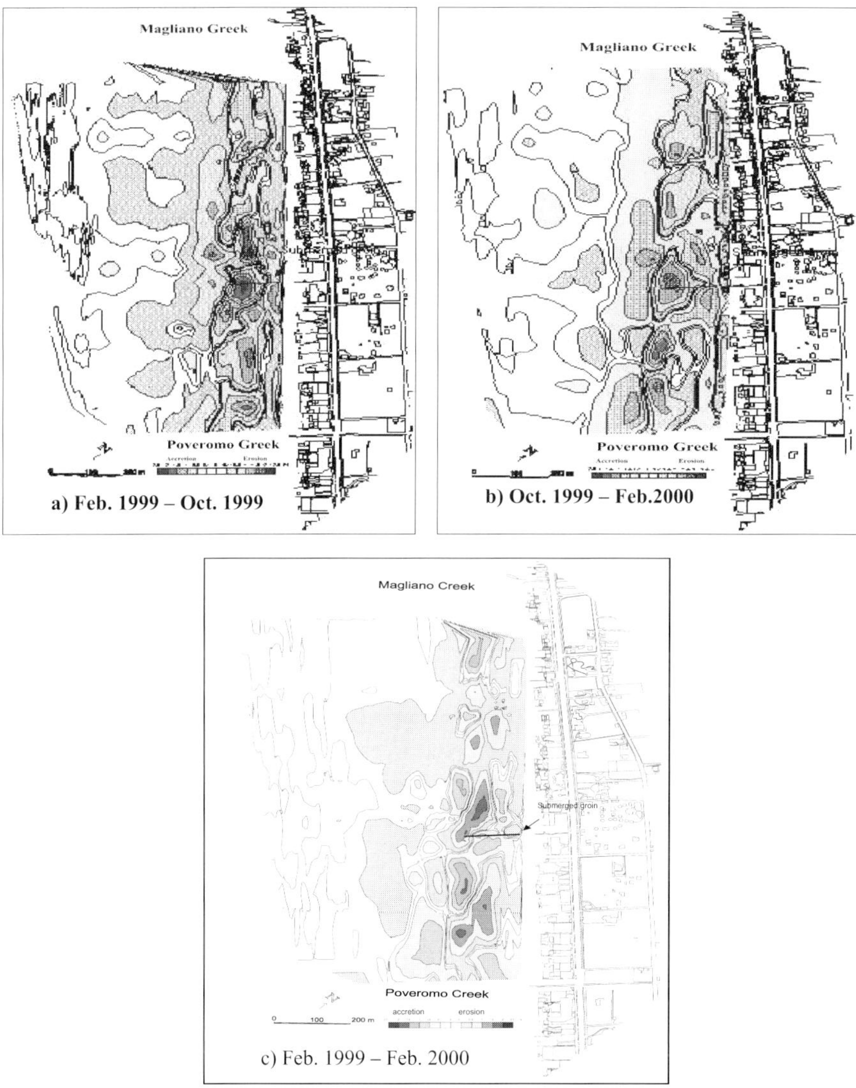

Fig. 7. Depth variations in the study area in the year following construction of the submerged groin

Results from the monitoring period (Feb. 1999 / Feb. 2000) indicate limited changes in beach morphology, diluted along the nearshore (Figure 7C). In the vicinity of the submerged groin, erosion reached values of approximately 2 m (vertical scouring of the sea bottom) especially at the submerged groin offshore tip (Figure 7C). On the other hand, deposition occurred downdrift of the submerged groin between the –2 m isobath and the shoreline (Figure 7C).

Figure 8 shows the shoreline evolution provided by the most representative bathymetric surveys. It shows three shoreline positions: the first (Feb. 1999) before the groin was constructed, the second (Oct. 1999) after the summer period and the third (Feb. 2000), one year after the first survey.

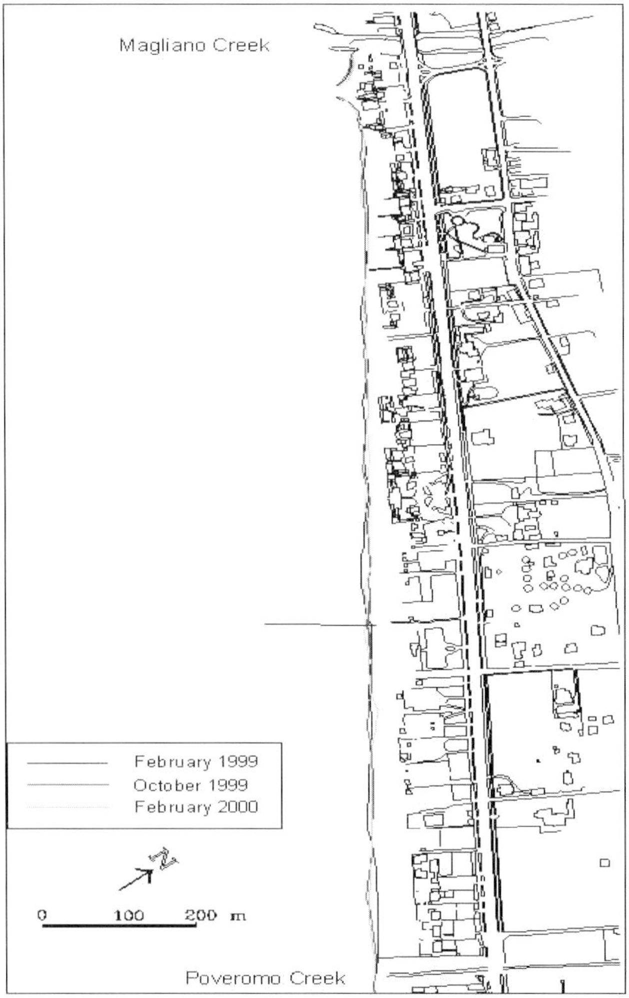

Fig. 8. Shoreline position in the study area during the monitoring period

5. Conclusions

Although the study area was monitored for only one year, useful information emerged which adds to our understanding of the behaviour of submerged groins. We have proved that submerged groins interact with nearshore dynamics accelerating natural processes, namely the seasonal cross-shore displacement of the bar-trough system. Submerged groin performance is closely related to actual length. In fact, in our case, since the submerged groin did not fully cross the bar system, strong currents concentrated at its head and created deep scouring, which could lead to the collapse of the structure. During low energy conditions, the submerged groin efficiently traps sediments inducing significant sediment deposition along its entire profile. The sedimentological impact of the submerged groin on the beach is negligible, although a slight increase in grain size was detected along the swash zone. The environmental impact is positive due to the fact that the groin is completely covered by sand both on the backshore and on the foreshore and near-shore region.

Acknowledgements

This study was supported by the MURST, Project "Analisi delle variazioni sedimentologiche e morfologiche delle piagge indotte da interventi di difesa e ricostruzione della fascia costiera" (Resp. G. Fierro) and by the GNDCI, Linea 2, U.O. 51 (Resp. G. B. La Monica).

REFERENCES

Albani D.: Le spiagge della Toscana settentrionale dalla foce del Fiume Magra ai Monti Livornesi. In *Ricerche sulle variazioni delle spiagge italiane. II - Le spiagge toscane*, C.N.R., Rome, 1940, 11–86.

Aminti P.: Studio su modello numerico per la costruzione di un setto sommerso in sacchi e ripascimento artificiale a Marina dei Ronchi. Unpublished technical report, Comune di Massa, 1998.

Aminti P. and Cappietti L.: Analisi del campo di moto nell'intorno di un pennello sommerso. IDRA 2000 – XXVII Convegno di Idraulica e Costruzioni Idrauliche, 2000, pp. 215–222.

Aminti P.L., Iannotta P. and Pranzini E.: Morfodinamica di un sistema costiero intensamente protetto: il litorale di Marina di Massa. Atti del Convegno: *Il rischio idrogeologico e la difesa del suolo*, Rome, 1-2 October 1998, Accademia Nazionale dei Lincei, 154, 1999, 263–270.

Berriolo G. and Sirito G.: Studio per la sistemazione del litorale di Marina di Massa. Atti del Convegno di studi per il riequilibrio della costa fra il Fiume Magra e Marina di Massa, Documento 2, Comune di Massa, 1977, pp. 16.

Cipriani L.E. and Pranzini E.: Evoluzione recente delle spiagge toscane. In: *Rapporto sullo Stato dell'Ambiente della Regione Toscana – 1998*, Regione Toscana, 1999, 77-85.

Folk R.L. and Ward W.C.: Brazos River bar: A study in the significance of grain size parameters, *Journal of Sedimentary Petrology*, 27, 1957, 3–26.

Gandolfi G. and Paganelli L.: Il litorale pisano – versiliese (Area campione Alto Tirreno). Composizione, provenienza e dispersione delle sabbie. Bollettino della Società Geologica Italiana, 94, 1975, 1273–1295.

Greenwood B. and Davidson-Arnott R.G.D.: Sedimentation and equilibrium in wave-formed bars: a review and case study, *Canadian Journal of Earth Sciences*, 16, 1979, 312–332.

Iannotta P.: Dinamica dei sedimenti del litorale posto a sud del porto di Marina di Carrara. Degree thesis, University of Florence, Italy, 1997, p. 98.

Pranzini E.: Cause naturali ed antropiche nelle variazioni del bilancio sedimentario dei litorali. *Rivista Geografica Italiana*, n.s. 1, 1995, 47–62.

ARTIFICIAL REEF METHODOLOGY FOR COASTAL PROTECTION USING SUBMERGED STRUCTURES: A NUMERICAL MODELING APPROACH

B. ONTOWIRJO[1] AND H.D. ARMONO[2]

[1] *Research Engineer Coastal Engineering Laboratory at the Agency for Assessment and Application of Technology (LPTP-BPPT) Jakarta - Indonesia.*

[2] *Graduate Student Faculty of Civil Engineering Queens University, Canada On leave from Faculty of Ocean Technology Sepuluh November Institute of Technology, Surabaya, Indonesia.*

Abstract

Waves induce flow around reef structures has been difficult to estimate, mainly due to complexity of porous structures and breaking waves conditions. An attempt has been made in this study to model the fluid domain and simplification of the existing structure barrier by a finite volume of fluid method for a three-dimensional Stokes model. The model is capable of simulating free-moving surface boundary conditions, the contribution of undertow, wave asymmetry and bottom boundary drift to sediment transport.

In addition to the present fluid model, a specially shaped submerged structure will be tested in order to observe its hydraulic capability to reduce the offshore waves energy as well as to provide a safe and productive environment for fish. It is expected that the flow created by the structure will be similar to flow around a natural reef so that the submerged structure when deployed may help to shelter marine creatures and accelerate reef reconstruction. The hydraulic parameters such as particle velocities, fluid flows, wave breaking and dissipation of wave energy in the vicinity of reefs are investigated. The results of a two-dimensional reef model using the finite volume method are presented and discussed. Comparison to laboratory measurement will also be discussed in this study.

1. Introduction

Artificial reefs are man-made structures that serve as shelter and habitat, a source of food, breeding area and shoreline protection. They are normally placed in areas with low productivity or where the habitat has been degraded (White et al, 1990). Coastal engineers dealing with the hydraulic or engineering aspects of artificial reefs have carried out related artificial reef studies. (e.g: Lindquist & Pietrafiesa 1989, Baynes & Szmant, 1989). Other studies have emphasized the utilization of reefs as breakwaters (Aono & Cruz, 1996, Hayakawa et al 1998, Kawasaki & Iwata, 1998) and reef reconstruction shelter (Armono, 1998).

Studies on artificial reefs using Finite Volume methods have been widely carried out. However, most of these reefs were permeable and had trapezoidal or rectangular shapes (Hayakawa et al 1998, Kawasaki & Iwata 1998), which caused tearing of fishing nets. Although effective in reducing the wave energy, in order to reduce the tearing of fishing nets, bottom-seated smooth-shaped reefs, such as cylindrical shapes, turtle blocks, and reef balls, were proposed as alternatives (Reefball Development Group Ltd., 1997).

C. L. Goudas et al. (eds.), Soft Shore Protection, 211-225.

This paper describes the flow field around either single or multiple hemispherical reefs obtained from numerical modelling. Two-dimensional computational models were developed and analyses carried out using the FLOW-3D software package to simulate the flow pattern in the vicinity of reefs. The numerical results obtained from the above analysis are discussed below.

2. Numerical Method

2.1 SOLA VOF

The computational fluid dynamic (CFD) software package FLOW-3D® (Flow Science, 1997) based on SOLA-VOF was used to model the flow field in the vicinity of reefs. The software was based on the fundamental laws of mass, momentum and energy conservation in which the finite difference method was applied to solve these equations.

The original development of the program was carried out in the Computational Fluid Dynamics Laboratory at the Los Alamos National Laboratory, New Mexico, in the early 1960s. The numerical algorithm used was called 'Solution Algorithm − Volume of Fluid (SOLA−VOF)', an extension of the SOLA method (Hirt et al, 1975, and Hirt et al, 1980). The obstacle geometries in the software were placed in the grids with fractional area volume obstacle representation (FAVOR) methods. By implementing this method, the geometry and grids were made completely independent of each other. These techniques generate smoothly embedded geometric features that were constructed from the program's pre-processor or imported from other CAD programs. This eliminated the 'stair-step' boundaries, which frequently occurred in the finite difference models.

FLOW-3D employs the staggered grid arrangements. All variables are located at the center of the cell volume, except for velocities, which are located at the cell faces. The flow region is divided into a mesh of fixed rectangular cells. When free surfaces of fluid interfaces are present, it is necessary to identify whether those cells are empty, contain a surface, or are full of water. The software considered a cell with an F value less than unity, but with no empty neighbour, as a full cell.

At the mesh boundaries, a variety of conditions may be set using the layer of fictitious cells surrounding the mesh. There are six types of boundary condition provided by FLOW-3D: symmetry plane (default), rigid wall, specified velocity, specified pressure, continuative, and periodic boundary condition. However, in this research only the first three were used.

As a rule, all rigid and free boundary surfaces are treated as free-slip boundaries (no tangential stresses on the surfaces) and referred to as symmetry plane boundary conditions. For a rigid wall boundary, the normal velocity is set to zero, and the tangential velocities can be set to any value by the wall shear stress model provided by the software for a free-slip type of wall. However, in the no-slip boundary condition, the tangential velocities in the fictitious cells are set to zero. Furthermore, in the specified velocity conditions, tangential velocities and normal velocities must be specified.

2.2 GOVERNING EQUATION

Hirt et al were the first to report on volume of fluid techniques in 1975 and extended them further in 1981. The 2-D fluid equations to be solved were the Navier-Stokes equations, viz.,

$$\frac{\partial u}{\partial t}+u\frac{\partial u}{\partial x}+w\frac{\partial u}{\partial z}=-\frac{\partial p}{\partial x}+g_x+\upsilon\left[u\frac{\partial^2 u}{\partial x^2}+w\frac{\partial^2 u}{\partial z^2}\right],$$

$$\frac{\partial w}{\partial t}+u\frac{\partial w}{\partial x}+w\frac{\partial w}{\partial z}=-\frac{\partial p}{\partial z}+g_z+\upsilon\left[u\frac{\partial^2 w}{\partial x^2}+w\frac{\partial^2 w}{\partial z^2}\right]. \qquad (1)$$

Since the ocean water was assumed to be incompressible, the incompressibility condition below must be satisfied, viz.,

$$\frac{\partial u}{\partial x}+\frac{\partial w}{\partial z}=0, \qquad (2)$$

However, to allow limited compressibility effects, the equation (2) above was replaced by the following expression (Hirt & Nichols, 1981),

$$\frac{1}{c^2}\frac{\partial p}{\partial t}+\frac{\partial u}{\partial x}+\frac{\partial w}{\partial z}=0. \qquad (3)$$

Essentially, the volume of fluid technique consisted of three components. Firstly, a VOF function F to locate the water surface was required. Within the water, F was equal to unity, and became zero outside the water surface. The governing equation was given as follows:

$$\frac{\partial F}{\partial t}+u\frac{\partial F}{\partial x}+w\frac{\partial F}{\partial z}=0. \qquad (4)$$

Secondly, an algorithm to track the surface as it moved and deformed within a computational grid was required. As noted by Hirt and Nichols (1981), for the equation (4) above, the flux of *F* moving with the fluid through a cell needs to be computed. However, standard finite-difference approximations would lead to a smearing of the F function and the interfaces would then lose clarity. Fortunately, the fact that F was a step function with values of zero or one permitted the use of a flux approximation that preserved its discontinuous nature. This approximation was referred to as a donor-acceptor method. The boundary conditions at the surface were applied as explained in the earlier section. Detailed techniques on applying previously mentioned boundary conditions have been given by Hirt et al (1981). Finally, the F function in the equation (4) above should be updated to give the new fluid surface.

3. Numerical Model

The reef was located on the ocean floor (see Figure 1) using a Cartesian two-dimensional coordinate system (x,z) from the toe of the beach (0 m) to a distance of 13.0 m, from the toe, in the upward direction (depending on the configuration of reefs). The applied beach was straight, with the toe located 50.0 m from the open sea or at x=0 with a 5% slope.

The computational grids used to cover the model (shown in Figure 1) consisted of 192 cells in a horizontal direction. Varying fluid cell volumes were used in the horizontal direction. Cells in the vicinity of reefs (between 0.0 to 13.0 m) were denser than in the other areas. In this area, 65 cells (each cell equal to 0.20 m wide, which gave nearly 20 cells within one hemispherical reef) were used to represent the reefs smoothly, while in the areas between −50.0 m to 0.0 and 13.0 m to 90.0 m, the width of each cell was equal to 1.0 m. The number of cells used in this area was 50 and 77, respectively. In the vertical direction, the cells totalled 30, over a height of 6.0 m. Both directions included, there was therefore a total of 5760. These grids were used to cover the physical size of the computational domain which was 140.0 m in length and 6 m in height as shown in Figure 1.

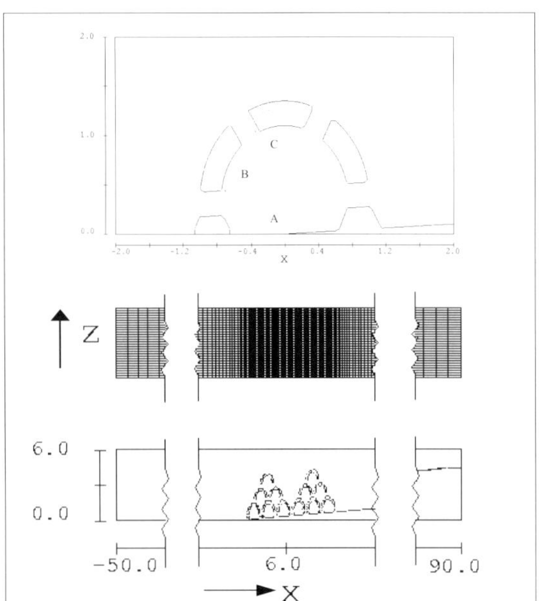

Fig. 1. Typical reef and applied grids

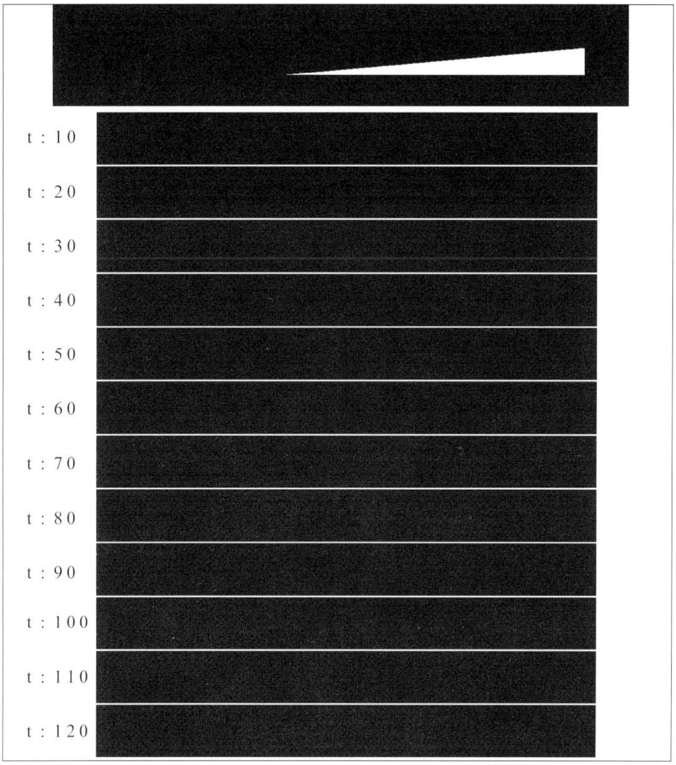

Fig. 2. Water profile without reefs

The reefs (2.0 m diameter), each having 4 openings, were located between 0.0 to 13.0 m from the toe, as shown in Figure 1. For the purpose of analysis, clusters of one, two, three, six, and twelve reefs were installed at a specified location. The three and six reef clusters were arranged in a triangular shape, while the twelve reef clusters were constructed with two triangles of six reefs. The velocity magnitudes within the bottom reef, at locations represented in the Figure by A, B, and C, were also examined.

For initial conditions, the water depth was set at 4.0 m, and the sea considered being at rest. A sinusoidal wave as previously considered by Richardson (1996) for a surf similarity model with 1.0 m height and 5 seconds period was applied for all reef configurations for the left boundary condition. Symmetry boundary conditions were used at the lateral boundaries and a wall (no-slip) boundary condition was used at the right computational domain.

The first step was to simulate the wave profile without any reef, and then compare the wave profile with that obtained from the installation of one, two, three, six, and twelve reef clusters. However, due to the limited space, only the results for three, six, and twelve reef clusters are given in the following pages.

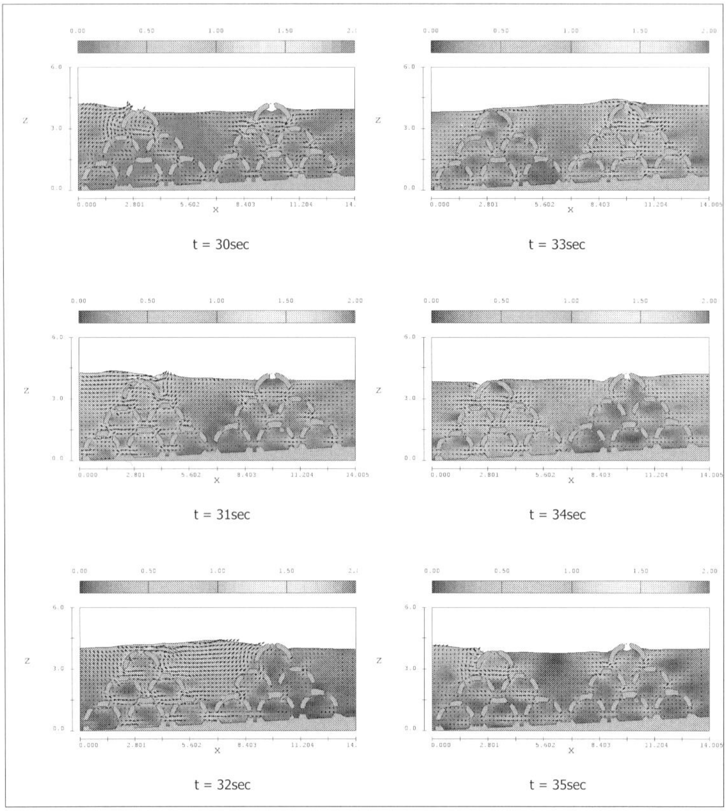

Fig. 3. Water profile and velocity magnitudes for a beach with twelve reefs at various times

4. Results and discussion

The model without reefs followed the surf similarity model that has been analysed previously using FLOW3D (Richardson, 1996). In this research for a 5% beach slope 1.0 m wave height and 5 seconds period, the surf similarity parameter ξ of the model was equal to 0.264. The breaking wave over the reefs could thus be characterized as the spilling and breaking type. The computed water surface profile agreed well with the results published in various sources for the spilling type of breaking waves (Gross, 1987 & Carter, 1988). The profiles for every 10 seconds are given in the Figure 2.

To check the sensitivity of the computed numerical values, the total number of cells in the computational domain were varied over the whole domain. The time history for the water profile and velocity magnitude on the water surface (x=-23.0 m, z=4.0 m) for the 12 reef configuration are presented in Figure 3. For the coarse grids, the number of horizontal cells in the vicinity of the reefs were 26, and for the normal and fine grids, they were 39 and 65, respectively. The difference between the normal and the fine grids was thus not significant for the water profile and velocity computation; however, it is acceptable. Use of finer grids is required for better convergence.

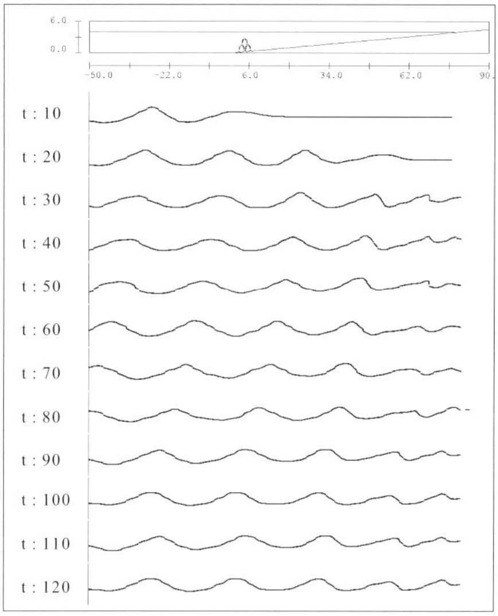

Fig. 4. Water profile with three reefs

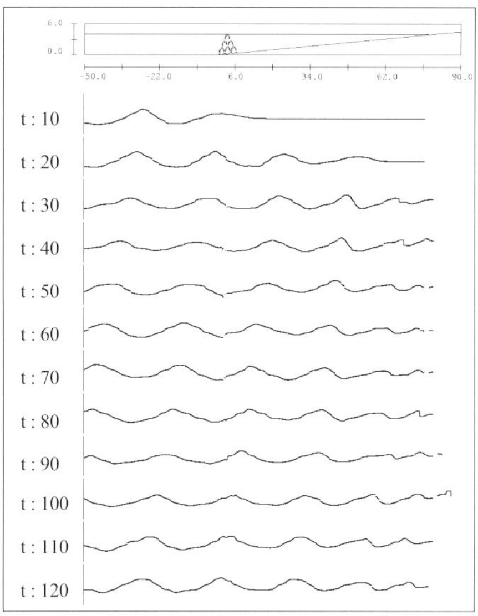

Fig. 5. Water profile with six reefs

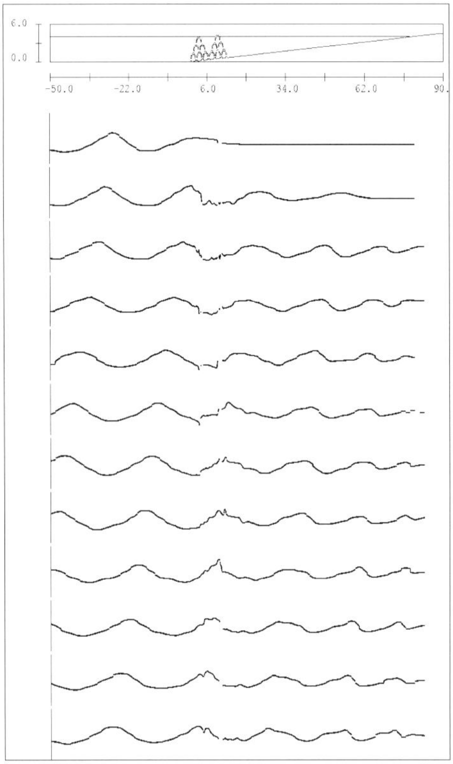

Fig. 6. Water profile with twelve reefs.

Typical water surface plots for models with three, six, and twelve reef clusters installed around 6.0 m from the toe of beach are given in Figures 4, 5, and 6 respectively. These figures show that the wave height is reduced when the number of reefs installed are increased. Below six reefs, the waves do not break over the reefs as shown in Figure 4. The waves travel over the reefs without any obstruction and hence do not break at the reef location; this is due to the fact that the reefs are placed deep below the water surface. However, as shown in Figures 5 and 6, where six and twelve reefs are installed, the waves break over the reefs.

In Figure 7, the water surface history plots at +70 meters from the toe of beach are compared before and after installation of the reefs. The dotted line represents the water surface without reefs, while the solid line represents the water surface time history when three, six, or twelve reef clusters are installed. The reductions are noticeable when the number of reefs installed is twelve. Since the model used for numerical analysis is only two dimensional, the cross flow generated in the lateral direction by the

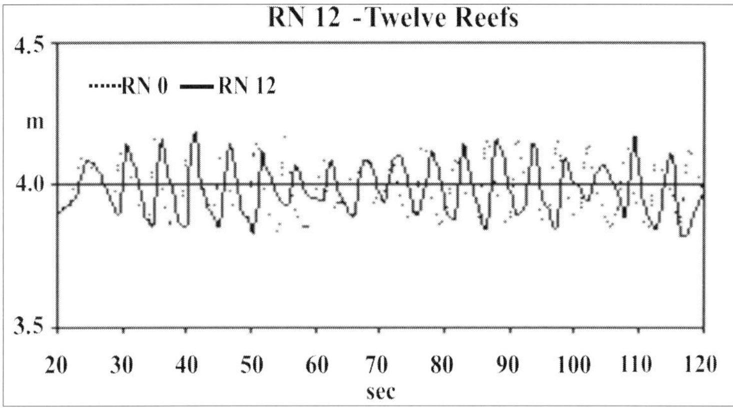

Fig. 7. Water profile time series at x=70 m

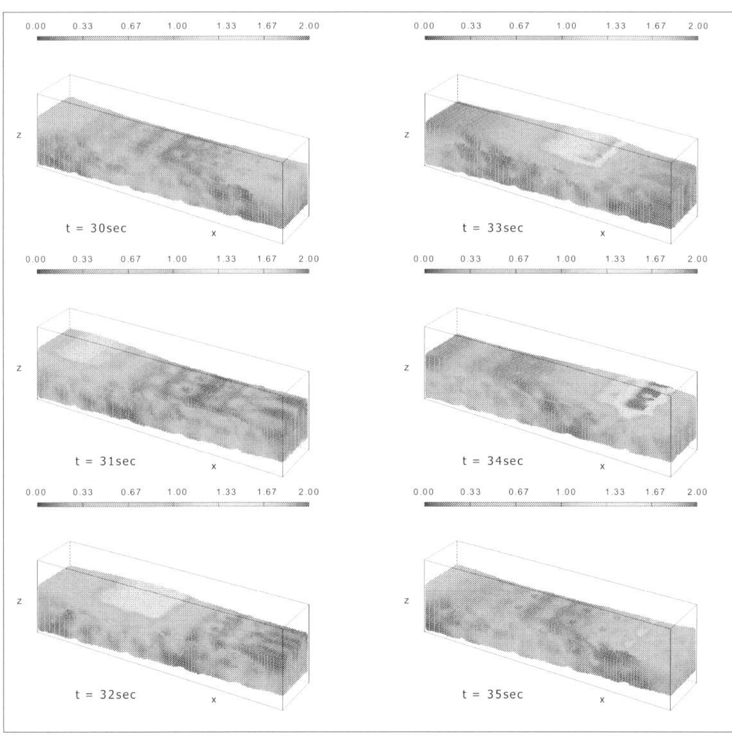

Fig. 8. 3-D profiles of velocity magnitudes for the beach with cross flow

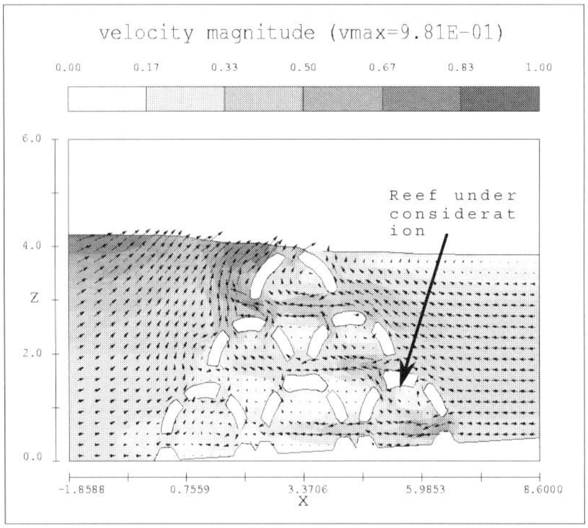

Fig. 9. Typical flow field around reefs

three-dimensional porous hemispherical reef (shown in Figure 8) is not properly modelled in analysis. If the three-dimensional nature of fluid flow is properly modelled, the wave basin between the reef and the coast would be more quiescent leading to the congregation of fishes, settling down of biomass and the development of productive aquaculture habitat. The two-dimensional model results show that the wave amplitudes are reduced inside the wave basin produced by reefs when the number of reefs is greater than six, even though the reduction is not significant.

As water moves through the reefs, the incoming wave energy is dissipated by turbulence; furthermore, pressure waves, which can be detected by fish, are produced as water exits through the holes on the top and sides of the hemispherical balls. Turbulent water, which exits/enters through the holes on the top and sides, moves upward/downward and modifies the incoming/outgoing wave field.

This movement of water is found to be effective in attracting fish swimming near the sea surface. Typical flow around the reef is given in Figure 9. The shading represents the varying velocity magnitudes in the vicinity of reefs. As seen in the figure the energy dissipation, due to turbulence around each structure, leads to almost quiescent local areas (at points A, B, and C, indicated in Figure 1) in and around the hemispherical balls where fish can settle down and spawn. The area where the velocities are low are represented by the light shading.

It should also be noted that the generated reduced fluid flow would lead to the deposition of sediments containing food organisms as well as the growth of various types of seaweed. This would attract bottom-feeding animals such as flatfish, sea urchins and lobsters. These benthic animals tend to congregate and populate the partially quiescent areas on the leeward side of these structures. The possibility of silting at the bottom of the hemispherical reef (due to very low velocities) must be judiciously minimized since it would inhibit the growth of benthic diatoms and prevent the attachment of seaweed spores.

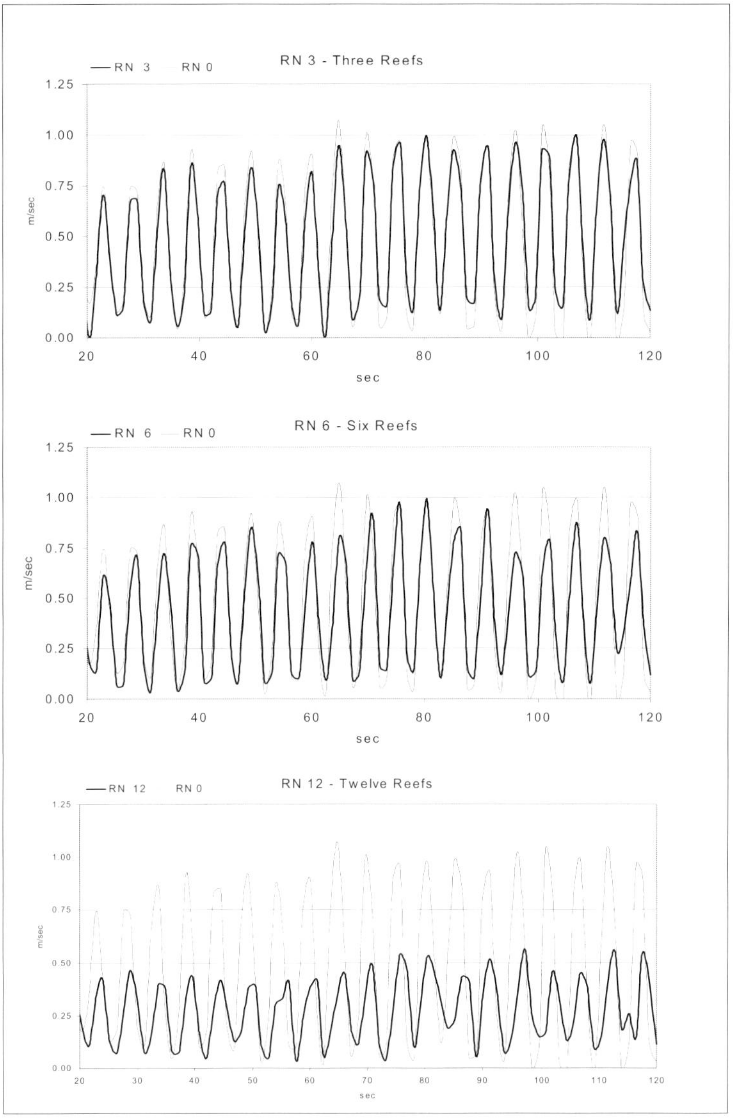

Fig. 10. Surface velocity magnitude time series at x=+15 m

Figure 10 gives the surface velocity variation within the basin (at x =+15.0 m) for three, six and twelve reefs. It will be seen that the velocities at this location (just near the reef on the shoreward side) decrease considerably for the twelve-reef cluster, leading to the settlement of benthic diatoms and congregation of fish. If the three-dimensional nature of flow is properly modelled, this would generate a wave motion

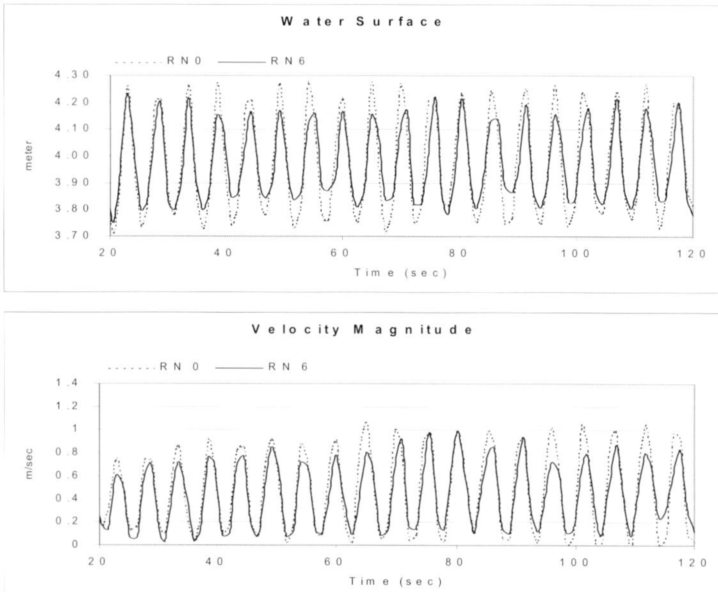

Fig. 11. Water surface profile and surface velocity magnitude time series for the six-reef clusters at x: +15.0 m

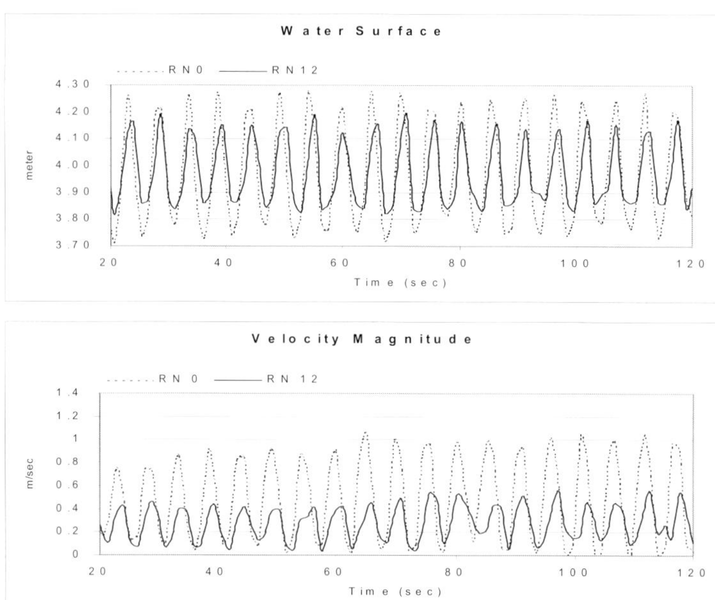

Fig. 12. Water surface profile and surface velocity magnitude time series for the twelve-reef clusters at x: +15.0 m

parallel to the reefs at this location. This transverse (to the reef) flow is expected to facilitate the motion of the fish in this direction, which would increase the productivity of the sea around this area.

Time series for velocity magnitudes before and after reef installation for the six and twelve reef clusters are shown in Figure 11 and 12, respectively. It will be seen that the water velocities decrease considerably at all three locations (A, B, and C) within the reef (indicated in Figure 9) facilitating the attachment of marine organisms and their subsequent growth, especially at locations B and C.

5. Conclusions

The following are this study's main conclusions:

1) The use of reefs below six units does not seem to reduce the wave height.
2) Use of a 12-unit reef seems to reduce the wave climate inside the basin considerably.
3) Provision of proper reef placements provides a conducive environment and suitable locations in which fish can congregate and spawn; in addition it also provides areas in which benthic diatoms and seaweed spores can develop and grow.

Further studies are being carried out to analyze the three-dimensional nature of fluid flow and to develop better numerical models. Moreover, additional studies are being carried out to investigate the influence of:

a) Combinations of reefs with and without holes
b) Reefs with a combination of normal and inverted forms
c) Reefs with external modifications

Symbols used

c: speed of sound in the water
g_x, g_z: gravity acceleration in x and z direction
k: wave number: $k = 2\pi / \lambda$
T: wave period
w: velocity components in z direction
λ: wave length
ω: wave angular frequ.: $\omega = 2\pi / T$
F: fluid function
H: wave height
p: pressure
u: velocity components in x directions
B: beach angle
υ: coeficient of kinematic viscosity
ξ: surf simil. par.: $\xi = (H / \lambda)^{-0.5} \tan\beta$

Acknowledgements

The numerical simulation in this paper was done by the second author at Memorial University, Newfoundland. The financial support provided by the Government of Indonesia is gratefully acknowledged. Acknowledgements are due to Dr. R. Seshadri, Dean of Engineering and Dr. M.R. Haddara, Associate Dean of Engineering, for the excellent computational facilities provided by the Faculty of Engineering and Applied Science, Memorial University, St. John's, Newfoundland.

REFERENCES

Ambrose, R.F., Swarbrick, S.L.: 'Comparison of Fish Assemblages on Artificial and Natural Reefs off The Coast of Southern California', *Bulletin of Marine Science*, Vol 44., No. 2, 1989, pp. 718–733.

Aono, T., Cruz, E.C.: 'Fundamental Characteristics of Wave Transformation Around Artificial Reefs', *Proceedings of the 25th International Conference of Coastal Engineering*, Chapter 178, 1996, pp. 2298–2311.

Baynes, T.W., Szmant, A.M.: 'Effect of Current on the Sessile Benthic Community Structure of an Artificial Reefs', *Bulletin of Marine Science*, Vol 44., No. 2, 1989, pp. 545–566.

Carter, R.W.G.: *Coastal Environtment: an Introduction to the Physical, Ecological and Cultural Systems of Coastlines*, Academic Press, London, ISBN 0-12-161855-2, 1988.

Fitzhardinge, R.C., Bailey-Brock, J.H.: 'Colonization of Artificial Reef Material by Coral and other Sessile Organisms', *Bulletin of Marine Science*, Vol 44, No. 2, 1989, pp. 567–579.

Flow Science, Inc.: FLOW-3D Excellence in Flow Modelling Software version 7.1, Los Alamos, NM, 1997.

Gross, M.G.: *Oceanography: a View of the Earth*, Prentice-Hall Inc., Engleword Cliffs, N.J., USA, ISBN 0-13-629692-0-01, 1987.

Hayakawa, N. et al: 'Numerical Simulation of Wave Fields around the Submerged Breakwater with SOLA-SURF Method', *Proceeding of the 26th International Conference of Coastal Engineering, Copenhagen, Denmark*, Paper No. 78, 1998, pp. 156–157.

Hirt, C.W., Nichols, V.D., & Romero, N.C.: 'SOLA: A Numerical Solution Algorithm for Transient Fluid Flows', Los Alamos Scientific Laboratory Report, Los Alamos Scientific Laboratory Report, LA-5852, 1975.

Hirt, C.W., Nichols,. V.D., & Hotchkiss, R.S.: 'SOLA-VOF: A Solution Algorithm for Transient Fluid Flow with Multiple Free Boundaries', *Los Alamos Scientific Laboratory Report*, LA-8355, 1981.

Hirt., C.W., & Nichols,B.D.: 'Volume of Fluid (VOF) Method for the Dynamics of Free Boundaries', *Journal of Computational Physics* 39, 1981, pp. 201–225.

Kawasaki, K and Iwata, K: 'Numerical Analysis of Wave Breaking Due to Submerged Breakwater in Three Dimensional Wave Field', *Proceeding of the 26th International Conference of Coastal Engineering, Copenhagen, Denmark*, Paper No. 79, 1998, pp. 158–159.

Lindquist, D.G., Pietrafesa, L.J.: 'Current Vortices and Fish Aggregations: the Current Field and Associated Fishes around a Tugboat Wreck in Onslow Bay, North Carolina', *Bulletin of Marine Science*, Vol 44., No. 2, 1989, pp. 533–544.

Pickering, H., Whitmarsh, D.: 'Artificial Reefs and Fisheries Exploitation: a Review of the 'Attraction versus Production' Debate, the Influence of Design and its Significance for Policy', *Fisheries Research*, Vol. 31, 1997, pp. 39–59.

Reef Ball Development Group, Ltd.: Internet Brochure., http://www.reefball.org.

Richardson, J.E.: 'Surf Similarity', *Flow Science Technical Note # 44*, FSI-96-00-TN44, Los Alamos, NM., 1996, 8, 1997.

White, A.T, et al: *Artificial Reefs for Marine Habitat Enhancement in Southeast Asia*, ICLARM Education Series 11, Philippines, ISBN 971-1022-83-4, 45, 1990.

SOFT PROTECTION USING SUBMERGED GROIN ARRANGEMENTS: DYNAMIC ANALYSIS OF SYSTEM STABILITY AND REVIEW OF APPLICATION IMPACTS

C.L. GOUDAS,[1] G.A. KATSIARIS,[2] G. LABEAS,[1] G. KARAHALIOS,[2] AND G. PNEVMATIKOS[2]

[1] *Dept. of Mathematics, University of Patras, Patras, Greece.*
[2] *Dept. of Physics, University of Patras, Patras, Greece*

Abstract

The submerged groin arrangement as a shore protection tool examined in detail by means of direct numerical integration by Kafousias et al, 2000, is examined in part I of this paper on the basis of the finite elements dynamic analysis in order to determine its capability to withstand the extreme loading of the 50–year return-wave conditions. The analysis performed using the ANSYS programme shows that the safety co-efficient value of the order of 3.0 is easily attainable under fifty-year return wave characteristics and hence there is no risk of failure in placing such groins in sites with extreme wave climates assumed here. All previous applications have verified this.

In part II, experimental systems of submerged groins, referred to as coastal protection and nourishment systems, or CPNS (patented arrangement), placed in various beaches in Greece, Italy and Egypt and surveyed at regular intervals after placement are reviewed. In all experiments the sites were not nourished and the systems were left to perform their protective and self-nourishment role on the basis of modification of the longshore current pattern. In all cases it was found that the CPNS systems prevented further recession of the shoreline. In one case (the site at Theologos, Rhodes), extensive self-nourishment (natural accretion) of sand was observed. To a lesser extent, accretion of sand was observed in all ten experiments conducted. The sites of the experimental placement of submerged groin systems are at the beaches at 1. Theologos, Rhodes (six-groin system of 40m each), 2. Theologos, Rhodes (ten-groin system of 40m each), 3. Theologos, Rhodes (two-groin system of 40m each), 4. Mytilini, Lesbos (ten-groin system of 40m each), 5. Perama, Lesbos (three-groin system of 40m each), 6. Vatera, Lesbos (nine-groin system of 40m each), 7. Enofyta, Viotia (ten groin system of 40m each), 8. Dilesi, Viotia (ten-groin system of 40m each), 9. Aghia Marina, Viotia (thirteen-groin system of 40m each), 10. Tsoukaleika, Achaia (three-groins of 20m each), 11. Rio, Achaia, (six-groins of 30m each), 12. Loutro, Corinthia (six-groin system of 40m each), 13. Xylocastro, Corinthia (six-groin system of 40m each), 14. Kiato, Corinthia (six-groin system of 40m each), 15. Tolo, Argolida (twenty-eight-groin system of 40m each), 16. Villa San Giovanni, Calabria, Italy (six-groin system of 40m each), and 17. Oasis Resort, Al Suckna (Red Sea), Egypt (twenty-eight groin system of 40m each).

Monitoring of all sites over a period of three years showed that in all cases the presence of the submerged groin systems had no adverse impact in the sense that in no case was shoreline recession observed. To a varying degree, accretion of sand or small size gravel occurred and in one case the accretion of sand was massive. All but one site (case XII) were not nourished. The nourished beach showed impressively stable change. The conclusion permitted by the monitoring results is that placement of groin

C. L. Goudas et al. (eds.), Soft Shore Protection, 227-260.

systems and nourishment constitutes a viable solution to beach rehabilitation and protection.

PART I

STABILITY OF CPNS

1. Introduction

Both the computation of stresses and of expected deformations and *in situ* experimental results have shown that highly inert seabed structures, such as the CPNS seabed groins (coastal protection and nourishment system), subject to 50-year return waves of up to 6m height or even larger, stay intact without any displacement or damage. Standard static analysis methods, such as analysis of stresses in transverse sections of such seabed groins, carried out by the Patras team are sufficient to validate the above statement. Nevertheless, the present detailed analysis, done on the basis of the Finite Element Method (FEM), was carried out in order to acquire a second independent verification. The physical dimensions of the said groins, which were dictated by functional requirements, namely the need to secure wave and current energy deflection toward the shore, lead to groins with a safety coefficient above 3.0 being designed.

This protection shore protection system, although not as yet tested in oceanic wave climates, seems, on the basis of the results, experimental and theoretical, capable of withstanding dynamic loads present in such environments and hence deserves *in situ* testing. It should be underlined that such testing refers more to the capability of the system to withstand the dynamic loads of ocean waves and much less its ability to protect the shore around the installation and its surrounding environment, an aspect which is also considered in this paper on a basis of a theoretical study (Goudas et al, 2000) and the results of applications of this arrangement in the Mediterranean and in the Red Sea.

The environmental impacts of CPNS, on the other hand, also based upon 15 *in situ* experiments, and covering all environmental parameters, physical and biological, demonstrate conclusively that this system has only beneficial impacts.

2. The CPNS System

As mentioned above, the CPNS system consists of a suitable number of long and rectilinear groins that are installed at suitable positions on the unexcavated seafloor. The alignment direction of the long axis of each, is perpendicular to the shoreline, unless other factors, e.g. strong permanent currents, dictate other choices. The principal semi-axes, a and b, of the almost elliptical cross-section, vary with the cross-section distance from the shoreline, while the major axis of length 2a is horizontal and the minor axis 2b is normal to the seafloor. The parametre a decreases with increasing distance from the shoreline and the marametre b increases with increasing seadepth. Figures 1 and 2 give a conceptual definition of the geometry of each groin-member of CPNS. The groins are installed at the shore by in situ filling with special concrete mix prefabricated long geotextile tubes. Their position on the seabed and distance offshore will vary according to site and is determined by research results.

3. Dynamic analysis by application of finite element methods

The finite element method was used by ISRTAM (Institute of Structures and Advanced Materials), which specializes in analyses and testing of composite structures.

The strength and static stability of the seabed groin design was studied for specific geometry and length, based on the assumption that the module will be resting upon an inclined seabed at a given depth characteristic of the installation sea area. A change to this parameter, with the rest fixed (geometry of the groin, inclination of the seabed, design wave characteristics, etc.), will effect the final results but not the static stability of the groin. The FEM method applied produced results that have demonstrated both the strength and static stability of the seabed groin design under conditions of a hundred-year return wave at any site.

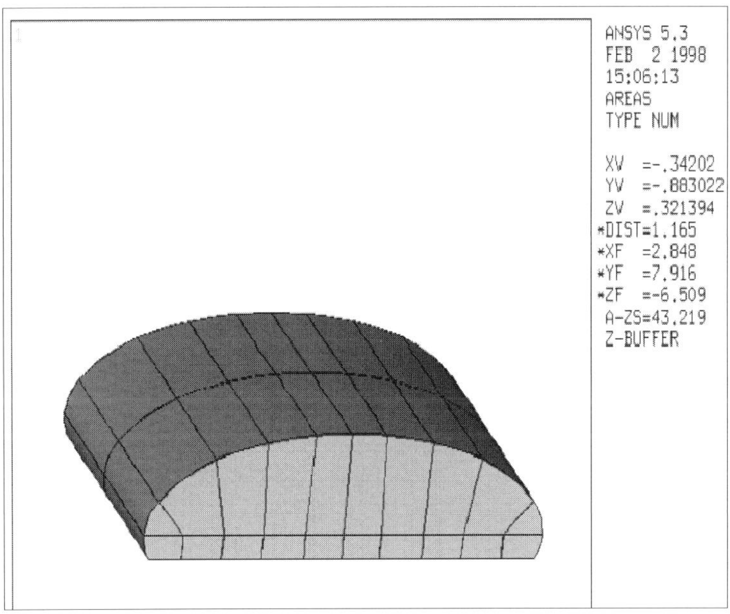

Fig. 1. Geometry of a seabed groin segment

4. Geometry and loads applied to the seabed groin

The groin segment examined was taken to be of a length of 20m and with a constant cross-section. The adopted cross-section is elliptical with semi-axes of 0.85m and 0.3 m, respectively. Increasing the total length and the semi-axes of the groin segment obviously leads to a stronger element, and hence the choice, which was geared to show that even the smallest and thinnest groin segment can withstand the maximum wave load likely to be experienced in this sea area. Figure 1 shows the perspective geometry of a 1m long segment.

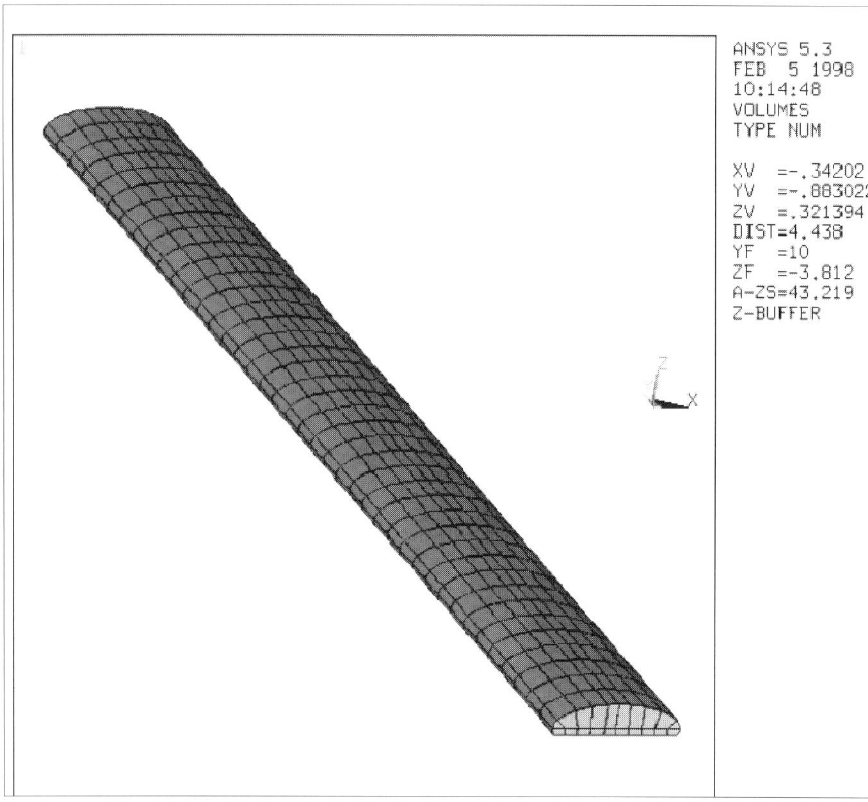

Fig. 2. Finite element model of the seabed groin

The groin is constructed of concrete type C16, not reinforced, but contained inside a strong geotextile cover. The strength of the geotextile is not considered, assuming that it performs as a skin of the concrete that prevent chipping. The mechanical properties of the concrete are:

Measure of Elasticity E = 30 Gpa,
Poisson Ratio $\nu = 0.25$
Breaking limit $\sigma_B = 20$ Mpa.
Seabed kinetic friction co-efficient $\eta = 0.8$.

The design groin segment is statically and dynamically as follows:

α) Its weight (the specific weight of C16 concrete is 2400 kg/m^3)

β) The wave static and dynamic load expressed through the total pressure p. The pressure applied upon all surface elements of the groin segment includes the hydrostatic of the variable unit water column and dynamic component due to the motion of the water. The unsteady flow caused by the waves produces cyclic time variation in this pressure. The time-dependent expression of the pressure p(x,y,z;t) at any groin surface point P(x,y,z) at any time instant t is obviously normal to this surface and is expressed by the formula:

$$p(x,y,z,t) = -\rho g z - r q^2 / 2 + \frac{\rho a c^2 k \cos[k(z+h)]}{\sinh(kh)} \sin[k(x-ct). \quad (1)$$

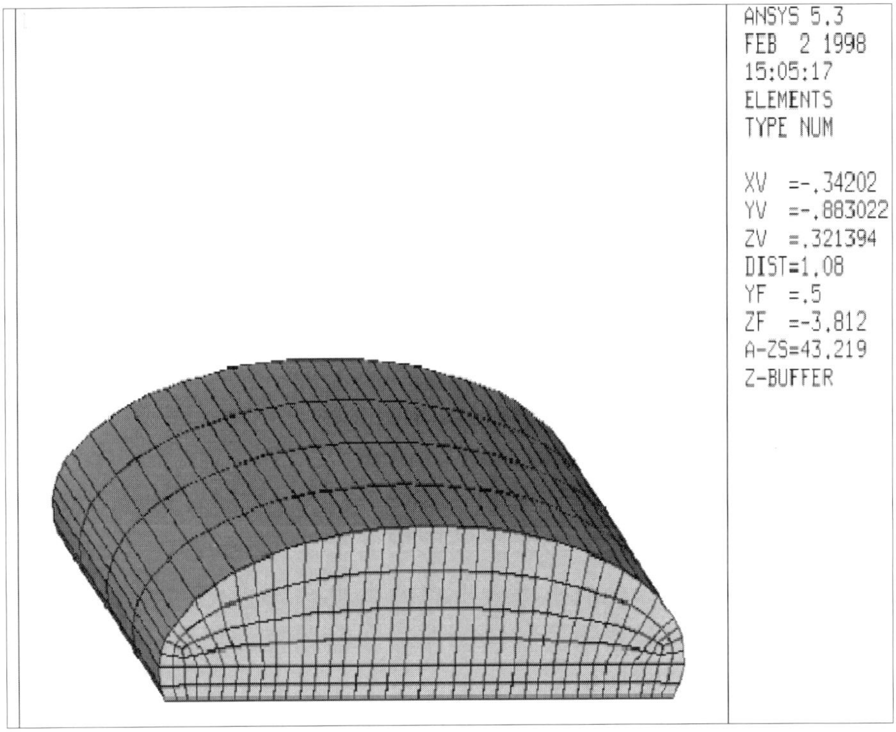

Fig. 3. Finite element model of a segment of the seabed groin

The first term of this expression gives the static pressure (time-independent term), here representing the mean hydrostatic pressure since z is zero at the mean sea level, taking into consideration the tide level at the time. The second term is the pressure component due to the kinetic energy of the unit volume of seawater. Although of a dynamic nature, this term is time-independent (assuming fixed wave characteristics), and hence part of the static load. The third term is purely time and position dependent and hence dynamic. The co-ordinate system Oxyz has its origin O resting on the MSL, the Oz axis directed along gravity and negative downward, and its Ox axis coinciding with the wave propagation direction.

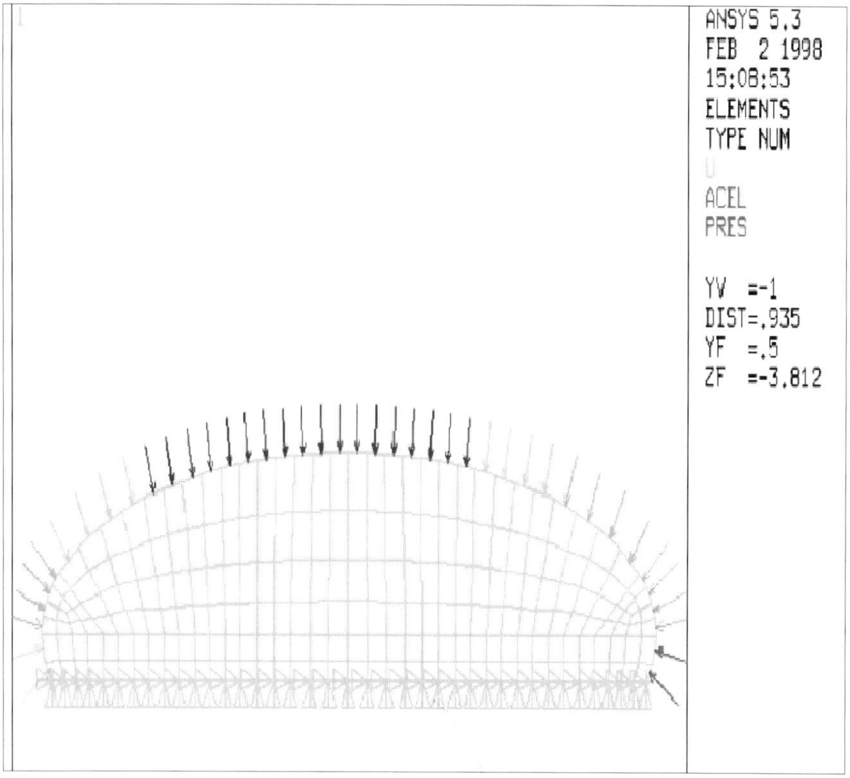

Fig. 4. Loading and boundary loading of the seabed groin

The notation used in eqn.1 is as follows:

ρ (sea water density) = 1030 kg/m^3 ,
g (acceleration of gravity) = 9.81m/s^2 ,
q = water molecule velocity,
h = water depth at the position of groin installation,
α = amplitude of wave motion = H/2 = half the wave height,
L and T = wave length and wave period,
k = 2π/L,
c = wave propagation velocity, and
t = time.

The numerical values taken for the above parameters are h = 4 m, α = H/2 = 3 m, L = 20 m, T = 5sec and c = 4 m/s.

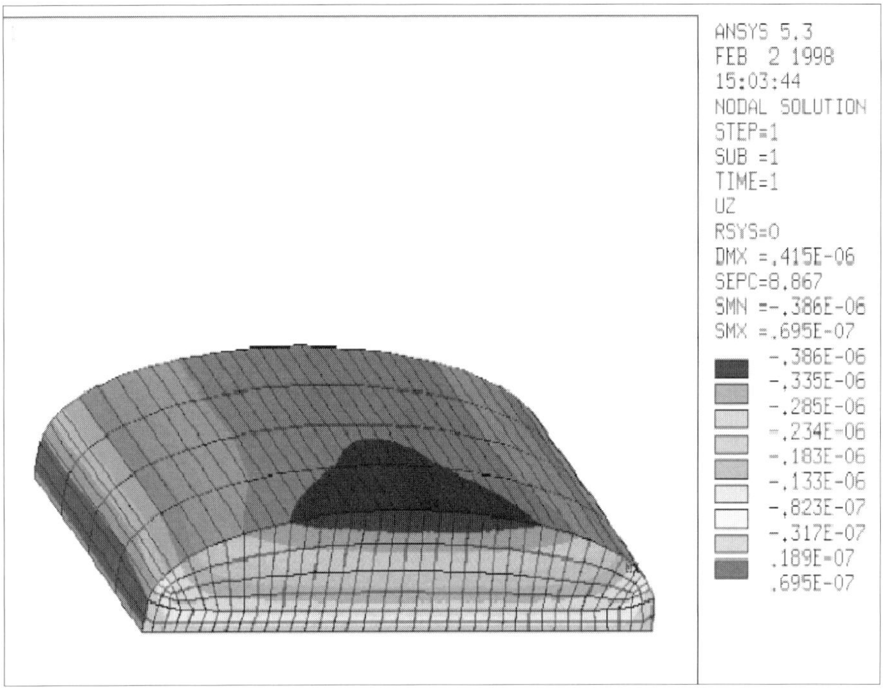

Fig. 5. Displacement in the z direction

5. FEM Model of the seabed groin

The seabed groin described above was partitioned into a finite element set (FES) using the computer program FES ANSYS.

5.1 GEOMETRIC MODEL.

Eight-node three-dimensional elements with three degrees of freedom per node were used for making the geometric model of the seabed groin.

The coding system SOLID 45 was employed for this purpose. The geometrical model consisting of 16 600 elements and 21 141 nodes is shown in Figure 2; Figure 3 shows a 1m long segment of the geometrical model. Repetition of this segment along the entire length of the seabed groin gives the geometrical FES used in full.

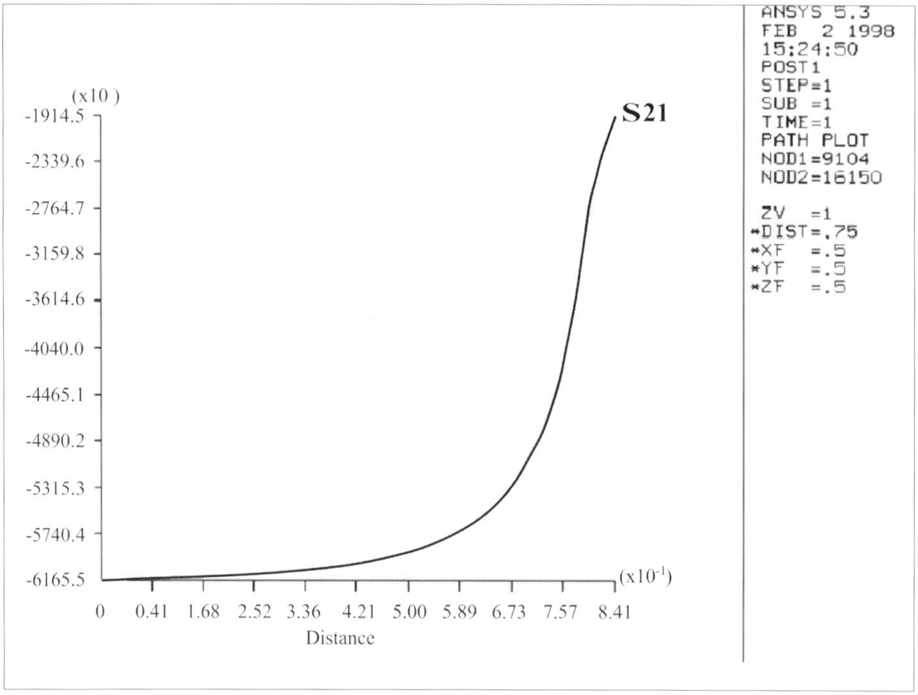

Fig. 6. Variation in the normal stress σ_z starting from the seating position of groin (z =−4m) up to its upper surface along the longitudinal axis of symmetry (x = 0)

5.2 LOADING, BOUNDARY CONDITIONS AND TYPE OF ANALYSIS

The wave load was considered as representative and corresponds to the angles of wave impact upon the seabed groin. More specifically, the angles between the wave propagation direction and the longitudinal groin axis were assumed to be 0°, 45° and 90°. For each of these wave impact directions, stress-strain analyses were performed, freezing the time parameter at t = 0sec., t = 1.25sec., and t = 2.5sec. Hence a total of nine static analyses were performed. The problem is solved by static analysis for specific values of the time parameter, and hence of the corresponding load, because the inertial loads of the groin contribute very little to the resistance and static stability.

For each of the nine cases considered, the pressure is computed using expression (1) for each surface finite element that is in contact with seawater. The corresponding force is applied to the element surface receiving it.

The seating surface of the groin upon the seabed is assumed to be of fixed position in all three spatial directions Hence the nodes upon this surface are locked. Figure 4 shows a groin cross-section with the loading and the boundary conditions of loading.

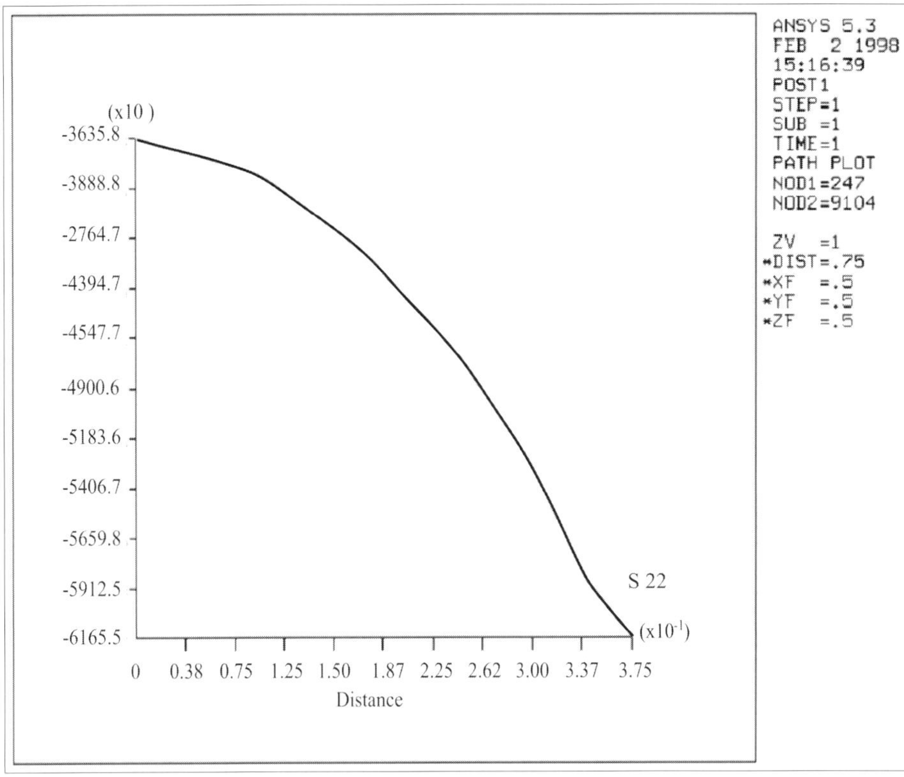

Fig. 7. Variation in the normal stress σ_z starting from the seating position of the groin (z = –4m) from the longitudinal axis (x = 0) up to x = –0.75m

6. Results of the analysis

The results obtained for the nine cases considered provide the strained and unstrained geometry of the groin due to the applied loading (see Figure 4). Figure 5 gives the vertical displacement (along the Oz axis) for the same segment of the groin shown in Figure 3.

The results obtained show that in all cases the greatest stresses developed near the base of the groin. Table 1 lists the equivalent maximum Von Mises stress and the structure safety factor N_α. It is interesting to note that the stresses developed are extremely small, a fact also attested by the safety factor N_α of the structure. This factor signifies how many times larger is the resistance of the structure to the maximum stress developed by the wave load.

Fig. 8. Example of full CPNS arrangement and one single groin (bottom) installed at the beach at Therma, Gulf of Gheras, Lesbos. The system consists of 10 seabed groins

For the worst cases as far as strength of the structure is concerned (No 7, Table 1) we show the normal stress σ_z and the corresponding Von Mises stress. Figures 6 and 7 show the normal stress σ_z in various parts of the structure. In particular, Figure 6 shows the variation in normal stress σ_z at the base (z=−4 m) as function of the groin height along the x-axis. Figure 7, on the other hand, shows the dependence of normal stress σ_z at the groin base level (z =−4) upon the distance from the central longitudinal trace (x=0) to the external longitudinal line (x = 0.75m).

As already mentioned, the analysis adopted the assumption, later established to be valid, that groin points in contact with the seabed are not displaced. From the analysis it was possible to calculate the friction force necessary for securing static stability of the groin. The maximum friction force that it was possible to develop along the directions x or y, for longitudinal or lateral displacements, is:

$$T_{max} = \eta \cdot F_z , \tag{2}$$

where η is the kinetic friction co-efficient between the groin's base and the seabed and Fz is the normal reaction force applied by the seabed upon the groin. This normal reaction is calculated from the analysis. The ratio of the maximum friction force that can be developed (computed from formula (2)) over the external force applied, under the condition that this external force does not cause slipping of the groin's base along all combinations of x and y directions, is identified as the safety factor N_o of the structure as regards static stability (slipping).

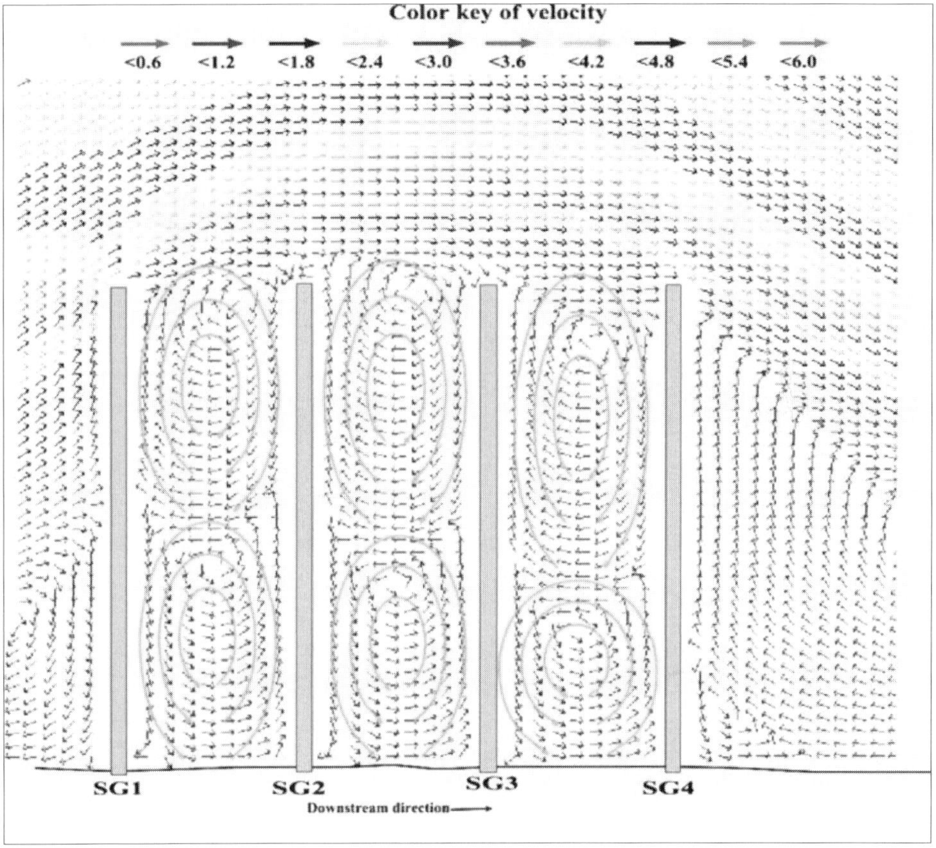

Fig. 9. The two-dimensional current flow in the presence of four seabed groins of 40 m length placed 20 m apart. Between consecutive groins are two vortices of comparable size and opposite polarity. It is again significant that the outer vortices rotates CCW and prevents escape of bedload. The CCW vortices reach the limits of the system and prohibit escape of bedload. The ratio L/M (L = 40 m, M = 2 m) is equal to 0.5. The case corresponds to Re=2640, u_o=0.5m/s, and max. velocity appearing in the entire field umax=5.98m/s

Table 2 lists for each of the nine cases considered the total reaction force components Fx, Fy, and Fz along the 3 space directions as computed by the FEM analysis, the maximum friction force T_{max} that can be developed, as given by formula (2), and the minimum safety factor N_o that safeguards slipping of the groin. This safety factor against slipping receives minimum value along the y direction, because the friction force required to face slipping along the x direction is smaller. This Table shows that even for the most disadvantageous case of a wave impact angle of 45° with the groin longitudinal axis, the safety factor N_o has a value of 3.53, implying that the required stability of the groin is secured and there is no risk of the groin slipping.

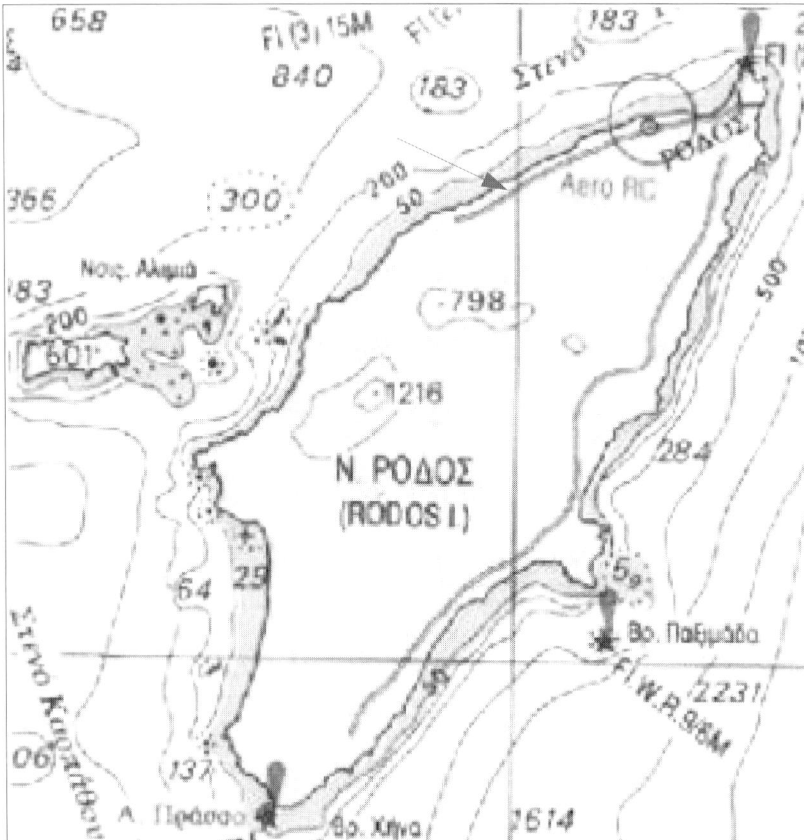

Fig. 10. Nautical chart of Rhodes. The arrow points to the site of projects 1, 2 and 3 and indicates the angle of attack of the prevailing waves

PART II

ENVIRONMENTAL IMPACTS OF CPNS

7. Introduction

CPNS systems were designed and installed on the beaches mentioned in the abstract using procedures to be briefly described in this part of the paper together with an analysis of the monitoring results and the broader environmental impacts. All of the projects, except those implemented outside Greece (Italy, Egypt), were contracted out to the University of Patras (the majority of them) or to a private company. However, the University of Patras team did the design work for all of the home and foreign projects.

Fig. 11. Theologos Beach at Rhodes, photographed from one of the five rock groins that have caused apparenty vast erosion of the seaside road. The site was a farm only seventy years before. The well structure bears witness to earlier use of land

The type of CPNS intervention is illustrated in Figure 8. The site of installation of the 10-groin system shown in this figure is Therma, Gulf of Gheras, Lesbos, in the NE part of the Aegean Sea in Greece. The groin type used in all of the projects are typical of those shown in this Figure (bottom picture) and they are constructed by casting concrete *in situ* in prefabricated geotextile bags of a suitable length (40 m in this case) and cross-perimeter (3.93 m in this case). The concrete cast is made from the inshore opening of the bag, which is sealed by thermal welding after the pre-estimated quantity of concrete (32 m^3 in this case) has been poured. The concrete in liquid condition prolonged through additives, reaches the offshore end of the bag and after solidification acquires an approximately elliptical form in which the cross-section semi-axes a and b at the state of equilibrium depend upon the seabed slope and the distance from the shoreline (in this case a varies from 1.65 m at the shoreline to 1.60 m at the offshore end, and b from 0.15 m at the shoreline to 0.45 m at the offshore end).

The geometry of the groin members of a CPNS and of the system itself is determined by a programme developed by the Patras team (see Goudas et al, 2000) which did successive trial runs for various values of the basic system parameters, namely the distance M between successive groins and the groin length L, leading to a longshore current flow pattern unable to affect longshore transport of the bedload resting between groins. The optimal flow pattern, akin to the seafloor and to the wave climate of the Therma, Lesbos site, is shown in Figure 9. The regime of water molecule velocities is that of permanent vortices, which obviously are not capable of displacing bedload from the seabed between pairs of successive groins. As pointed out by Goudas et al (2000) multiple vortices are probably safer than a single vortex per pair of successive groins. This is merely an empirical criterion, based on the obvious argument of enhanced confinement of the corresponding bedload within the area of circular water motion hence preventing it from reaching the offshore limit of the protected area and the risk of escape and exposure to longshore currents that prevail there. Stronger, however, is the argument that all CPNS design work based upon this criterion was found to be effective in terminating bedload transport away from the protected site.

TABLE 1

Maximum equivalent stress and safety factor N_α against failure of design loading

Case No Studied	Angle of groin to Wave Direction (°)	Phase Time t (sec)	Maximum Equivalent Stress (Pa)	Safety Co-efficient Against Failure (N_α)
1	0	0	60,058	333
2	45	1.25	48,199	415
3	90	2.5	60,043	333
4	0	0	60,410	331
5	45	1.25	60,410	331
6	90	2.5	60,410	331
7	0	0	60,444	331

TABLE 2
Reaction force components in the directions x,y,z, maximum possible friction force T_{max} and safety co-efficient against slipping N_0 for the design loading examined

Case no Studied	Angle of groin to wave direction (°)	Phase time t (sec)	Reaction force components			Max. frict. force	Safety co-eff. against failure (N_o)
			F_x (KN.)	F_y (KN.)	F_z (KN.)	T_{max} (KN.)	
1	0	0	308,8	11,6	1403,3	1122,6	3,64
2	45	1.25	257,1	9,7	1190,6	952,5	3,71
3	90	2.5	309,7	11,7	1403,0	1122,4	3,62
4	0	0	317,7	12,2	1403,3	1122,6	3,53
5	45	1.25	317,7	14,4	1403,5	1122,8	3,53
6	90	2.5	317,8	12,4	1403,3	1122,6	3,53
7	0	0	315,4	11,9	1428,3	1142,6	3,62

Fig. 12. The site of Figure 11 photographed 5 months after placement of the CPNS. The instigated sedimentation induced by the system has partly covered the still visible well top. Over 5 000 m^3 was deposited by the sea on the protected site

In general, the projects described in this section were designed and constructed in the period 1994 to 1998 and hence have been in operation for insufficient time for conclusive evidence covering all possible environmental impacts upon the site under protection and the adjacent shores. It should also be stressed that none of the experiments has been completed since a period of as many as 10 years is allowed for testing the ability of the system to induce natural sedimentation, a property that was predicted and verified to be active in all cases, but with varying overall effect. As one would expect, this property depends upon the available bedload resource in the broader site area, and the period of 10 years before providing nourishment was necessary in order to avoid unnecessary expenditure.

Fig. 13. The beach of Project 1 one year after placement of CPNS. The destroyed road and the rock surface groin that caused the destruction are visible and so is the seabed groin system that induced sedimentation of the beach. The system covered both sides of the rock groin

However, since two to six years have elapsed since the projects were commenced, some of the CPNS impacts may be considered conclusive and generally accepted, though to avoid claims of bias on the part of the Patras team which recommended, designed, implemented and monitored these experimental projects, or on the part of third parties, any conclusions in this paper should be independently verified. As supporting evidence in favour of the validity of these conclusions is the fact that the project owners, mostly local authorities, have not filed any complains, except for the low rate of sedimentation induced by some of the systems.

With the above proviso in mind, the following conclusions in relation to the cost, design, implementation, impacts and behaviour of the CPNS projects can be drawn:

a. All experiments were conducted in Mediterranean wave-tidal-current climate and the conclusions in relation to the CPNS capability to prevent erosion and induce sedimentation concern this climate.

b. CPNS is a low cost shore protection measure, ranging from 5% to 25% of the cost of a hard protection project for the same beach length.

c. The total project time up to installation (data collection, design, and implementation) is very short, ranging from 3 to 6 months per 1 km of beach length.

d. All experiments verified that CPNS systems interrupted any erosion process as soon as they were installed.

e. All experiments verified that CPNS systems did not to increase the rate of erosion of downstream beaches.

f. The induced natural sedimentation has not been very satisfactory, particularly on beaches with poor sand resources, though this can be remedied by nourishment.

g. Depending upon the beach profile, CPNS systems may be visible or invisible to the viewer, but will be invisible after nourishment.

h. The capability of CPNS systems to retain nourishment materials, a property most important in all projects involving unprotected nourishment throughout the world, has to be tested.

i. CPNS systems encourage sea vegetation, shellfish and fish. In all of the experiments, the fish school populations increased. However, in one beach with large number of sea urchins before CPNS installation, the entire troop disappeared soon after installation.

j. The installation of CPNS systems is smooth and fast with a recorded installation time of 5 days per km of beach. The bag prefabrication is also fast and casting *in situ* takes about 1h/groin.

k. The effectiveness of CPNS systems in oceanic climates is only conjectured, since they have not been tested.

l. CPNS is recommended for protecting long spans of eroding beaches at a feasible cost.

In the following section each project will be briefly described in terms of environmental conditions, design, implementation, and impacts.

7.1. RHODES – PROJECT 1

Site location and characteristics.

The site is located at Theologos.

The beach is aligned along a SW-NE direction making an angle of about π/6 with the local parallel, thus permitting the strong N and NE weather to determine the wind-wave climate of the site. Details of this climate based upon measurements made over are recorded in the *Wind and Wave Atlas of the North-Eastern Mediterranean Sea* (Athanasoulis et al, 1992). This source was used in all CPNS studies carried out at the Greek sites.

Fig. 14. Site of Project 2, the beach on Theologos Rhodes, photographed in summer 1994. The arrow pointing to a boat marks the shoreline position one year later (see Figure 15)

The tidal climate of the site is recorded in the 4th edition of *Elements of Tide of Greek Harbours*, 1991, published by the Hydrographical Service of the Greek Navy, Athens, Greece.

This project was assigned to the University of Patras team, together with Projects 2 and 3, due to the rapid erosion caused by the placement of four rock groins that resulted in the destruction of the alongshore road. The condition of the site is shown in Figure 11, while Figure 12 shows the same shore during the phase of the sedimentation induced by the installed CPNS.

The 100-year return wave characteristics, computed on the basis of classic formulae for the predominant wind direction (NW) and for the 100 km fetch of the site, are as follows:

$$H_{max} = 4.59, L = 32.31, T = 4.55.$$

For the design of the model we also used

$$u_x = \text{longshore current velocity} = 0.5 \text{ m/s},$$

(the upstream direction – NE - is that viewed by the photographer of Figure 11), and

maximum tidal amplitude = 0.45 m.

For details of the design, see Goudas et al (1993).

7.2. RHODES – PROJECT 2

Site location and characteristics.

The site is the beach at Theologos, adjacent to the site of Project 1 (see Figure 10).

The CPNS design was based upon the same parameter values used for Project 1.

Fig. 15. Site of Project 2, shown in Figure 14, is shown here one year (1995) after installation (1994) of CPNS. An impressive change is evident although the groins are still visible (arrows)

7.3. RHODES – PROJECT 3

Site location and characteristics.

The site is the beach at Theologos, next to the sites of Project 1 and 2 (see Figure 10).

The CPNS design was based upon the same parameter values used for Project 1.

The CPNS installed consisted of only 2 seabed groins.

For details of the design see Goudas et al (1993b).

It needs to be stated at this point that these successful initial applications of CPNS (projects 1, 2, and 3) is not to suggest that the natural nourishment induced by the system is always successful or will be sufficient for other sites, since the bedload that will sediment the site of intervention depends on the available resources at the upstream part of the site. As a rule, sites exposed to erosion for long periods have poor resources and hence the sedimentation caused by a CPNS will be insufficient and nourishment is needed to supplement the protected area. The CPNS protected site, on the other hand, has not yet been tested for its ability to retain nourishment material. What cannot be disputed is the conclusion drawn from all experiments, that the sites protected by CPNS do not erode further, although accretion can be from extensive to poor.

7.4. LESBOS – PROJECT 4

Site location and characteristics.

The site — Therma, after its natural hot springs — is the beach of the main town of Mytilini on Lesbos Island, and specifically the southern shore of the Gulf of Gheras. The island and site have a very long history.

The 100-year return wave characteristics, computed on the basis of classic formulae for the predominant wind direction (NE) and for a fetch of 15 km of the site, are as follows:

$$H_{max} = 1.78,\ L = 12.50,\ T = 2.83.$$

For the design of the model we also used

u_x = longshore current velocity = 0.3 m/s, (the upstream direction is SW)
maximum tidal amplitude = 0.23 m.

The CPNS installed consisted of 10 seabed groins.

For details of the design see the project study by Goudas et al (1994).

The site immediately after installation is shown in Figure 8. The system put a complete stop to the erosion. The small and unsatisfactory sedimentation is evident. The site requires nourishment. It is predicted that the ability of the site to keep nourishment material is very high and that the beach will remain in excellent condition for many years.

7.5. LESBOS – PROJECT 5

Site location and characteristics.

The site is the beach at Perama on Lesbos Island, and specifically at the southeastern end of the northern shore of the Gulf of Gheras.

The 100-year return wave characteristics, computed on the basis of classic formulae for the predominant wind direction (NW) and for a fetch of 100 km of the site, are as follows:

$$H_{max} = 1.78, L = 12.50, T = 2.83.$$

For the design of the model we also used

u_x = longshore current velocity = 0.3 m/s, (the upstream direction SW)
maximum tidal amplitude = 0.23 m.

The CPNS installed consisted of 10 seabed groins.

For details of the design, see the project study by Goudas et al (1994a).

The CPNS installed consisted of 3 seabed groins.

The system put a complete stop to the erosion. Small and unsatisfactory sedimentation is evident. The site requires nourishment. It is predicted that the ability of the site to keep nourishment material is very high and that the beach will remain in excellent condition for many years.

7.6. LESBOS – PROJECT 6

Site location and characteristics.

The site is the beach at Vatera, on Lesbos, and specifically the northwestern end of the northern shore of the island.

The 100-year return wave characteristics, computed on the basis of classic formulae for the predominant wind direction (NW) and for a fetch of 194 km of the site, are as follows (see Goudas et al, 1996):

$$H_{max} = 6.40, L = 45.50, T = 5.40.$$

u_x = longshore current velocity = 0.5 m/s, (the current direction is to S) and
maximum tidal amplitude = 0.23 m.

The CPNS installed consisted of 9 seabed groins. The system has put a complete stop to the erosion. Satisfactory sedimentation is evident.

7.7. PREFECTURE OF VIOTIA, MUNICIPALITY OF ENOPHYTA – PROJECT 7

Site location and characteristics.

The site is a beach in the municipality of Neophyta, prefecture of Viotia.

The 100-year return wave characteristics, computed on the basis of classic formulae for the predominant wind direction (NW) and for a fetch of 33,5 km of the site, are as follows (see Goudas et al, 1996):

H_{max} = 2.64, L = 19.01, T = 3.49,
u_x = longshore current velocity = 0.25 m/s, (alternating direction – NE and SW),
maximum tidal amplitude = 0.51 m.

The CPNS installed consisted of 10 seabed groins.

The site is shown in Figure 16 immediately after the installation of the CPNS system. The system put a complete stop to the erosion. Small and unsatisfactory sedimentation is evident. The site requires nourishment. It is predicted that the ability of the site to keep nourishment material is very high and that the beach will remain in excellent condition for many years.

Fig. 16. Site of Project 7, after the installation of a CPNS. The impact is obvious. However, nourishment at a future phase is necessary

7.8. PREFECTURE OF VIOTIA, MUNICIPALITY OF DILESI – PROJECT 8

Site location and characteristics.

The site is a beach in the municipality of Dilesi, prefecture of Viotia, only 1 km from that of Project site 7.

The CPNS design was based upon the following site condition data (see Goudas et al, 1996):

H_{max} = 2,64, L = 19,01, T = 3,49,
u_x = longshore current velocity = 0.25 m/s, (alternating direction – NE and SW),
maximum tidal amplitude = 0.51 m.

The CPNS installed consisted of 10 seabed groins.

The system put a complete stop to the erosion. Small and unsatisfactory sedimentation is evident. The site requires nourishment. It is predicted that the ability of the site to keep nourishment material is very high and that the beach will be in excellent condition for many years.

7.9. PREFECTURE OF VIOTIA, MUNICIPALITY OF AGHIA MARINA – PROJECT 9

Site location and characteristics.

The site is the shore of Aghia Marina, Gulf of S. Evoikos, near the historic Marathon battle plain.

The 100-year return wave characteristics, computed on the basis of classic formulae for the predominant wind direction (SE) and for a fetch of 30,0 km of the site, are as follows (see for more details, Goudas et al, 1995):

H_{max} = 2.52 m, L = 17.72 m, T = 3.37 s,
u_x = longshore current velocity = 0.25 m/s,
(alternating direction – NE and SW),
maximum tidal amplitude = 0.51 m.

The CPNS installed consisted of 13 seabed groins.

The system put a complete stop to the erosion. Small and unsatisfactory sedimentation is evident. The site requires nourishment. It is predicted that the ability of the site to keep nourishment material is very high and that the beach will be in excellent condition for many years. The grain size of the nourishment material was specified in a special study since the site is exposed to violent winds that move fine grain sand landwards far from the shore.

7.10. PREFECTURE OF ACHAIA, COMMUNITY OF TSOUKALEIKA – PROJECT 10

Site location and characteristics.

The site is on the shores of the Gulf of Patras, about 20 km away from Patras, capital of western Greece.

The 100-year return wave characteristics, computed on the basis of classic formulae for the predominant wind direction (NW) and for a fetch of 49,0 km of the site, are as follows (see also Goudas et al, 1994):

$$H_{max} = 3.20 \text{ m}, L = 22.54 \text{ m}, T = 3.80 \text{ s},$$
$$u_x = \text{longshore current velocity} = 0.25 \text{ m/s}, \text{(Direction NW)},$$
$$\text{maximum tidal amplitude} = 0.10 \text{ m}.$$

The CPNS installed consisted of 3 seabed groins.

The system put a complete stop to the erosion. Small and unsatisfactory sedimentation is evident. The site requires a greater number of groins and more nourishment. It is predicted that the ability of the site to keep nourishment material is very high and that the beach will be in excellent condition for many years.

Fig. 17. The beach of Project 10 five months after the installation of a 3-seabed-groin CPNS

7.11. PREFECTURE OF ACHAIA, MUNICIPALITY OF RIO – PROJECT 11

Site location and characteristics.

The site on the shores of Rio, Gulf of Patras, only two km from the University of Patras campus.

The CPNS design was based upon the following site condition data (see Goudas et al, 1996):

$$H_{max} = 4.29 \text{ m}, L = 30.07 \text{ m}, T = 4.39 \text{ s},$$
$$\text{fetch} = 87 \text{ km}$$
$$u_x = \text{longshore current velocity} = 0.25 \text{ m/s}, \text{(Direction NW)},$$
$$\text{maximum tidal amplitude} = 0.51 \text{ m}.$$

The CPNS installed consisted of 6 seabed groins.

The system put a complete stop to the erosion. Small and unsatisfactory sedimentation is evident. The site requires extension of the CPNS to protect the rest of the beach and road (part of this road is now permanently closed to traffic) as well as nourishment. It is predicted that the capability of the site to keep nourishment material is very high and that the beach will be in excellent condition for many years.

Fig. 18. The beach of Rio one year after the installation of a 6 seabed groin CPNS. After establishing confidence in the system, the Municipality repaired and widened the road

7.12. PREFECTURE OF CORINTHIA, MUNICIPALITY OF XYLOCASTRO – PROJECT 12

Site location and characteristics.

The site is the beach of Loutro, an exclusive tourist resort on the southern coast of the Gulf of Corinth. It is close to Sykion and Corinth, towns famous from classical times.

The CPNS design was based upon the following site condition data (see Goudas et al, 1996):

$$H_{max} = 3.5 \text{ m}, L = 25.0 \text{ m}, T = 4.0 \text{ s},$$
$$\text{fetch} = 58.0\text{km}$$
$$u_x = \text{longshore current velocity} = 0.25 \text{ m/s, (Direction NW)},$$
$$\text{maximum tidal amplitude} = 0.30 \text{ m}.$$

The CPNS installed consisted of 6 seabed groins.

The designed CPNS has not yet been fully implemented, since of the 51 seabed groins scheduled for placement, only 6 were installed. However, the erosion has completely stopped. Small and unsatisfactory sedimentation is evident. The site requires the designed system to be fully implemented and nourishment carried out. It is predicted that the capability of the site to keep nourishment material is very high and that the beach will be in excellent condition for many years.

Fig. 19. The beach of Xylocastro after installation of a six-seabed-groin CPNS, a small part of the system designed for total protection. The results, far from satisfactory, indicate small progress and the need to implement the design in full

7.13. PREFECTURE OF CORINTHIA, MUNICIPALITY OF XYLOCASTRO – PROJECT 13

Site location and characteristics.

The site is the beach at Xylocastro, an exclusive tourist resort on the southern coast of the Gulf of Corinth. It is close to Sykion and Corinth, towns famous from classical times.

The CPNS design was based upon the following site condition data (see Goudas et al, 1995):

$$H_{max} = 3.5 \text{ m}, L = 25.0 \text{ m}, T = 4.0 \text{ s},$$
$$\text{fetch} = 58.0\text{km}$$
$$u_x = \text{longshore current velocity} = 0.25 \text{ m/s}, \text{(Direction NW)},$$
$$\text{maximum tidal amplitude} = 0.30 \text{ m}.$$

The CPNS installed consisted of 6 seabed groins.

The designed CPNS is again not yet fully implemented, since of the 25 seabed groins scheduled for placement only 6 were installed. Nevertheless, the system has put a complete stop to the erosion. Small and unsatisfactory sedimentation is evident. The site requires completion of the designed system and nourishment. It is predicted that the capability of the site to keep nourishment material is very high and that the beach will be in excellent condition for many years.

7.14. PREFECTURE OF CORINTHIA, MUNICIPALITY OF SYKION – PROJECT 14

Site location and characteristics.

The site is the beach at Kiato, municipality of Sykion, and is adjacent to the site of Project 13.

The CPNS design was based upon the following data relating the site conditions (see Goudas et al, 1993):

$$H_{max} = 3.5 \text{ m}, L = 25.0 \text{ m}, T = 4.0 \text{ s},$$
$$\text{fetch} = 58.0\text{km}$$
$$u_x = \text{longshore current velocity} = 0.25 \text{ m/s, (Direction NW)},$$
$$\text{maximum tidal amplitude} = 0.30 \text{ m}.$$

The CPNS installed consisted of 6 seabed groins.

The system brought the erosion to a complete halt. Small and unsatisfactory sedimentation is evident. The site requires nourishment. It is predicted that the capability of the site to keep nourishment material is very high and that the beach will be in excellent condition for many years. The grain size of the nourishment material was specified in a special study since the site is exposed to violent winds that move fine grain sand landwards far from the shore.

7.15. PREFECTURE OF ARGOLIDA, MUNICIPALITY OF ASSINI – PROJECT 15

Site location and characteristics.

The site is the shores of Tolo, an exclusive tourist resort on the southern coast of the Gulf of Argolida. It is close to Nafplion and Epidauros, towns famous from classical times.

The CPNS design was based upon the following site conditions data (see Goudas et al, 1995):

$$H_{max} = 7.27 \text{ m}, L = 51.06 \text{ m}, T = 5.72 \text{ s},$$
$$\text{fetch} = 250{,}0\text{km}$$
$$u_x = \text{longshore current velocity} = 0.35 \text{ m/s, (Direction NW)},$$
$$\text{maximum tidal amplitude} = 0.30 \text{ m}.$$

The CPNS installed consisted of 26 seabed groins.

The system put a complete stop to the erosion. Small and unsatisfactory sedimentation is evident. The site now only requires nourishment. It is predicted that the capability of the site to keep nourishment material is very high and that the beach will be in excellent condition for many years.

Fig. 20. Two successive photographs of the same part of project site 15 indicating positive progress as far as protection is concerned but pour sedimentation impact of the 26 groin CPNS installed

7.16. CALABRIA ITALY, MUNICIPALITY OF VILLA SAN GIOVANNI – PROJECT 16

Site location and characteristics.

The site is the beach of Villa San Giovanni, known as Porticello Beach, overlooking the Strait of Messina.

The CPNS design was based upon the following site condition data (see Goudas et al, 1995):

H_{max} = 6.42 m, L = 45.17 m, T = 5.38 s,
fetch = 195 km, (straight-line distance between Porticello and G. Policastro)
u_x = longshore current velocity = 0.25 m/s,
(alternating direction – NE and SW), maximum tidal amplitude = 0.51 m.

The CPNS installed consisted of 13 seabed groins.

The system put a complete stop to the erosion. Small and unsatisfactory sedimentation is evident. The site requires nourishment. It is predicted that the capability of the site to keep nourishment material is very high and that the beach will be in excellent condition for many years. The grain size of the nourishment material was specified in a special study since the site is exposed to violent winds that move fine grain sand landwards far from the shore.

7.17. OASIS RESORT AT AIN SUKHNA ON THE RED SEA, EGYPT – PROJECT 17

Site location and characteristics.

The site is the beach of Oasis Village Resort, Al Hegaz, near the town of Ain Sukhna on the Red Sea.

The CPNS design was based upon the following site condition data (see Goudas et al, 1998):

H_{max} = 4.45 m, L = 31.32 m, T = 4.48 s,
fetch = 93 km
u_x = longshore current velocity = 0.25 m/s, (alternating direction – NE and SW),
maximum tidal amplitude = 0.75 m.

The CPNS installed consisted of 13 seabed groins.

The system put a complete stop to the erosion. Figures 21 and 22 show the site before and after installation of the 26 groin CPNS, while Figures 23, 24, and 25 show the monitoring results. The sediment added to the seabed after installation varies from 0.20 m to 0.50 m. This experiment, in spite of the satisfactory results and the great improvement in the beach of the Oasis Resort on the Red Sea, was conducted under difficult conditions and the groin casting did not have the desired professionalism. Nevertheless, the experience was a valuable one, since it has paved the way for future applications. It is predicted that the capability of the site to keep nourishment material is very high and that the beach will be in excellent condition for many years.

Fig. 21. The beach of Al Hegaz Oasis Resort at Ain Sukhna on the Red Sea. The earlier protection scheme, maintained until 1998, had to be made 'softer'

Fig. 22. A wide-scale photo of the beach of Figure 21 as it looked in September 1999

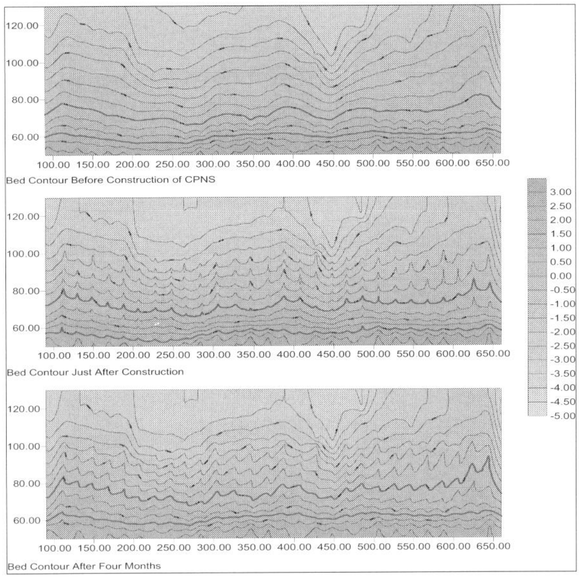

Fig. 23. Seabed contours based upon monitoring data before, immediately after, and four months after placement of the 26 groin CPNS, indicate the major positive impact upon this beach of the Red Sea

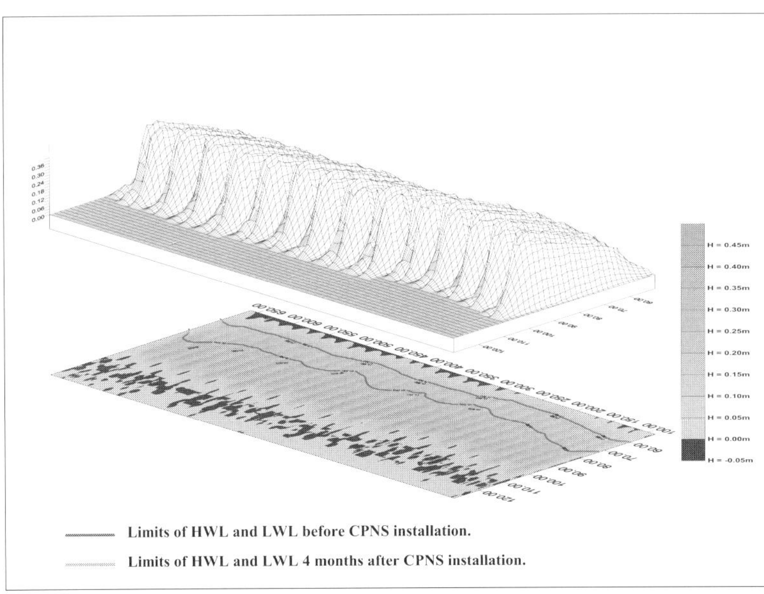

Fig. 24. Sediment level distribution 4 months after installation of CPNS at the Red Sea site

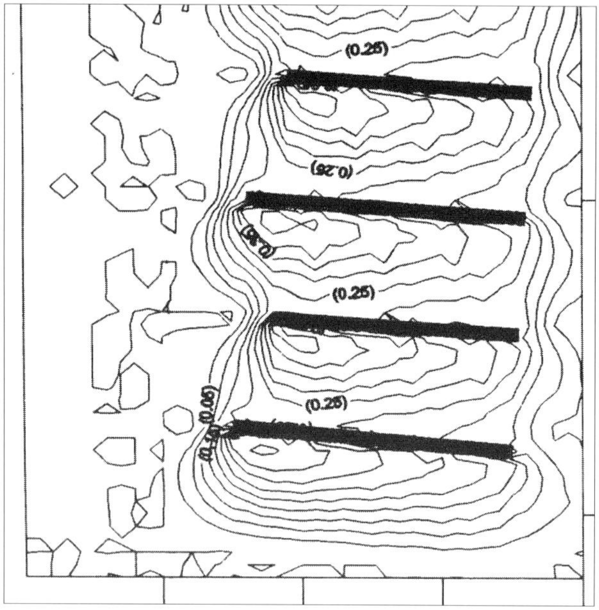

Fig. 25. Sediment differential (added) distribution above original seabed level 4 months after installation of CPNS at the Red Sea site. These results can be compared to the computed seawater vortices (Figure 9) induced by the groins

8. Conclusions

a. FEM was able to provide precise stress-strain relationships and the static stability of the seabed groins.

b. The safety co-efficient N_α of the groin as regards its strength is found to be at least 3.31 and hence there is no failure risk whatsoever, implying that there is no necessity for changes in the dimensions of the material constitution of the groins. The designer is free to determine the dimensions of groins on the basis of performance factors other than strength and static stability, so long as the new dimensions are bigger than the ones assumed in the present analysis.

c. The 17 CPNS systems installed under Mediterranean and Red Sea conditions verified the theoretical prediction in relation to the water flow pattern based upon an extensive numerical solution of the Navier-Stokes equations. Specifically, the placement of low height-groins aligned perpendicularly to the shoreline modifies the linear pattern of longshore currents, making them rotational. The formation of permanent vortices, single or multiple, render inoperable the known property of longshore currents to transport bedload.

d. The 17 *in situ* experiments showed that coastal erosion is unfailingly stopped by CPNS.

e. The capability of CPNS to induce natural sedimentation, although observed to some extent in all cases, depends upon the availability of the resource. As a rule, eroded beaches lack satisfactory sediment amounts to support self-nourishment.

f. The ability of CPNS to stop or at least slow down substantially the sediment transport process on nourishment materials away from the beach of placement, is only predicted theoretically but not proven.

REFERENCES

Hydrographic Service of the Greek Navy: *Elements of Tide of Greek Harbours*, 4th edition, published by the Hydrographic Service of the Greek Navy, Athens, Greece, 1991 (in Greek).

Athanassoulis, G. A., and Skarsoulis, E. K.: *Wind and Wave Atlas of the North-Eastern Mediterranean Sea*, Laboratory of Ship and Marine Hydrodynamics, National Technical University of Athens, Greece, 1992.

Goudas, C., Katsiaris, G., Pnevmatikos, G.: *Design of CPNS for the Beach of the Community of Theologos, Rhodes, and Study of Environmental Impacts, Project 1*, Studium of Mechanics, University of Patras, Patras, Greece, 1993 (in Greek).

Goudas, C., Katsiaris, G., Pnevmatikos, G.: *Design of CPNS for the Beach of the Community of Theologos, Rhodes and Study of Environmental Impacts, Project 2*, Studium of Mechanics, University of Patras, Patras, Greece, 1993a (in Greek).

Goudas, C., Katsiaris, G., Pnevmatikos, G.: *Design of CPNS for the Beach of the Community of Theologos, Rhodes and Study of Environmental Impacts, Project 3*, Studium of Mechanics, University of Patras, Patras, Greece, 1993b (in Greek).

Goudas, C., Katsiaris, G., Pnevmatikos, G.: *Design of CPNS for the Beach of the Municipality of Kiato, Prefecture of Corinthia, and Study of Environmental Impacts*, Studium of Mechanics, University of Patras, Patras, Greece, 1993c (in Greek).

Goudas, C., Katsiaris, G., Pnevmatikos, G.: *Design of CPNS for the Beach of the Therma, Municipality of Mytilini, Lesbos, and Study of Environmental Impacts*, Studium of Mechanics, University of Patras, Patras, Greece, 1994 (in Greek).

Goudas, C., Katsiaris, G., Pnevmatikos, G.: *Design of CPNS for the Beach of the Community of Perama, Lesbos, and Study of Environmental Impacts*, Studium of Mechanics, University of Patras, Patras, Greece, 1994a (in Greek)

Goudas, C., Katsiaris, G., Pnevmatikos, G.: *Design of CPNS for the Beach of the Community of Tsoukaleika, Prefecture of Achaia, and Study of*

Environmental Impacts, Studium of Mechanics, University of Patras, Patras, Greece, 1994b (in Greek).

Goudas, C., Katsiaris, G., Pnevmatikos, G.: *Design of CPNS for the Beach of the Municipality of Enophyta, Prefecture of Viotia, and Study of Environmental Impacts*, Studium of Mechanics, University of Patras, Patras, Greece, 1995 (in Greek).

Goudas, C., Katsiaris, G., Pnevmatikos, G.: *Design of CPNS for the Beach of the Community of Aghia Marina, Prefecture of Viotia, and Study of Environmental Impacts*, Studium of Mechanics, University of Patras, Patras, Greece, 1995a (in Greek).

Goudas, C., Katsiaris, G., Pnevmatikos, G.: *Design of CPNS for the Beach of the Municipality of Xylocastro, Prefecture of Corinthia, and Study of Environmental Impacts*, Studium of Mechanics, University of Patras, Patras, Greece, 1995b (in Greek).

Goudas, C., Katsiaris, G., Pnevmatikos, G.: *Design of CPNS for the Beach of the Community of Tolo, Prefecture of Argolida, Study of Environmental Impacts*, Studium of Mechanics, University of Patras, Patras, Greece, 1995c (in Greek).

Goudas, C., Katsiaris, G., Pnevmatikos, G.: *Design of CPNS for the Beach of Porticello, Municipality of Villa San Giovanni, Calabria, Italy, and Study of Environmental Impacts*, Studium of Mechanics, University of Patras, Patras, Greece, 1995d.

Goudas, C., Katsiaris, G., Pnevmatikos, G.: *Design of CPNS for the Beach of the Community of Vatera, Lesbos, and Study of Environmental Impacts*, Studium of Mechanics, University of Patras, Patras, Greece, 1996 (in Greek).

Goudas, C., Katsiaris, G., Pnevmatikos, G.: *Design of CPNS for the Beach of the Municipality of Schematari, Prefecture of Viotia, and Study of Environmental Impacts*, Studium of Mechanics, University of Patras, Patras, Greece, 1996a (in Greek).

Goudas, C., Katsiaris, G., Pnevmatikos, G.: *Design of CPNS for the Beach of the Community of Rio, Prefecture of Achaia, and Study of Environmental Impacts*, Studium of Mechanics, University of Patras, Patras, Greece, 1996b (in Greek).

Goudas, C., Katsiaris, G., Pnevmatikos, G.: *Design of CPNS for the Beach of the Community of Loutro, Municipality of Xylocastro, Prefecture of Corinthia, and Study of Environmental Impacts*, Studium of Mechanics, University of Patras, Patras, Greece, 1996c (in Greek).

Goudas, C., Katsiaris, G., Pnevmatikos, G.: *Design of CPNS for the Beach of Oasis Resort Tourism Company, Community of Ain Sukhna, Red Sea, Egypt, and Study of Environmental Impacts*, Studium of Mechanics, University of Patras, Patras, Greece, 1998.

'BACK TO THE BEACH': CONVERTING SEAWALLS INTO GRAVEL BEACHES

PIERLUIGI AMINTI,[1] LUIGI E. CIPRIANI[2] AND ENZO PRANZINI[3]

[1] *Dipartimento di Ingegneria Civile - Università degli Studi di Firenze. Via di S. Marta, 3 - 50137 Firenze, Italy.*
[2] *Regione Toscana - Dipartimento delle Politiche Territoriali e Ambientali. Via di Novoli, 26 - 50127 Firenze, Italy.*
[3] *Dipartimento di Scienze della Terra - Università degli Studi di Firenze. Via Iacopo Nardi, 2 - 50132 Firenze, Italy.*

Abstract

During the 20th century, 5 km of hard structures (seawalls, detached breakwaters and groins) were built along 2.5 km of coast to stabilise the shoreline at Marina di Pisa (Tuscany, Italy). These reflective structures increased nearshore erosion and now a water depth of 7 m deep is found at the offshore toe of the detached breakwaters. Wave reflection pushes sediment flux offshore, inducing downdrift erosion. A project to reduce hard structures, based on their conversion into gravel beaches, is proposed here. The project layout was tested on a 2D physical model and proved to be effective as a coastal protection measure. Furthermore, it is less expensive than the maintenance costs of the present structure and produces a 30 m wide gravel beach, so offering economic benefits because of its recreational value.

1. Introduction

Italian coastlines began experiencing beach erosion during the second half of the nineteenth century, and nowadays more than 50% of the national shoreline is threatened by erosion (CNR, 1997; D'Alessandro and La Monica, 1999). The first response to this process was hard shore protection, favoured by the vicinity of the mountains and the fact that rock quarrying is a traditional industry. Consequently, rocky boulders are cheap and available everywhere and their use proliferated in the 1960s and 1970s. In addition, little was known about marine sand dredging for beach nourishment at the time due to the lack of fluvial and estuarine navigation. Today there are approximately 330 km of hard structures along our coastlines, drastically modifying coastal landscapes, inducing downdrift erosion, affecting sea water quality and circulation and, in many cases, not even providing efficient beach preservation.

Soft shore protection methods, such as beach nourishment with sand have recently been adopted in several projects, but they are not applicable where older structures such as seawalls, groins and breakwaters have dramatically changed the original coastal morphology. In these cases, water depth in front of the structures suddenly increases: as a consequence, incoming waves usually break over the structures, dissipating their energy on the structure itself and inducing scouring at the toe. Hypothetical beach nourishment with sand-like material would require an extremely large amount of sediment in order to restore a more gentle nearshore profile and thus at a cost rarely affordable for local and regional administration. In addition, ordinary maintenance of rock rubble/boulder constructions is extremely expensive, but

C. L. Goudas et al. (eds.), Soft Shore Protection, 261-274.

necessary to protect urban and industrial settlements, coastal infrastructures and recreational areas.

Retention of these 'archaeostructures' clashes with the soft shore protection strategies recently adopted by national and regional administration. However, the skill and know-how for new solutions in these extreme cases is limited, the economic requirements are huge and the long time needed before reaping the advantages is at odds with the political life span, which needs short-term projects and results visible in 4 to 5 years.

'Back to the beach' is our slogan: hard structures have to be removed and converted into soft solutions, that is, gravel beaches. This process will require a relatively long period including intermediate phases and long rests to allow the environment to evolve gradually towards its new and more natural configuration.

2. Beach erosion and coastal protection at Marina di Pisa

Marina di Pisa is a seaside resort built on the southern wing of the Arno River delta (central Tuscany) (Figure 1) in the early 1800s. The Arno River delta has built up over the last 25 centuries, as a consequence of inland hill and mountain deforestation performed since pre-Etruscan times (Pranzini, 1984). The shoreline advanced more than 7 km at the river mouth during this time, while a 5 km-wide beach ridge and band of dunes formed alongshore from Livorno to Forte dei Marmi (46 km) (Figure 1).

Fig. 1. Map of the area under discussion

Changes in land-use within the basin, and the practice of river bed quarrying, together with dam construction, have reduced the river sediment load from the estimated $5 \cdot 10^6$ m^3/yr in the seventeenth century to less than $2 \cdot 10^6$ m^3/yr in the last 50 years. Beach erosion began in the mid 1800s at the delta apex and gradually expanded to adjacent beaches (Pranzini, 1989). In the meantime the expanding Marina di Pisa was confronted by shoreline retreat.

At first, groins were built, though they did not prevent the houses on the sea front being demolished by waves in the early twentieth century. Detached breakwaters and seawalls were built immediately after World War II and gradually expanded southward, running behind the erosion, though they were themselves one of the causes of this phenomenon. There were no settlements on the northern side of the delta and the shoreline was left free to retreat: as a consequence, erosion led to the loss of 1 300 m of land in a century, producing a sharp asymmetry at the Arno River mouth. At Marina di Pisa the coastline is now artificially stabilized for 2.5 km by a continuous seawall and by 10 detached breakwaters (Figure 2); in addition, a few groins divide the protected coast into five cells of different sizes. As a result, over 5 km of hard structures are defending 2.5 km of coastline.

Fig. 2. Seawall and detached breakwaters at Marina di Pisa

Mean water depth between the seawall and the detached breakwater is approximately 2 m, while on the offshore side of the breakwaters the water reaches a depth of approximately 7 m. The nearshore profile is almost horizontal with the breakwater for several hundred meters, although it is slightly convex, peaking at approximately 120 m from the breakwater (Figure 3). Farther offshore, the mean nearshore slope is approximately 0.4% down to the –15 m isobath. The nearshore profile, with its convex shape, demonstrates the offshore migration of sediments. As a consequence of this nearshore morphology, wave energy dissipation is very limited along the nearshore profile and the waves break directly over the structure. Sea bottom scouring is strong at the toe of the structure and maintenance costs high. In addition, wave reflection pushes the southward longshore drift coming from the Arno River offshore, so the fluvial sediments do not efficiently feed the downdrift beaches.

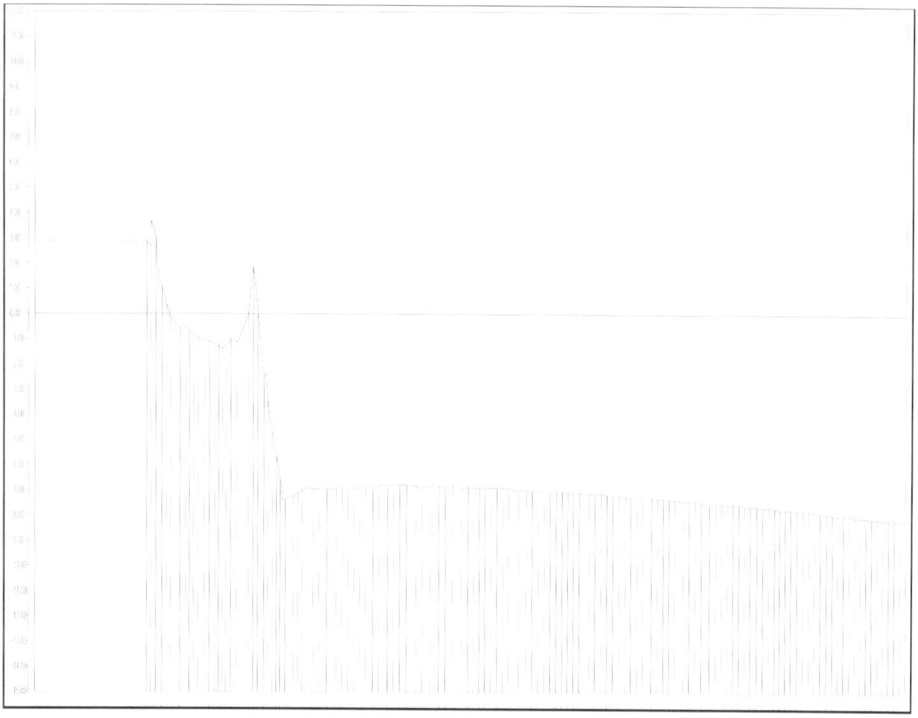

Fig. 3. Beach profile at Marina di Pisa.

Nevertheless, for the inhabitants of Marina di Pisa, these hard structures provide reassurance during severe storms, although the sea-front promenade is often closed due to overwash and a certain amount of damage is expected. Toe scouring induces the collapse and lowering of the breakwaters, which require frequent maintenance in the form of adding boulders to the tops of the breakwaters.

The study relating to a general plan for shore protection in the Pisa jurisdiction area revealed that these hard structures were no longer sustainable, but at the same time any proposed soft defence for the adjacent beaches was destined to fail due to the negative impact of the old works. The main task is to reduce wave reflection along the detached breakwaters in order to allow longshore sediment transport to reach the shoreline and feed the beaches to the south. This should also favour sedimentation in front of the breakwaters and thus anticipate wave shoaling and breaking. All these factors must be obtained without reducing the protection afforded by the seaside promenade and buildings.

3. A back to the beach project: presentation and discussion

While the members of the commission (P. Aminti, G. Berriolo, G. De Filippi, J. Oneto and E. Pranzini), appointed by the Pisa county administration to plan a new strategy for coastal conservation, were discussing this topic, the national agency for coastal defence (Genio Civile Opere Marittime) were about to raise two breakwaters (n. 6 and n. 7 from north) to a height of 3.80 m above the mean sea level, at a cost of approximately 3 billion Italian lira (1.5 million euro). This work was delayed and a different solution, aimed at reducing the reflective structures along the coast, was tested with a physical model. The project is based on lowering the detached breakwaters to mean sea level, and absorbing the overtopping wave energy with a gravel beach placed in front of the present seawall (Figure 4).

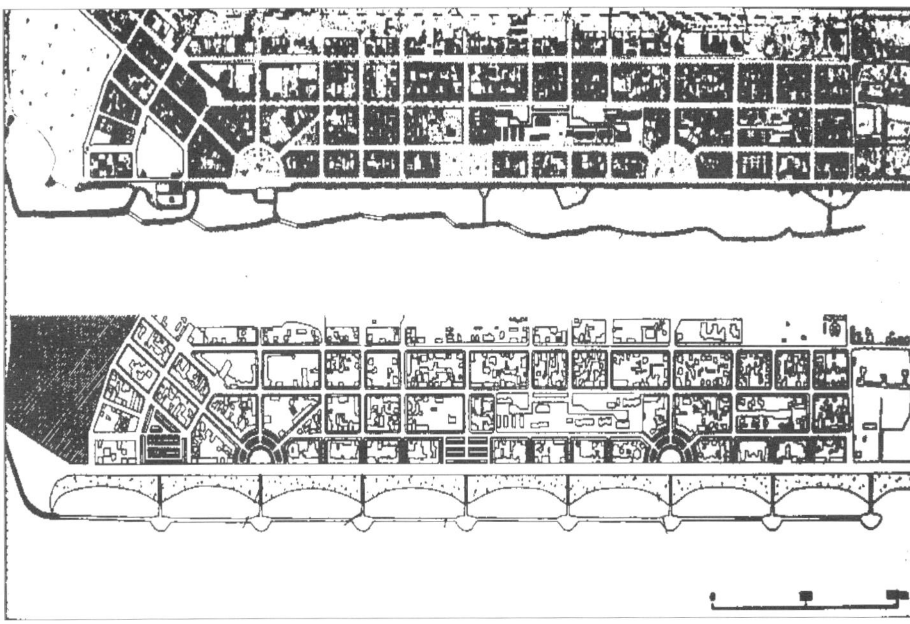

Fig. 4. Present (up) and proposed (down) coastal defences at Marina di Pisa

In the present situation, the breakwater berm is about 2 m above mean sea level and the transmitted wave height in the protected area is about 30% of the significant wave height under extreme conditions. Lowering the breakwater to mean sea level increases the transmission coefficient to 42% under the same incoming waves.

Energy transmitted behind the breakwater can be efficiently absorbed by coarse sediments, which have proved to be very stable both in nature and on artificially nourished beaches, thanks to their size, permeability and porosity (Pacini et al., 1997, 1999). These characteristics allow water to infiltrate during uprush, and return to the sea via sub-surface flow, eliminating the tractive forces acting during the backwash responsible for grain removal from the swash-zone. All this favours settlement of grain carried on the berm crest by on-shore fluxes, and the formation of a high berm (Orford, 1977; Orford & Carter, 1985). This behaviour was also confirmed by the available numerical modes (van Hijum & Pilarczyk, 1982; Van der Meer, 1998), although they probably overestimate the berm height which forms in gravel and boulder beaches.

One problem which still must be solved is the compatibility between the fine sediments of the nearshore and the coarser ones artificially dumped on the beach. On natural ‘mixed sand and gravel beaches’, the coarsest grains are found (Miller & Zeigler, 1958), where the last wave breaks, depositing fine sand in front of it, with the sudden reduction in grain size often occurring along a sharp line. The tendency of gravel to move onshore in mixed sediment beaches was also observed during recent beach nourishment performed in Sardinia (Pacini et al., 1997, 1999). In this case, fill material – gravel produced from hard rock – was pushed offshore three times with a bulldozer, forming a platform 30 cm below the mean sea level and each time the waves immediately brought it back to the shore. This helped to clean the material of its silt particles and provided preliminary rounding of the grains.

Another aspect that needs to be studied in more detail is the longshore mobility of gravel along sandy beaches.

The gravel filling performed at Marina di Cecina (Tuscany) south of the last groin along the coast is interesting in this respect. After four years, the material was found 15 km to the south! Similar results can also be observed at Lido di Policoro (Basilicata), where gravel and boulders discharged onto the unprotected shore are moving toward a fine sand beach downdrift.

All these factors suggest that while offshore dispersion of the filling will be very limited, strong lateral fluxes must be taken into account. Our project foresees a reduction in this longshore transport through the construction of short groins connected to the artificial reef by a submerged groin (Figure 4).

Cross-shore beach profiles were studied in gravel beaches with wave channel experiments at the Delft Hydraulics Laboratory (Van Hijum & Pilarczyk, 1982; Pilarczyk & De Boer, 1983; Van der Meer, 1988) and at the Wallingford Laboratory (Powell, 1990), producing parametric models of equilibrium profiles. Numerical models also allow the study of gravel beach profiles (CIRIA CUR, 1983). All these studies relate to beaches formed solely by gravel, reaching a depth where the breaking waves cannot cause any changes; no data exists for profiles where gravel lies over a fine sand bed.

In Italy, wave channel experiments aimed at supporting coastal protection projects have been done to study cross-shore beach evolution after artificial coarse nourishment has been carried out (Aminti, 1988; De Santis & Ruol, 1988). Although these experiments proved the stability of the model beach, we only have a small amount of data relating to natural beaches defended through coarse sediment nourishment (Berriolo, 1999; Pacini et al., 1999).

4. The physical model

To test the effectiveness of the project, a 2D physical model of the Marina di Pisa coast was performed on a scale of 1:25 at the Civil Engineering Department of the University of Florence. Here, a wave channel 50 m long, 0.8 m large and 0.8 m deep was used, equipped with an oleodynamic piston-type wave generator with a random input signal to reproduce a JONSWAP-type spectrum of pre-determined values of wave height $H_{1/3}$ and period $T_{1/3}$. A graduated tank collected the water overtopping the seawall or the beach thus allowing evaluation of the efficiency of the coastal protection.

In our tests, each wave was run for 45 minutes, representing 4 hours in nature according to the Froude law. Five sensors measured wave parameters along the channel as shown in Figure 5, and sampling at 20 Hz was processed both in the temporal (zero-up-crossing) and frequency (spectral analysis) domain. Generator performance was checked at gauge 1, whereas at gauge 2 the waves reaching the structures were measured (Table 2). Gauges from 3 to 5 produced data on wave height at the beach toe (Figure 6), while two piezometers allowed measurement of the sea level near the coast and the set-up in the protected area (Figure 7).

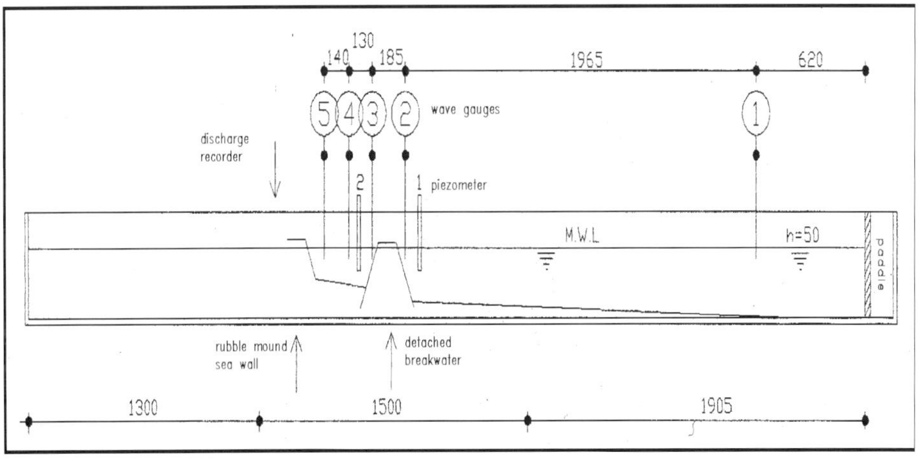

Fig. 5. Diagram of the wave channel used for the physical model of the coastal defences at Marina di Pisa

The main parameters to be considered when setting up a physical model of a beach are the Froude number and Dean's parameter (Dean, 1973; Gourlay, 1980; Vellinga, 1986)

$$N_0 = \frac{H_0}{VT}, \tag{1}$$

with H_0 = wave height in deep water, T = wave period, V = sediment settling velocity in still water.

These parameters allow size and density of the sediments to be included in the same relationship.

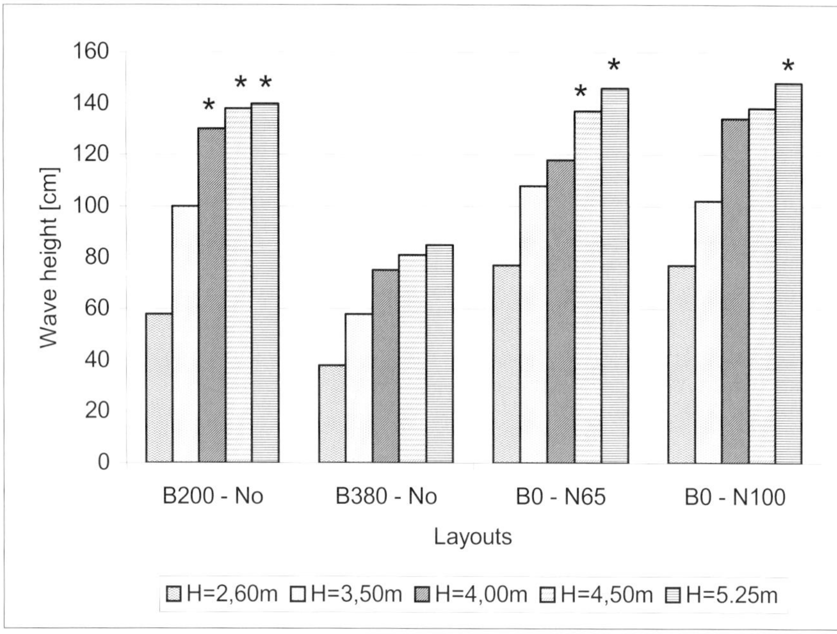

Fig. 6. Wave height at the beach toe for different waves for the different tested layouts $(H_{1/3})_3$

Table 1. Sediment physical characteristics in the prototype and the model at 1:25 scale.

Material	Density (g/cm^3)	D_{50} in the model (mm)	Settling velocity in the model (m/s)	Scaled settling velocity (m/s)	D_{50} in the prototype (mm)
Sand	2.65	1.20	0.12	0.60	16

Table 2. Wave parameters at location 1 and 2 (see Figure 5).

H_0 [m]	$(H_{1/3})_1$ [m]	$(H_{1/3})_2$ [m]	$(T_{1/3})_2$ [s]
2.60	2.34	1.89	6.3
3.50	3.36	3.00	7.7
4.00	3.81	3.32	8.4
4.50	4.25	3.45	8.9
5.30	4.98	3.69	9.95

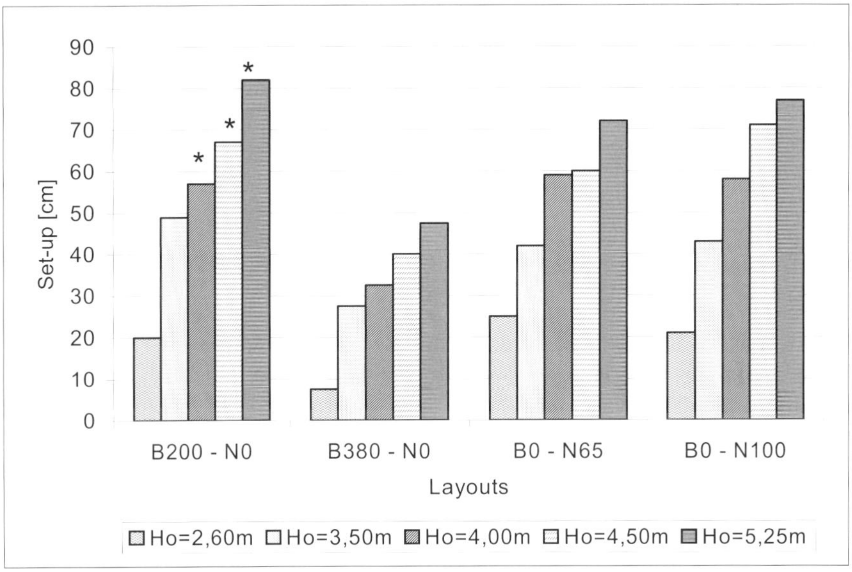

Fig. 7. Set-up values at the fill toe for the various layouts under different waves

The present configuration of the breakwater was tested first (B200) and then tested with its height raised to +3.8 m as proposed for the maintenance works (B380); no nourishment (N0) was simulated at this stage.

Later, an artificial gravel beach was constructed and the detached breakwater lowered to the mean sea level (B0). The equivalent of 65 m^3/m (N65) and 100 m^3/m (N100) nourishment were tested. Although sediments of various size were tested, our data here only refer to 16 mm equivalent grains (1.20 mm in the model; Table 1). Some considerations relating to the other tested materials are reported in the conclusions to this paper. As far as wave climate is concerned, tests were carried out with increasing H_{0S} values, up to 5.30 m (Table 2), which correspond to a wave with a return period of 30 years. The offshore wave conditions have been reproduced in 12.5 m deep water $(H_{1/3})_1$ and measured at a depth of 7 m in front of the breakwater $(H_{1/3})_2$.

In the bidimensional model it was not possible to measure the sediment loss produced by the flow occurring between adjacent breakwaters due to the different mean sea levels. Set-up values for the area between the seawall and the detached breakwater can help in estimating this effect.

5. Results

5.1 WAVE HEIGHT IN FRONT OF THE ARTIFICIAL BEACH

Significant wave height $(H_{1/3})_{3,4,5}$ at the beach base was related to the transmission coefficient of the breakwater and the reflection index of the work near the beach. The former was strictly determined by the height of the breakwater above the sea level (Van der Meer, 1992). On a low reflectivity beach, the increase in wave height produced by lowering the breakwater is only partially balanced by the reduction in wave reflection by the nearshore structures, and wave heights in the protected zone are quite similar in any layout with low offshore breakwaters (Figure 6).

5.2 SET-UP

The extreme set-up value behind the detached breakwater (Figure 7) was 83 cm when the latter was in its present configuration, with the top at 2.0 m above m.s.l., and fell to 48 cm with a 3.5 m high breakwater, when levels were balanced by filtration through the rocks. Low breakwaters were more easily overtopped, but offshore fluxes more efficiently restored the water levels and set-up was reduced (70-75 cm). This data can be considered the maximum values, since 3-dimensional processes were not simulated and longshore fluxes could render the prototype very different to the 2D model. During severe storms, the waves were actually observed to pass over the detached breakwater, running over a higher sea level and easily reaching the coastal road.

5.3 OVERTOPPING VOLUMES

In the present configuration (B200 N0) overtopping occurred when Hs reached 4.0 m (Table 3, Figure 7), and when the waves were strong, the coastal road was completely flooded (2.07 l/s/m). Raising the detached breakwater to 3.8 m above m.s.l. (B380 N0) overtopping did not occur under any wave condition. This was largely to be expected, since behind other recently raised breakwaters, no problems with the coastal road have been reported, not even during severe storms.

When the breakwater was lowered to the mean sea level (B0), a fill volume of 65 m^3/m (N65) prevented overtopping by waves with Hs lower than 4.5 m. In extreme conditions (Hs = 5.30 m) waves reached the seawall and reflection occurred, preventing the berm crest from forming: a concave profile directly connected to the seawall evolved. With a 100 m^3/m fill, a large berm was formed (equivalent to 30 m in nature), with a crest well distant from the seawall preventing road flooding in all sea states tested. The highest waves, however, did overtop the berm, but the water was absorbed by the backshore, which rose in elevation.

6. Discussion and conclusion

A gravel beach proved to be effective in reducing the overtopping of the present seawall and inundation of the coastal road. This was also the case with a significant lowering of the detached breakwaters. As observed in natural beaches (Orford & Carter, 1985), coarser sediments stay on the nearshore and no mixing occurs with the fine sand constituting the present shore-face. Marked grain size discontinuity occurs, at the step base (Figure 8).

Table 3. Overtopping, for each layouts, under different wave conditions.

Layout	Hs [m]	Overtopping
B200 N0	4.00	0.10
B200 N0	4.50	0.43
B200 N0	5.30	2.07
B380 N0	5.30	0.00
B0 N65	4.50	0.04
B0 N65	5.30	0.14
B0 N100	5.30	0.00

Sometimes the sand was seen to slightly cover the gravel: this could choke up the pores and reduce beach permeability and porosity. As this phenomenon, together with grain size, is a conclusive characteristic, this aspect should be analysed in detail.

Our results prove that it is possible to lower the detached breakwaters along our coast, with a direct gain in nearshore water quality. Gravel beaches could substitute seawalls, even without an external breakwater, as proved in other tests performed by the authors (Aminti et al., in prep.). Additional tests, not reported here, show that there is no significant gain in efficiency by changing the grain size, as long as we are inside the gravel–pebble range.

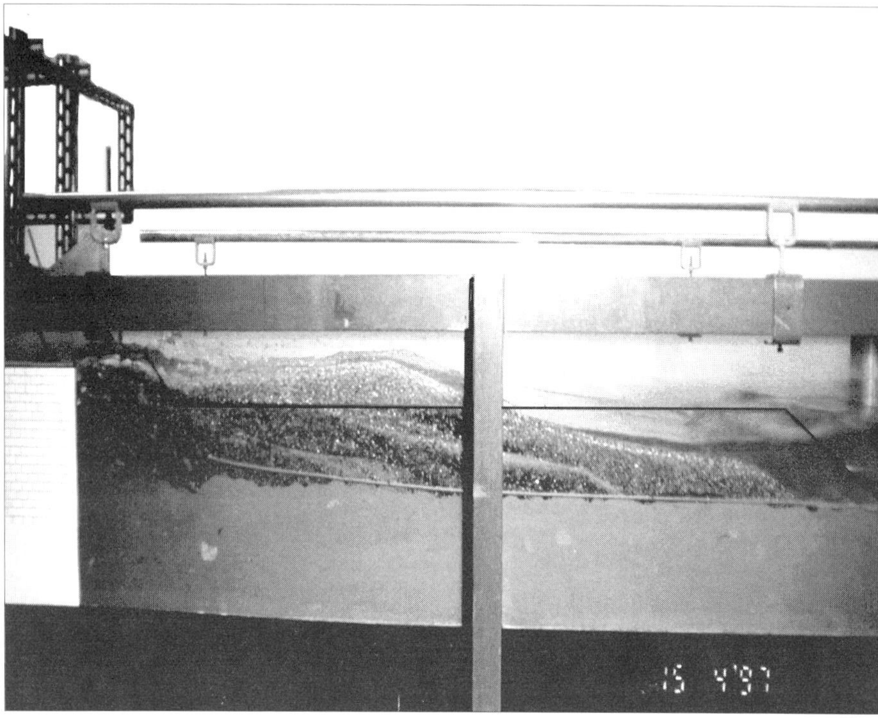

Fig. 8. Gravel beach formed by nourishment in the physical model. The horizontal line shows the top surface of the material laid initially; the beach profile was shaped by Hs = 5.30 m waves.

These results convinced the national agency to approve our project and order a cost analysis. The new project cost was estimated to be 2 200 million Italian lira (1.1 million euro), which is approximately 30% less than the cost of raising the breakwater (3 000 million Italian lira; 1.5 million euro). A 30-meter-wide dry beach is also obtained, which, in a site developed for tourism in Italy, is far more valuable than the cost of construction. Lowering the detached breakwater should reduce near-shore water pollution, and the reduction in wave reflection should reduce offshore dispersion of longshore drift. The forecast nearshore accretion should induce wave shoaling and energy dissipation. In the future, this could allow lowering of the detached breakwater even further, or reducing of beach grain size. Work on this project will begin in the near future, together with detailed monitoring. If the results prove to be positive, the entire sea-front side of the town will be modified in the forthcoming years (Figure 9).

Fig. 9. New look of the coast at Marina di Pisa according the project of hard structure reduction

Acknowledgements

This study was supported by the MURST, Project *Analisi delle variazioni sedimentologiche e morfologiche delle piagge indotte da interventi di difesa e ricostruzione della fascia costiera*, (Resp. G. Fierro) and by the GNDCI, Linea 2, U.O. 51 (Resp. G. B. La Monica).

REFERENCES

Aminti P.: Indagine sperimentale sulla protezione delle spiagge con ripascimenti artificiali. Conv. Idraulica e Costruzioni Idrauliche, L'Aquila, 1988, 562–575.

Berriolo, G.: L'intervento di riequilibrio della spiaggia di Fondi-Sperlonga, *Studi costieri*, 1, 1999, 33–41.

CIRIA CUR: *Manual on the use of rock in coastal and shoreline engineering*, 1983.

CNR: *Atlante delle spiagge italiane*, C.N.R., S.El.Ca., Firenze, 1997.

D'Alessandro, and La Monica, G.F.: Rischio per erosione dei litorali italiani. *Atti dei Convegni Lincei*, 154, 1999, 251–256.

De Santis, M. and Ruol, P.: *Studio sperimentale su un particolare intervento per la salvaguardia delle spiagge in erosione*, Ist. di Costruzioni Marittime e Geotecnica, Padova, 1988.

Dean, R.G.: Heuristic model of sand transport in the surf zone, *Conference on Engineering Dynamics in the Coastal Zone*, Institute of Engineers, Australia, 1973, p. 208–214.

Gourlay, M.R.: Beach profiles, process and permeability, Research Report no. CE14, Department of Civil Engineering, University of Queensland, St. Lucia, 1980, pp. 36 .

Miller, R.L., and Zeigler, J. M.: A model relating dynamics and sediment pattern in equilibrium in the region of shoaling waves, breaker zone, and foreshore. J. Geology, 66, 1958, 417–441.

Orford, J.D.: A proposed mechanism for storm beach sedimentation. Earth Surf. Proc. 2, 1977, 381–400.

Orford, J.D., and Carter, R.W.G.: Storm-generated dune armouring on a sand-gravel barrier system, Southeastern Ireland. Sed. Geol., 42, 1985, 65–82.

Pacini, M., Pranzini, E., and Sirito, G.: La ricostruzione della spiaggia di Cala Gonone. Studi costieri, 1, 1999, 43–55.

Pacini, M., Pranzini, E., and Sirito, G.: Beach nourishment with angular gravel at Cala Gonone (Eastern Sardinia, Italy). *Medcoast '97*, Qawra, Malta, 11-14 Nov., 1997, 1043–1058.

Pilarczyk, K.W., and den Boer, K.: Stability and profile development of coarse materials and their application in coastal Engineering, *International Conference on Coastal and Port Engineering in Developing Countries*, Colombo, 1983, 43-60.

Powell, K.A.: Predicting short term profile response for shingle beaches. *HR Wallingford Report* SR219, 1990.

Pranzini, E.: A model for cuspate river delta erosion. 6th Symp. on Coastal and Ocean Management/ASCE. Charleston, SC. *Coastal Zone '89*, 1989, 4345–4357.

Pranzini, E.: Bilancio sedimentario e evoluzione storica delle spiagge. *Il Quaternario* 7, 1994, 197–202.

van der Meer, J.: Rock Slopes and gravel beaches under wave attack. Delft Tech. Univ., publication no. 396, 1988.

van der Meer, J.: Conceptual design of rubble mound breakwaters. *Proceedings of the Short Course on Design and Reliability of Coastal Structures*, Venice, October 1992, pp. 17–64.

van Hijum, E., and Pilarczyk K.W.: Gravel beaches: equilibrium profile and longshore transport of coarse material under regular and irregular wave attack. Delft Hydr. Lab., publication no. 274, 1982.

Vellinga, P.: Beach and dune erosion during storm surges. *Delft Hydraulic Communication*, N 372, 1986.

MICRO–SCALE DYNAMICS OF SAND TRANSPORT IN THE PRESENCE OF LOW-HEIGHT SUBMERGED GROIN ARRANGEMENTS

C. L. GOUDAS[1], G. KATSIARIS[2], G. KARAHALIOS[2], G. PNEVMATIKOS[2]

[1] *Dept. of Mathematics, University of Patras, Patras, Greece*
[2] *Dept. of Physics, University of Patras, Patras, Greece*

Abstract

The micro-scale dynamics of bedload transport in the surf zone with or without submerged groins placed on the seabed are discussed on the basis of flow energy dissipation using the analytical solution of the equation of flow within the boundary layer formed immediately above the seabed.

The calculation of energy dissipation within the boundary layer formed at the seabed is carried out assuming a two-dimensional unsteady flow of sea water past a horizontal and porous seabed, through which a small amount of suction at a constant velocity takes place, while the free-stream velocity (i.e., the velocity of sea water above the seabed but outside the velocity boundary layer) results from a constant part and an oscillatory component of given frequency. The analytical solution shows that the velocity boundary layer formed near the seabed has a thickness insensitive to large changes in the wave frequency of oscillation. Moreover, the dissipated flow energy is as much as about 40% of the kinetic energy of the water moving inside this boundary layer. On the other hand, a number of *in situ* experiments carried out by installing arrangements of submerged groins resting upon the seabed sites show alignment between groin direction and the direction normal to the wavy sand formations created on sandy seabeds. The same experiments also showed that the amplitude and the wavelength of these sandy waves is distinctly larger in the neighbourhood of seabed groins.

The numerical solution of the two-dimensional flow within a seabed area with installed submerged groins, such as those shown in Figure 1, and the deposition results of the *in situ* experiment shown in Figure 2, indicate remarkable agreement between the predicted and the experimental result.

1. Introduction

The net effect of the interaction between coastal processes and the shore is a short-term quasi-periodic fluctuation in the beach state and particularly the shoreline position, as well as a secular change evidenced by long-term or secular recession or advancement in the position of the latter. The secular shoreline recession, often accompanied by unfavorable change in the material constituents onshore and nearshore, such as loss of sand and the dominance of sizeable rocks, are phenomena of some concern. The international community's alarm at such phenomena, often associated with destruction of man-made nearshore structures and loss of rural land, has lead to numerous theoretical, experimental *in vitro* and *in vivo* projects aimed at developing schemes for defensive stabilisation, at least for the local onshore-offshore environment.

C. L. Goudas et al. (eds.), Soft Shore Protection, 275-287.

Classical and novel shore defense schemes, some classified as 'hard' and some as 'soft' shore protection methods, are qualified on the basis of at least three criteria:

1. *Effectiveness in defending the shore they where designed for and installed*
2. *Active life expectancy of the arrangement and its overall effectiveness*
3. *Impact upon the environment and especially upon the adjacent shores*

Other criteria, such as cost, technical feasibility, aesthetic implications and so on may play some role, but passing the above three criteria has to be examined first.

The 'soft' shore protection system known as the Coastal Protection and Nourishment System (CPNS) tested theoretically (Goudas et al, 2000) for its large-scale impact upon the sediment transport within the strips of seabed defined by installed seabed groins requires further mathematical treatment in the area of small-scale dynamics relating to the effect of the groin parameters, especially their geometry and orientation and the sea water kinetic energy present within the velocity boundary layer that is transferred to every bedload grain thus producing the initial conditions for their motion. In such an analysis, an assumption has to be made regarding the percentage of this energy that is transferred to the seabed material and the part that is transformed into heat.

2. The CPNS system

The CPNS system consists of long, shallow and rectilinear groins that are placed on the unexcaveted seafloor. The direction of their long axis is perpendicular to the shoreline, unless other factors such as strong permanent currents, dictate other choices.

An example of the system's layout is shown in photograph 1, taken at Therma, Gulf of Gheras, Lesbos, Greece. Photograph 2, shows a single groin unit placed at the usual direction in relation to the shoreline, of almost elliptical cross-section, with axes varying, in general increasing with increasing distance from the shoreline. It was made using a long prefabricated geotextile tube that was first placed flat on the seafloor with the help of ballast and anchors, then stretched in the alignment shown in the photograph and finally filled through a hole at its onshore end with a special concrete mix in liquid condition. As is evident, the groins thus made are sitting on the seabed and dive under sea surface at a distance from the shoreline as defined by the special research and design work made for each project.

The research unit aims in particular to define the characteristics of the seabed groin-system on the basis of data collection pertaining to the weather, sea water movements (wave, current, tidal, etc.) and the way they interact with inshore-offshore water motion boundaries, the morphology of these boundaries (topography), and composition of the seabed material. The data must include the following:

1. The total number of groins to be developed and installed
2. The exact position of the inshore origins and the offshore terminal points of each seabed groin and hence its directions as well as its relative direction with respect to the impacting seabed currents of whatever origin

3. The physical dimensions (total length, shapes of cross sections and longitudinal section, total volume and weight) of the groin
4. The dynamic behavior of each groin under the action of extreme wave load and the behavior of the CPNS as a whole under such loads
5. A description of the materials (geotextiles, special concrete, etc) and equipment to be used and of the procedures to be followed during the prefabrication (thermal welding, attachment of ballast metal bands, sealing of the offshore ends) of the geotextile tubes, and finally the procedures of placement and tensioning, of casting and filling of the tubes with special concrete mix, and of the final sealing of the inshore end of the groin
6. Data relating to monitoring of the system and evaluation of the results

3. Hydrodynamic effect of CPNS

The general impact of the CPNS systems upon the movement of water masses is as complicated as the movement itself. However, for water masses below the upper surface of the seabed groins, the effect, through deflection, is 'polarization' of particle movement expressed by the predominance of particle velocities resting upon vertical planes parallel to the axes of the seabed groins. Evidence of this is not only the low size of the grain sedimentation on the shore but also the wavelike seabed formations with the principal direction parallel to the groin axes (see photographs 3 and 4). This observation has been made in all (17) *in situ* experiments conducted in shores with varying wave-current climates.

Despite the experimental nature of the evidence, it will be adopted in the mathematical formulation of the problem of energy transfer from the water to the seabed within the small size velocity boundary layer formed immediately after the seafloor. Assuming also that no slip motion occurs (i.e. the water particle velocity on the seafloor is zero), water kinetic energy is dissipated inside the thin velocity boundary layer at a rate reaching 40% of the total kinetic energy of this water layer per unit of time. This will be shown later by direct solution of the Navier-Stokes equations formulated for this case.

Returning to the critical assumption of water motion 'polarization' due to the presence of the CPNS, it needs to be said that the direction of water motion can be toward or away from the shore but essentially parallel to the groin system axes. Water set-up during storms and the development of rip currents close to the seafloor with offshore direction interfere with the wave-generated and density currents, which in general lead to the shore. The presence, however, of the CPNS diminishes the level of the water set-up to a degree depending upon the length and vertical elevation above the seafloor of the groin system and hence lowers of the erosive effect of the rip currents. As one would also expect, the system absorbs a good percentage of wave energy and hence beneficially decreases the energy that reaches the shore. The net effect of CPNS, as evidenced by material transfer toward the shore and as observed in all experiments conducted, justifies the assumption, made in the solution given in this paper that the dominant water mean current velocity near the seafloor is directed toward the shore.

Bearing in mind the above two assumptions as well as the assumption that on the seabed the water particle motion component caused by the wave motion (taken as being regular, single and travelling waves), the hydrodynamic problem will now be formulated and the energy dissipation and mass transportation calculated.

4. Hydrodynamic model

The problem of total energy dissipation within the boundary layer formed at the seabed is formulated as follows. We consider the unsteady two-dimensional flow of an incompressible fluid (e.g., seawater) past the seabed, which we take as infinite, horizontal and porous, permitting a small amount of suction at a constant velocity. The fluid motion near the seabed, but outside the boundary layer, results from a constant velocity that represents the current water motion, and an oscillatory component that represents the linear oscillatory wave motion near the seabed. An unavoidable assumption is that the oscillatory component has the frequency of a single wave. The assumptions made here simulate, but only approximately, the seawater motion, taking place within the surf zone and close to the seafloor. The theoretical motion of water particles, beside the steady current component, is cyclical near the sea surface and linear oscillatory close to the seabed (see e.g., Shaw, 1982, pp. 62–63). The wave driven and density currents, provide the steady flow component assumed above. The equations of motion of the fluid under the assumptions set will now be formulate.

5. Equations of flow and solution

We consider the unsteady two-dimensional flow of an incompressible viscous fluid (seawater) past an infinite, horizontal porous seabed, through which a small amount of suction at a constant velocity occurs. The Cartesian coordinate system Oxyz (sea Figure 1) has its origin O upon the seabed and the axes Ox and Oy along and perpendicular to the seabed, respectively. The Oz axis is normal to the Ox axis and rests on the seabed. The fluid properties are considered constant. Under these assumptions, the physical variables are functions of the space coordinate y and the time t only, and consequently the basic equations governing the fluid motion are:

$$\partial u'/\partial t' + v'\partial u'/\partial y' = \partial U'/\partial t' + \nu\partial^2 u'/\partial y'^2, \tag{1}$$

$$\partial v'/\partial y' = 0, \tag{2}$$

where u′ and v′ are the velocity components along the Ox and Oy axes, respectively; ν the kinematic viscosity coefficient and U(t) the free-stream velocity.

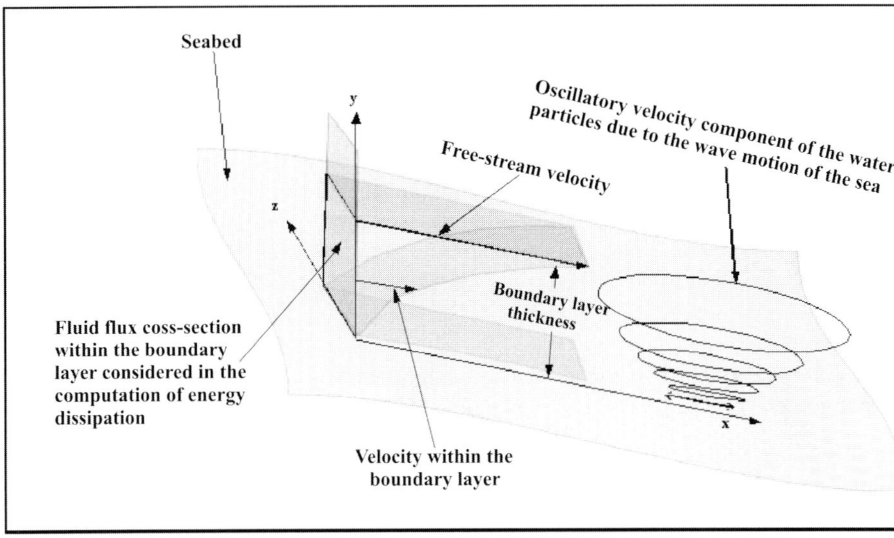

Fig. 1. The energy dissipated near the seabed is computed on the basis of the width of the boundary layer and the velocity profile inside it up to the free-stream

Since the seabed is assumed to be porous and through it suction with uniform velocity occurs, Eq. (2) integrates to:

$$v' = -v_o, \ (v_o > 0). \tag{3}$$

The negative sign indicates that the suction velocity is toward the interior of the seabed.

Now with the aid of the following non-dimensional transformations:

$$y = (v_0/\nu)y', \ t = (v_0^2/4\nu)t',$$

$$u = u'/U_0, \quad U = U'/U_0, \tag{4}$$

and Eq. (3), the Eq. (1) becomes

$$\partial u/\partial t - 4\partial u/\partial y = dU/d\,t + 4\partial^2 u/\partial y^2, \tag{5}$$

where U_0 is the mean value of $U(t')$.

The appropriate boundary conditions to the problem are:

$$u = 0 \text{ at } y = 0, \ \ u = U(t) \text{ as } y \to\infty. \tag{6}$$

The solution of Eq. (5) for the boundary conditions Eq. (6) will be obtained under the assumption that there exists a mean steady flow upon which the unsteady flow is superimposed. Then, in the neighborhood of the seabed, we represent the velocity field as:

$$u(y,t) = u_0(y) + u_1(y,t), \tag{7}$$

and now, following Watson (1958), we also assume

$$U(t) = 1 + F(t), \tag{8}$$

where $u_0(y)$ is the mean velocity while $u_1(y,t)$ the corresponding unknown function of y and t to be determined; finally, $F(t)$ is an arbitrary function of time. Watson (1958) has obtained the solution of Eq. (5) under the above boundary conditions and with the help of the two-sided Laplace transform technique for different forms of free-stream velocity. We discuss here only the case of oscillatory motion (added to the fixed current velocity), i.e., when the free-stream velocity oscillates with given wave frequency, and take it to be of the form:

$$F(t) = \varepsilon e^{i\Omega t}, \tag{9}$$

where $\Omega = 4\nu\omega/v_o^2$; ω is the dimensional frequency of oscillations and ε a small constant quantity <<1.

Under the above assumptions, the solution is

$$u(y,t) = 1 - e^{-y} + \varepsilon e^{i\Omega t}(1 - e^{-hy}), \tag{10}$$

where

$$h = h_r + ih_i = \tfrac{1}{2}[1 + (1 + iW)^{1/2}]$$
$$= \tfrac{1}{2} + \tfrac{1}{2}\{\tfrac{1}{2}[(1 + W^2)^{1/2} + 1]\}^{1/2} + i\{\tfrac{1}{2}\{\tfrac{1}{2}[(1 + W^2)^{1/2} - 1]\}^{1/2}\}. \tag{11}$$

Denoting by M_r and M_i the quantities

$$M_r = 1 - \exp(-h_r y)\cos(h_i y), \qquad M_i = \exp(-h_r y)\sin(h_i y), \tag{12}$$

where h_r and h_i are given by Eq. (11), then Eq. (10) reduces to

$$u(y,t)=1-e^{-y}+\varepsilon[M_r\cos(\Omega t)-M_i\sin(\Omega t). \tag{13}$$

6. Calculations and discussion

Water motion close to the seabed, particularly within the boundary layer, results in dissipation of its energy on account of its own internal viscosity and the friction caused by the boundary. A percentage of boundary layer kinetic fluid energy is transformed into heat, while the remainder is expounded in seabed mass transport, if the seabed composition permits such displacements.

The total energy dissipated due to the presence of the boundary (seabed) is calculated as follows: assuming fluid motion through a section of size $dydz$ (see Figure 1) with velocity $u(y)$ within the boundary layer, then the elementary mass is given as

$$dm = \rho u(y)dydz, \tag{14}$$

while its energy dE is given as

$$dE = \tfrac{1}{2}dmu^2(y) = \tfrac{1}{2}\rho u^3(y)dydz. \tag{15}$$

Then the total kinetic energy E of the fluid moving within the boundary layer is

$$E = \tfrac{1}{2}[\rho \int_0^{\delta} u^3(y)dy]dz, \tag{16}$$

where δ is the thickness of the boundary layer.

If the boundary (seabed) were absent, the energy E_o would be

$$E_o = \tfrac{1}{2}[\rho \int_0^{\delta} U_o^3 dy]dz = \tfrac{1}{2}(\rho \delta dz U_o^3). \tag{17}$$

Thus the energy loss caused by the presence of the boundary (seabed) is

$$E_o - E = \tfrac{1}{2}\{\rho dz \int_0^{\delta} [U_o^3 - u^3(y)]dy = \tfrac{1}{2}\{\rho dz[\delta U_o^3 - \int_0^{\delta} u^3(y)dy]\}. \tag{18}$$

To calculate the energy dissipated within the boundary layer, we first calculate the thickness δ of this layer using Eq. (13) for the velocity and computing $u(y)$ from the seabed outwards and solving the equation

$$|u(y,t) - u(\infty,t)| \leq 10^{-2}u(\infty,t). \tag{19}$$

The minimum value of y, satisfying Eq. (19) for any value of t?[0,T], is δ. Solving Eq. (19) for y by successive approximation methods of second order while giving fixed values to t, implies calculation of $\partial u(y,t)/\partial y$, which can be readily made from expression Eq. (13).

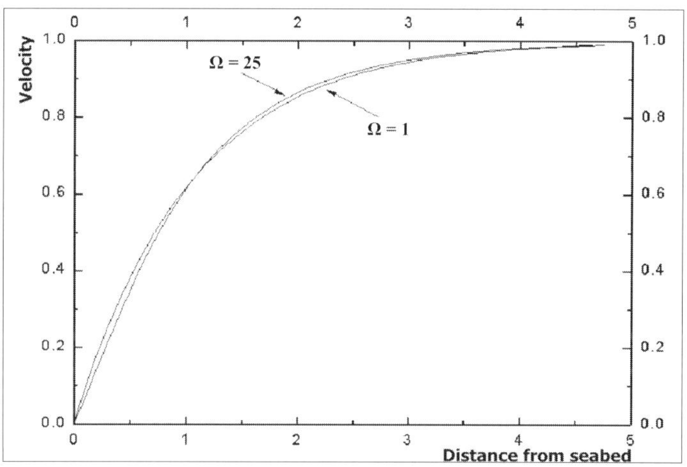

Fig. 2. Boundary layer thickness against distance from seabed (non-dimensional units)

The above calculations were made for the following values of the non-dimensional parameters involved, e.g., $v_o=0.001$, $\rho=1.03$, $v=0.001006$, $T=5$, $U_o=50$, $\varepsilon=0.2$, $\Omega=1$, 5 and 25.

For these values of the parameters, we found that (See Table I):

$\delta \approx 4.70$, *for* $\Omega = 1$,

$\delta \approx 4.65$, *for* $\Omega = 5$,

$$\delta \approx 4.60, \ \ for \ \ \Omega = 25.$$

This result shows that the velocity boundary layer thickness is not sensitive to big changes in the parameter Ω.

TABLE I
VELOCITY, ENERGY LOSS AND ENERGY LOSS % OF TOTAL WITHIN THE BOUNDARY LAYER

	Ω=1			Ω= 5			Ω=25		
y	u	E_{loss}	$\%E_{loss}$	u	E_{loss}	$\%E_{loss}$	u	E_{loss}	$\%E_{loss}$
0.00	0.00	0.00	1.00	0.00	0.00	1.00	0.00	0.00	1.00
0.25	0.21	0.13	1.00	0.20	0.13	1.00	0.17	0.13	1.00
0.50	0.38	0.25	0.98	0.36	0.25	0.98	0.35	0.25	0.99
0.75	0.51	0.37	0.95	0.49	0.37	0.96	0.49	0.37	0.96
1.00	0.62	0.47	0.92	0.60	0.48	0.93	0.61	0.48	0.93
1.25	0.70	0.56	0.87	0.69	0.57	0.89	0.70	0.57	0.88
1.50	0.76	0.64	0.83	0.75	0.65	0.84	0.77	0.64	0.83
1.75	0.81	0.70	0.78	0.81	0.71	0.79	0.83	0.70	0.78
2.00	0.85	0.76	0.73	0.85	0.77	0.75	0.87	0.75	0.73
2.25	0.89	0.80	0.69	0.89	0.81	0.70	0.90	0.79	0.68
2.50	0.91	0.83	0.65	0.91	0.85	0.66	0.92	0.82	0.64
2.75	0.93	0.86	0.61	0.93	0.87	0.62	0.94	0.85	0.60
3.00	0.94	0.88	0.57	0.95	0.89	0.58	0.95	0.87	0.56
3.25	0.96	0.90	0.54	0.96	0.91	0.54	0.96	0.89	0.53
3.50	0.97	0.92	0.51	0.97	0.92	0.51	0.97	0.90	0.50
3.75	0.97	0.93	0.48	0.98	0.93	0.48	0.98	0.91	0.47
4.00	0.98	0.94	0.45	0.98	0.94	0.46	0.98	0.92	0.44
4.25	0.98	0.94	0.43	0.99	0.95	0.43	0.99	0.92	0.42
4.50	0.99	0.95	0.41	0.99	0.95	0.41	0.99	0.93	0.40
4.75	0.99	0.95	0.39						

In Figure 2 the fluid velocity u is drawn against the distance y from the seabed within the boundary layer, for $\Omega = 1$ and 25, while in Figures 3 and 4 the dependence of energy loss (E_{loss}) and energy loss percent $(E_{loss}\%)$ against the distance from the seabed is given. The remark made earlier about the non-sensitivity of the boundary layer thickness to large changes in Ω, obviously also goes for E_{loss}. An essential observation concerning E_{loss}, i.e., energy dissipated within the boundary layer, is that it rises to about 40% of the total kinetic energy of seawater inside the boundary layer. Other conclusions are self-evident from Table I and hence any further discussion about them would seem redundant.

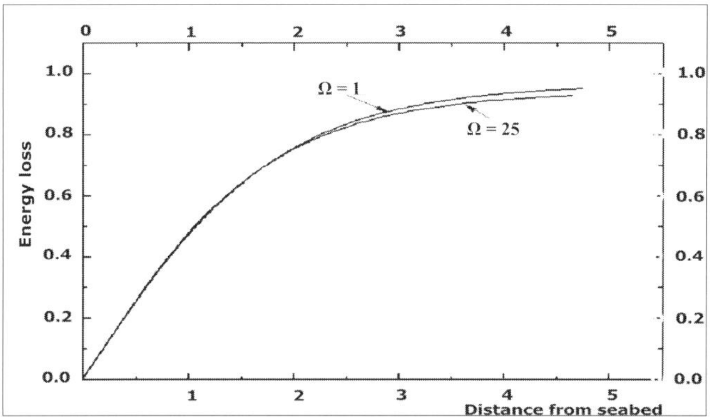

Fig. 3. Energy loss against vertical distance from seabed (non-dimensional units)

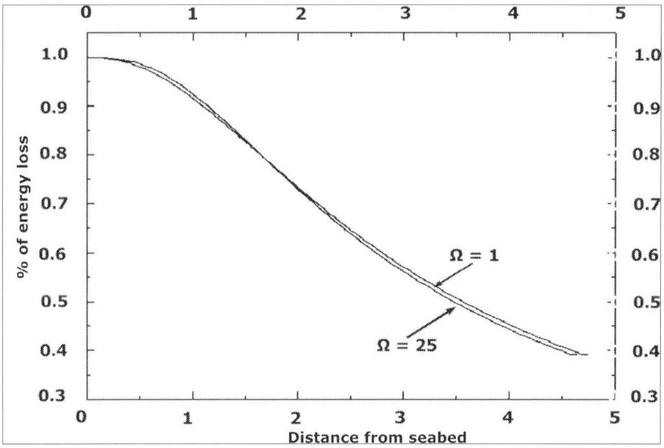

Fig. 4. Percent of total energy loss within the boundary layer against distance from seabed (non-dimensional units)

The energy dissipated within the boundary layer is turned into thermal energy and transportation of movable particles (sand) available on the seabed. In order to determine the amount of E_{loss} affecting mass (sand) transportation, more data as well as experiments are required. In the absence of sand it is evident that all E_{loss} is turned into heat. If the seabed has an abundance of sand, observations show that movement of sand (littorals) in large quantities takes place, a fact implying that a sizeable portion of E_{loss}

is used for mass displacement. In such cases, as observations show, the oscillatory component of fluid motion results in the formation of sand wave features on the sandy seabed, while the steady water current is responsible for the linear displacement of sand. Due to the presence of steady water currents, the wavy sand formations on the seabed are not symmetrical, or of a sinusoidal shape but acquire sharp descending surfaces immediately after the crests.

7. Wavelength of sand waves

In the calculation of energy dissipation given earlier we made the assumption that near the boundary (seabed) the water motion was horizontal and the velocity resulted from a constant component (current) and an oscillatory component (wave motion near the seabed). The oscillatory component is responsible for the appearance of sand waves usually observed upon the seabed. The length of the sand waves in the event of regular water waves is approximately equal to twice the amplitude of the oscillatory water motion near the seabed. This amplitude and wavelength we shall compute here on the basis of the linear theory for progressive waves (see e.g. Shaw, 1982, pp. 173–180).

The velocity potential *f* of the single progressive wave fulfills Laplace's Equation

$$\partial^2\phi/\partial x^2 + \partial^2\phi/\partial y^2 = 0, \tag{20}$$

which upon treatment gives the separable solution

$$❷ = -(ac)\cosh[k(y + h)]\cos[k(x - ct)]/\sinh(kh), \tag{21}$$

where a is the amplitude of the wave motion at the surface, c the phase velocity defined in terms of the surface wave parameters by the expression $c = L/T$, with L the surface wavelength and T the wave period, κ the wave number $(\kappa = 2\pi/L)$, y the distance below MSL $(y = 0$ at the sea level and negative toward the seabed), and h the seadepth.

The horizontal oscillatory velocity component $u_1(y,t)$ appearing in Eq. (7) is given by the expression

$$u_1(y,t) = \partial\phi/\partial x=(ac)\cosh[\kappa(y + h)]\sin[\kappa(x - ct)]/\sinh(\kappa h), \tag{22}$$

and near the seabed ($y \approx -h$)

$$u_1(0,t) = (ac)\sin[k(x - ct)]/\sinh(kh). \tag{23}$$

Hence, the length of the sand wave L_s, if it is assumed to be the same as that of the water particle oscillation limits near the seabed, is

$$L_s = 2a/\sinh(\kappa h) = H/\sinh(2\pi h/L), \tag{24}$$

and hence the relationship between the water wave length at the surface and the sand wave length at depth h is

$$L = 2\pi h/\sinh^{-1}(H/L_s), \tag{25}$$

where H is the wave height and L the wave length, both computed at the surface. An alternative derivation of formula (25) is given by Shaw (1982).

Photographs 1 and 2 clearly show the presence of such a wavy seabed. They are taken from a field of seabed groins deployed for the purpose of deflecting currents and waves toward the shoreline. The deflected water movement resulted in fluid energy dissipation and to some extent, sand displacement. Photographs 3 and 4 show sand waves that are both long and high. The sand wave crests are usually parallel to the CPNS groins in all seabed areas close to groins, while the direction of sand displacement is clearly shown by the sharp slope of the sand wave crests which always occurs on the crest side closer to the shore.

Photographs 1 and 2. Sand wave formations produced by water wave action reaching the seabed as oscillatory motion. The presence of the groins causes the sandy waves to adopt the same main direction as the groins, in this case toward the shore. The photographs show that the lengths and heights of the sandy waves are greater near the groins. The direction of sand wave propagation is obvious.

The sand wave length in photographs 1 and 2, taken at a depth $h = 3m$, are $L_{1s} = 0.25m$, $L_{2s} = 0.45m$, and $L_{3s} = 0.65m$, respectively. By evaluating formula (25) and assuming wave heights $H = 2m$ and $H = 1.5m$, we find that the corresponding surface wave lengths responsible for their formation are

$L_1 = 6.79$m, $L_2 = 8.57$m, and $L_3 = 10.23$m, for wave height H = 2m, and

$L_1 = 7.56m$, $L_2 = 9.82m$, and $L_3 = 11.98m$, for wave height $H = 1.5m$.

These surface wave lengths required for the formation of sand waves of the same lengths upon the seabed as the ones observed are compatible with the wave climate of a site in which the wave lengths observed never exceed 20m. Two of the authors (C. Goudas and G. Katsiaris) observed this after a storm in December 1996 (photographed by the team's diver).

The seabed groin system (CPNS) has been used in 17 cases of coastal erosion and three countries (Greece, Italy and Egypt). The least that can be said from monitored results is that they are simply positive for the corresponding beaches, though most of the results have been very successful. No side effects have ever been observed. The impact of the system is obvious in the typical before and after photographs 3 and 4.

The environmental acceptability of this 'soft' shore protection method, as well as the low cost of implementing it, make it a method suitable for widespraed application.

Photographs 3 and 4. A beach at Theologos, Rhodes Island, transformed by a CPNS installed in 1994 (bottom photo received in 1995)

REFERENCES

Cebeci, T. and Bradshaw, P.: *Momentum Transfer in Boundary Layers*, New York: McGraw-Hill Book Co., 1977.

Davies, R.A. Jr.: *Coasts*, N. Jersey: Prentice Hall, Simon and Schuster, 1996.

Elliot, G. and Caratti, G. (editors): *European Wave Energy Symposium*, Edinb. National Engineering Laboratory Exec. Agency, Scotland, 1993.

Fischer, C., Jirka, G.H. and Kaldenhoff, H.: 'Wave Energy Dissipation in the Wake of Submerged Horizontal Plate', *4th International Conference on Computing in Civil and Building Engineering*, Tokyo, 1991.

Goudas, C.L., Katsiaris G.A., Kafousias, N., Massalas, C., Pnevmatikos, G., Xenos, M., and Tzirtzilakis, E.: 'Longshore Current Modification near the Boundary by Seabed Groin Arrangements: a Numerical Approach', in this volume.

Mollison, D.: Wave Energy Losses in Intermediate Depths, *App. Ocean Research*, **5**, 1983, 234–237.

Sarmento, A.J.N.A. and Falcao, A.F. de O.: Wave Generation by an Oscillating Surface-Pressure and its Application in Wave Energy Extraction, J. Fluid Mech., 150, 1985, 467–485.

Schlichting, H.: *Boundary Layer Theory* (Seventh English Edition), New York: McGraw-Hill Book Co., 1987.

Shaw, R.: *Wave Energy – a Design Challenge*, Chichester, England: Ellis Horwood Limited, 1982.

Watson, H.: 'A Solution of the Navier - Stokes Equations Illustrating the Response of Laminar Boundary Layer to a given Change in the External Stream Velocity', *Quart. J. Mech. Applied Maths*., 11, 1958, 302–325.

Wieghardt, K.: Uber einen Energie-satz zur Berechnung Laminar Grenzschichten, *Ing. - Arch*., 16, 1948, 231–242.

MANAGEMENT OF MIXED SEDIMENT BEACHES

B. LÓPEZ DE SAN ROMÁN-BLANC,[1] J. S. DAMGAARD,[1] T. T. COATES[1] AND P. HOLMES[2]

[1] *HR Wallingford Howbery Park, Wallingford. Oxon OX108BA, blb@hrwallingford.co.uk*
[2] *Civil & Environmental Engineering Dpt. Imperial College. London SW7 2AZ*

Abstract

Mixed grain beaches include sediment sizes ranging over three orders of magnitude from fine sand (100mm), through gravels (2mm – 64mm) right up to small boulders (>256mm) (Coates & Damgaard, 1999). This type of beach is frequent around the world, including the Mediterranean. Sediment distributions may vary across the beach profile, along the shore and with depth below the beach face, as well as with time. Depending on these distributions, the beach response to wave conditions will vary.

Re-nourishment schemes using materials that are significantly different to the indigenous ones in terms of size and distribution have also produced mixed beaches all over the world. Nourishment schemes are often used in preference to hard defences, particularly in areas that are valued for their coastal recreation.

While the transport processes involved in the creation of sand beaches are being extensively researched throughout the coastal engineering community, little international effort is directed to gravel beaches and even less to mixed sediments beaches. Understanding and confidently predicting the behaviour of all types of beaches under different conditions and time periods is important to beach management and coastal defence.

This paper discusses the most characteristic mixed beach processes and then describes ongoing research at HR Wallingford on predicting beach responses, including a discussion of the ANEMONE suite of models.

1. Characteristics of mixed sediment beaches

Mixed sediment beaches may take a variety of forms. Their distribution can be bi-modal with concentrations of fine sand and medium gravels, or can be fully mixed with a constant grading from sand to cobbles as found along New Zealand's south east coast (Single & Hemmingsen, 1999). The sediment distribution can vary across the foreshore, along the shore, vertically down from the surface and over time periods ranging from hours to seasons. Two basic foreshore types have been identified:

- complex beaches typical of meso and macro tidal conditions, with a lower sand platform and a steep coarse-grained upper beach, with an area of mixing at the boundary
- fully mixed beaches typical of micro and meso tidal conditions, with a variable cross-shore sediment distribution ranging from a higher percentage of sand across the lower beach to predominantly coarse gravel and cobbles along the storm crest.

C. L. Goudas et al. (eds.), Soft Shore Protection, 289-299.

Either of these types can be found along the toe of cliffs or as a protective ridge in front of low-lying flood plains. Understanding these beach types can be made more complex by the presence of an impermeable clay or bedrock sub-strata, low-density shell fragments or structures such as groynes or breakwaters. Management problems associated with mixed-grain beaches depend on the nature and value of the backshore zone, and may include storm or long-term erosion and overtopping or breaching leading to flooding.

2. Methodology of study of mixed sediment beaches

Management of these complex beaches requires an ability to predict short-term beach response to storms and long-term changes in response to developments along the beach. Such a predictive ability is necessary to reduce the risk to life and property, the cost of beach maintenance and the impact on the natural coastal environment. In particular, tools are needed to predict:

- Crest cut back, wave run up, wave overtopping, back face instability and breaching due to storm events
- Longshore drift rates along open beaches and around structures or headlands, under both short-term storm conditions and under long-term average conditions
- The effects of recharging beaches with material that is significantly different to the indigenous sediment in terms of size and distribution.

2.1 EXISTING PREDICTIVE METHODS

Predictive methods are available to assist shoreline management, including physical and numerical models. Physical models are divided into flume and wave basin models, each providing valuable information. Numerical approaches are broadly divided into parametric models, often derived from observed results in the laboratory or from field measurements, and physics-based models which attempt to account explicitly for the main physical processes active across the foreshore. Each of these approaches has its own weaknesses and strengths, but none are able to simulate all of the important and complex processes influencing mixed beaches.

Available facilities and costs usually limit the use of physical models for shoreline management. Mobile bed modelling at small scales is further limited by the necessity to work out a compromise between the requirements of the different processes that must be simulated. Although acceptable approaches to scaling narrow-graded gravel beaches have been used for over 25 years (Powell, 1990), small-scale modelling of beaches with a mixture of sand and gravel is limited because of the incompatibility of having both fractions within the same model.

Existing numerical models are also limited in many ways. Most have been derived for sand beaches and have then been extended to include a wider range of grain sizes. The main problems with respect to extending these models to mixed beaches are set out in Coates and Mason (1998) and López de San Román Blanco et al (2000), as follows:

- Assuming a simplistic description of beach sediment, usually defining the complete beach by a single D50 value or another simple parameter
- Assuming that beaches do not vary across-shore, along-shore, vertically or over time
- Assuming an impermeable surface and ignoring flows within the beach face (infiltration and exfiltration)
- Assuming a simple threshold of motion based on the defined grain size

Longshore predictive tools in use at present, both those based on numerical simulation of appropriate physical processes (Damgaard et al., 1996) and those based on equations such as the CERC formula (Brampton & Motyka 1984 or Damgaard & Soulsby 1996) provide reasonable results. In contrast, cross-shore numerical models have had limited success – mainly due to a lack of knowledge of the governing physical processes and/or an inability to model the processes adequately. Hence no process-based model is available to predict the response of a coarse-grain beach to a given hydrodynamic forcing. At present, the best of the coarse-grained models is probably the parametric profile model of Powell (1990) known as SHINGLE, based on extensive laboratory flume tests and verified against field data.

2.2 ONGOING RESEARCH

HR Wallingford is currently carrying out a 3-year research project, funded by the UK Ministry of Agriculture, Food and Fisheries, in order to develop predictive tools for mixed beaches. The methodology of the project is three-fold:

- Characterisation of important processes occurring in mixed beaches
- Physical modelling, both of the processes controlling the basic physics and full-scale modelling to simulate mixed beaches (avoiding scaling problems)
- Development of the ANEMONE wave transformation and run up suite of models to include the effects of a permeable beach face in the swash zone

3. Mixed Beach Processes

3.1 PERMEABILITY

The most important characteristic of coarse and mixed sediment beaches is that they are permeable, so that the flow within the beach is significant enough to affect the flow above the surface. The permeability of beach material is related to its particle size, shape and sorting. Coarse gravel has large pore spaces allowing water to flow relatively easily through the matrix. If the material has a wide-size distribution including sand, then many of the large spaces are filled by smaller particles, hence the flow is restricted and the permeability reduced.

Fig. 1. Mixed beach on Andros Island, Greece

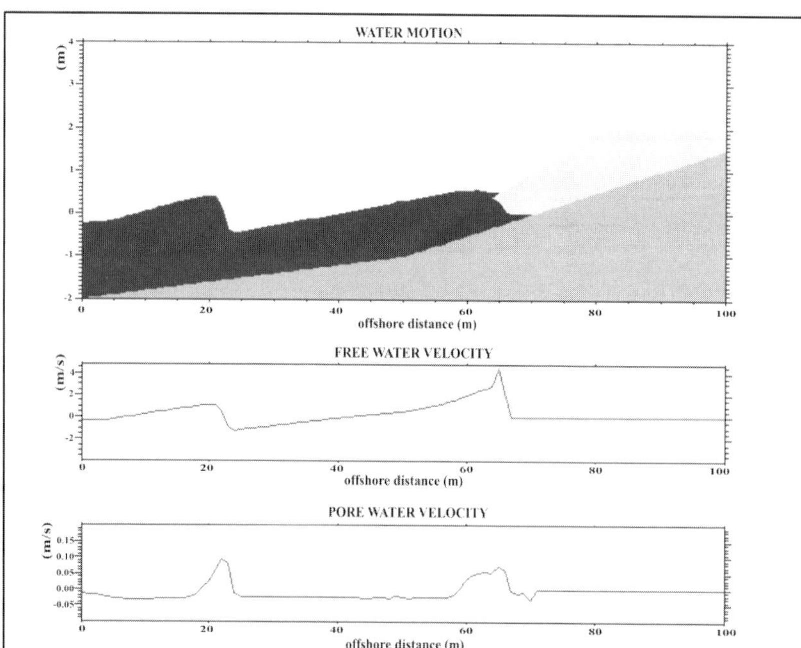

Fig. 2. Diagram of OTTP-1d model run on an impermeable beach with a permeable gravel berm (initial dry beach)

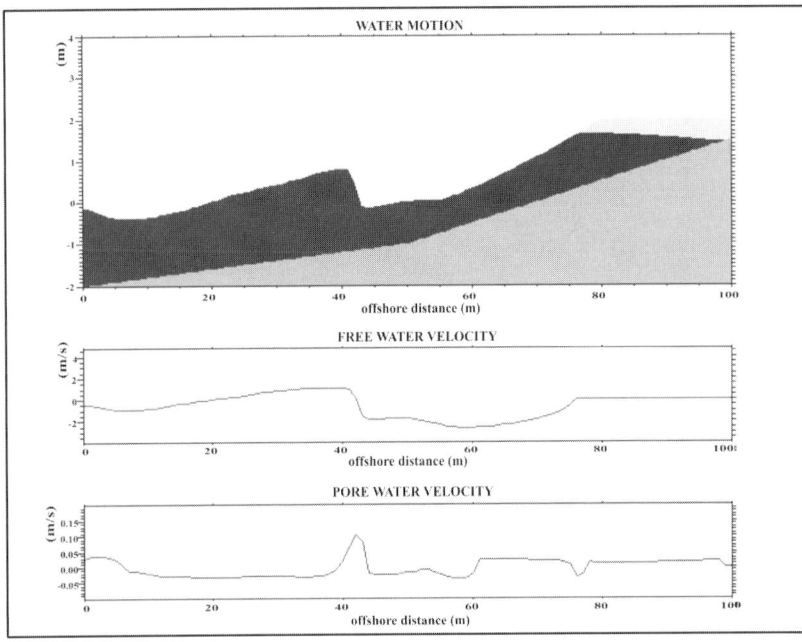

Fig. 3. Diagram of an OTTP-1d model run on an impermeable beach with a permeable gravel berm (final fully saturated beach)

3.2 ASYMMETRY OF WAVE PROFILE

The region of the beach which is intermittently covered by water as waves run up and down is called the swash zone. In this region, the exchange of fluid and momentum between the free water and the pore water can have a significant effect on the flow field. As a wave runs up the beach it may carry a large amount of water at a high speed. This can result in a significant transport of beach material further up the beach. On an impermeable beach, the following backwash of water will balance the mass flux of the run-up. This backwash rush is likely to produce sediment transport down the beach, which is similar in magnitude to the up-rush related transport, resulting in a small, wave-averaged net sediment transport rate. On a permeable beach, water may drain into the bed on the up-rush, thus reducing the shoreward limit of the run-up. Water may continually drain into the bed during the up-rush, and during much of the following backwash, resulting in a reduction of flow in the backwash and hence the potential for seaward sediment transport may be greatly reduced.

3.3 THRESHOLD OF MOTION OF SURFACE MATERIAL

If it is assumed that the probability of entrainment for each fraction in a mixed sediment bed depends only on the absolute grain size, then for a given bed shear stress, when grain sizes decrease, the driving forces will increase and therefore the transport rate will also increase. This assumption ignores the effects of the relative grain size within a non-uniform bed. The relative size effect accounts for the hiding and exposure of graded sediments due to the surrounding particles. According to this principle, the mobility of the fine grains decreases due to the presence of larger grains, whereas the mobility of coarse grains increases in the presence of finer grains. Some authors (e.g. Day, 1980) define a critical diameter, DA, so that if the diameter of the particle is smaller than DA, the particle is shielded from the flow and the initial motion requires a greater tractive force than suggested by its diameter. When the diameter is bigger than the critical one, the particle is more exposed and less tractive force is required to begin movement. This critical diameter has been found to vary inversely with grading (√(D84/D16)), although the physical explanation for this is not known.

Several authors (e.g. Wiberg & Smith, 1987; Wilcock & Southard, 1988; Kunhle, 1993) have studied theoretically and experimentally the threshold of motion of mixed sediments within alluvial channels, rivers or streams. Although their findings are important, care must be exercised in extrapolating their results to the swash zones, as the hydrodynamics and time scales are completely different.

The three major findings from the river studies are:

- Most sizes in a mixed size bed apparently begin to move at about the same shear stress, known as equal mobility.
- 'Winnowing' or natural vertical sorting occurs when most (or all) sizes in a mixture are at least occasionally in motion as the finer grains fall into positions evacuated by coarser grains (and not vice-versa).
- The Shields curve is reasonably well fitted by the D50 of the mixture of the sediments.

The major drawbacks when extrapolating these findings into the swash zone are:

- Some of the processes are not yet well understood, resulting in apparently contradictory results among researchers.
- The mixed sediments considered are not always characteristic of the mixed sediments found along the beaches.
- The time-scales of fluvial transport are very different to beach transport
- Oscillatory motions are not taken into account.

3.4 SEEPAGE AND GROUNDWATER FLOW

Flow in and out of the beach face due to waves propagating on a permeable beach induces pressure gradients within the sediment bed. Both these pressure gradients and the flow itself have an influence on the motion of particles, the threshold of motion, the boundary layer and on the effective strength of the bed as a whole.

In the presence of seepage (upwards or downwards), the critical ratio of the forces changes, as there is an additional force on the particles. Oldenziel and Brink (1974), Baldock and Holmes (1998) and Nielsen (1992), suggested that the Shields parameter should be modified in view of the fact that the submerged weight force of bed particles was reduced by an upward seepage.

Several experiments with induced seepage have been carried out, most of them with regard to alluvial channels, rivers or streams. The results to date are quite contradictory: Watters and Rao (1971), Willets and Drossos (1975) and Rao and Sitaram (1999) found that suction or inwards seepage decreases the stability of the bed particles and therefore increases their mobility, and injection or outwards seepage decreases the stability, decreasing their mobility. On the other hand, Oldenziel and Brink (1974), Richardson et al. (1985) and Nagakawa et al. (1988), found the opposite, though more intuitively likely results, with suction increasing the stability and injection decreasing it.

These contradictory results can be explained by analysing the stability of the bed, which is a function of both the magnitudes of the hydrodynamic forces acting on the particles and on their resistive forces. Suction or downward seepage increases both the resistive and hydrodynamic forces, whereas injection, blowing or upward seepage reduces them both. Depending on the relative influence of the seepage in one or the other forces, the stability of the bed will increase or decrease. The influence of the seepage on the lift and drag forces is two-fold:

- Direct – adding a dynamic force in the direction of the seepage velocity
- Indirect – altering the main flow conditions and therefore altering the forces on the particle.

This last influence can be observed in the change of angle of wave attack of the main flow onto the bed particle, by flushing the dead water out or into and therefore exposing the particle surface and by altering the wake of the particle. All these effects will depend on the bed and flow configuration, as well as on the direction of the seepage.

When the effect of a permeable bed on the incipient motion of bed particles is evaluated, it is apparent that not only does the seepage force have to be considered, but also the alteration of the boundary layer characteristics by the flow out or into the bed.

Other characteristics of beaches to consider are the capillary fringe and air entrapment.

4. Methodology – ANEMONE Models

As mentioned above, existing numerical sediment transport models are of limited applicability to mixed beaches as most of them have been derived for sand beaches and then extended to include a wider range of grain sizes. For this reason the models make inappropriate assumptions about the permeability of the beach and/or attempt to represent the entire beach as having a single nominal grain size.

Rather than attempting to modify an existing model, HR Wallingford is using the most recent nearshore wave transformation models from the ANEMONE suite as the foundation for developing physics-based beach response models that will include the effects of a porous beach with variable sediment distributions. The initial work will concentrate on cross-shore storm response using a model called OTTP-1d, which is currently being validated against field data and undergoing further development of its sediment transport capabilities. Detailed information about the hydrodynamics can be found in Dodd (1998), while the ANEMONE suite of models is described in Dodd et al (2000) and the initial development of OTTP-1d is summarised in Peet et al (1999).

OTTP-1d can simulate wave run-up, overtopping and regeneration both above and within a permeable beach. In more concrete terms, the model can provide predictions of wave heights and associated depth-averaged velocities in the inner surf zone; the time variation of the moving shoreline (which includes run-up and run-down); and overtopping rates and volumes (average and instantaneous). It also provides velocities and depths within permeable beaches/structures, and overtopping rates resulting from overtopping over and through the structure/beach. The program can model these quantities on arbitrary bathymetries and sections, as long as bottom slopes are not too large. In theory, horizontal water motions should be much greater than vertical ones for the model to work well and provide good predictions; in practice the model has been shown to work on slopes up to 1:1.

The wave breaking process is simulated 'naturally' in the model. This means that the numerical waves will steepen and form bores, which are commonly seen in the inner surf zone of natural beaches and on steeper, man-made structures. Therefore, the model may be expected to work best where waves do actually propagate as bores. However, for other types of breaker the model has been shown to work well in a bulk sense (i.e., in terms of transport of mass and momentum), even though it cannot reproduce the details of, say, a plunging breaker.

4.1 MODEL EQUATIONS IN FREE REGION

The model equations are the standard non-linear shallow water (NLSW) equations (also known as the finite-amplitude shallow water equations), which in flux-conservative form are:

Conservation of mass:

$$\frac{\partial d}{\partial t} + \frac{\partial (d U)}{\partial x} = q. \tag{1}$$

Conservation of momentum:

$$\frac{\partial (dU)}{\partial t} + \frac{\partial (dU^2 + \frac{1}{2} g d^2)}{\partial x} = gd \frac{\partial h}{\partial x} - \frac{f_w}{2} U \left|U\right| + qq_x. \tag{2}$$

The equations are derived by depth-integrating the Navier-Stokes equations and are valid under the assumption that horizontal velocities are typically much greater than vertical velocities, which is so in shallow water.

5. Pore-water flow

Mass and momentum is exchanged between the free-water region (flow above the bed) and the pore-water region (flow within the beach). When the beach is completely covered by water the effects of this exchange are not felt strongly by the flow field.

OTTP-1d describes the free-water flow in terms of depth-averaged velocities and local water depths. A consistent approach is used for describing the additional flow within the permeable bed. Pore-water velocities Up are depth-averaged over some specified permeable layer thickness dp. Depth-averaged flow modelling neglects vertical accelerations and all pressure is assumed to be hydrostatic. The pressure gradient is balanced by the friction, inertia and convective terms.

The conservation of mass for the permeable layer is given by:

$$\frac{\partial d_p}{\partial t} + \frac{1}{n}\frac{\partial \left(d_p U_p\right)}{\partial x} = -\frac{q}{n}. \tag{3}$$

The conservation of momentum is found by expressing the pressure as purely hydrostatic:

$$\begin{aligned} \frac{1+c_A}{ng}\frac{\partial U_p}{\partial t} + \frac{1}{n^2 g}U_p\frac{\partial U_p}{\partial x} + \frac{\partial (d + d_p - h_p)}{\partial x} &= \\ - aU_p - bU_p\left|U_p\right| - \frac{1}{n^2 g d_p} q q_x. \end{aligned} \tag{4}$$

These equations are solved within the model in two different ways:

- When the permeable bed is full and is covered with fluid in the free-water region it cannot change its layer thickness, hence dp remains locally constant.
- When the permeable layer is not full, or if it is full but is not covered by free-water flow with different friction and extra permeability terms. In this region the exchange flux q is governed by infiltration when the free-water region overlies an area of permeable bed which is not full, and exfiltration or seepage which occurs when the phreatic surface tries to extend above the bed due to the absence of overlying free-water.

These equations are solved by the finite volume method (for details see Dodd, 1998). Some results are shown in Figures 2 and 3.

6. Conclusions

Although the proper management of coarse and mixed-grain beaches (both natural and nourished) is of interest to many countries around the world, the available predictive tools are still inadequate for reliable predictions of the short- and long-term responses of these types of beaches. This is an important issue and the gaps in our present knowledge and predictive capability can only be addressed by extensive research.

Due to the complexity of the governing physical processes, the problems can only be solved by applying a combination of the tools available:

- To study and learn from existing field data

- To develop physically sound process descriptions
- To compare developed theories with large-scale physical model experiments
- To develop numerical models capable of simulating the significant physical processes.

HR Wallingford is undertaking this research through a three-year programme funded by the UK Ministry of Agriculture, Fisheries and Food. Together with the ongoing development of the HR Wallingford ANEMONE suite of models, this work will provide coastal zone managers with improved predictive tools for the management of coarse and mixed sediment beaches.

REFERENCES

Baldock, T.E. and Holmes, P.: Seepage effects on sediment transport by waves and currents. *International Conference on Coastal Engineering*, ASCE Paper no. 135, 1998.

Brampton, A.H., and Motyka, J.M.: Modelling the plan shape of shingle beaches, *Proceedings of the POLYMODEL 7 Conference*, 1984, Sunderland Polytechnic, UK.

Coates, T.T. and Daamgard, J.S.: Towards improved management of mixed grain beaches, *Proceedings of the HYDRALAB workshop,* Hannover, 1999, 69–73.

Coates, T.T., and Mason, T.: Development of predictive tools and design guidance for mixed beaches: scoping study. *HR Wallingford Report TR 56*, 1999, Wallingford, UK.

Damgaard, J.S. and Soulsby, R. L.: Longshore bed-load transport. *Proceedings of the* 25th *International Conference on Coastal Engineering*, ASCE, 1996, 3614–3627, Orlando, USA.

Damgaard, J.S., Stripling, S. and Soulsby, R.L.: Numerical Modelling of Coastal Shingle Transport, *HR Wallingford Report TR 4*, 1996, Wallingford, UK.

Day, T.J.: A study of the transport of graded sediments. HR Wallingford Report IT190, April 1980.

Dodd, N.X.: Numerical model of wave run-up, overtopping and regeneration, *Journal of Waterway, Port, Coastal & Ocean Engineering*, 124, 1998, (2), 73–81.

Dodd, N.X., Peet, A.H. and Coates, T.T.: ANEMONE: An overview of the advanced non-linear engineering suite of models for the nearshore environment *HR Wallingford Report SR 553*, March 2000.

Kuhnle R.A.: Incipient motion of sand-gravel sediment mixtures, *Journal of Hydraulic Engineering*, 119, 1994, 1400–1415.

López de San Román Blanco, B., Coates, T.T., Peet, A.H. and Damgaard, J.S.: Development of predictive tools and design guidance for mixed sediment beaches, *HR Wallingford Report TR 102*, 2000, Wallingford, UK.

Nagakawa, A., Tsujimoto, T. and Murakami, S.: Effect of suction or injection through a stream on bed load transport process, *Proceedings of the International Conference on Fluvial Hydrology*, 1988.

Nielsen, P.: Coastal bottom boundary layers and sediment transport, *Advanced Series on Ocean Engineering*, volume 4. World Scientific, 1992.

Oldenziel, D.M. and Brink, W.E.: Influence of suction and blowing on entrainment of sand particles, *Journal of the Hydraulics Division*, HY7, 1974, 935–949.

Peet, A.H., Dodd, N.X. and Coates, T.T.: Development of OTTP-1d, *HR Wallingford Report TR 96*, 1999, Wallingford, UK.

Powell, K.A.: Predicting short-term profile response for shingle beaches, *HR Wallingford Report SR 219*, 1990, Wallingford, UK.

Rao, A.R. and Sitaram, N.: Stability and mobility of sand-bed channels affected by seepage, *Journal of Irrigation and Drainage Engineering*, 12 (1), 1999, 370–379.

Richardson, J.R., and Richardson, E.V.: Inflow seepage influence on straight alluvial channels, *Journal of Hydraulic Engineering ASCE*, 111 (8), 1985, 1133–1147.

Single, M.B., and Hemmingsen, M.: Mixed sand and gravel barrier beaches of South Canterbury, New Zealand, in *Ecology and Geomorphology of Coastal Shingle*, Packman, JR, et al (eds), Conf. Proc., University of Wolverhampton, UK 1999 (in preparation).

Watters, G.Z. and Rao, M.V.P.: Hydrodynamic effects of seepage on bed particles, Journal of the Hydraulics Division ASCE, 97 (HY3), 1971, 421–439.

Wiberg, P.L. and Smith, J.D.: Calculations of the critical shear stress for motion of uniform and heterogeneous sediments, *Water Resources Research*, 23, 1987, 1471–1480.

Wilcock, P.R. and Southard, J.B.: Experimental study of incipient motion in mixed-size sediment, *Water Resources Research*, 24 (7), 1988, 1137–1151.

Willetts, B.B. and Drossos, M.E.: Local Erosion caused by rapid force infiltration, *Journal of the Hydraulics Division*, ASCE, 101 (HY12), 1975, 1477–1488.

COST–BENEFIT ANALYSIS AND SOME ENGINEERING CONSIDERATIONS IN SHORE PROTECTION CARRIED OUT IN THE LAZIO REGION, ITALY

RAIMONDO BESSON AND PAOLO LUPINO

Regione Lazio, Dipartamento Opere Publiche, Via Capitan Bavastro 110, 00154 Roma, Italy

Abstract

After about 25 years of experience relating to shore protection with rock groins, sand-sack submerged groins, seawalls, submerged breakwaters and so on, the regional administration of Lazio (coast line of 350 km) has done a careful analysis of the costs and benefits of these systems compared to those of the nourishment system. The results of this analysis demostrate that if the cost of the sand is less than about \$10/m^3, for this coast, the nourishment system is more suitable. These findings, along with several environmental considerations (less physical impact, no erosion propagation, etc.), has prompted the administration to develop a new program of research oriented towards finding extracting sites in the sea (sand at low cost) and to finding appropriate dredging, replenishment and maintenance technology.

In 1999, the region of Lazio started the Ostia Beach (Rome) replenishment project involving about 1 million m^3 of sand for nourishing 3,5 km of coast and now is starting the Anzio Beach (Rome) project involving about 0,5 million m^3 of sand for nourishing 1,5 km beach. The experience has led us to consider various other engineering options such as profile design, replenishment technology, dredging technology, computing methods and so on.

For example if the existing seabed profile is without a sand bar or very poor ones (compared to the nearest equilibrium profile), we also have to restore this sand volume if we want a relative stable beachline. With this kind of design project, we may find that for 1 metre of coastline advancement we need much more than 6-8 m^3 (advancement ratio) of sand than the shift profile method suggests. In the Anzio Beach project, in the most critical zones, even A_d= 20-25 m^3 for each metre of coastline advancement has been reached.

Furthermore, when a contractor commences the replenishment, he cannot be expected to deposit the sand according to the new equilibrium profile. A 'setting-profile' that can built onto from the existent beach and that contains the same volume of sand as projected is necessary. This kind of approach is very useful for computing work. When the A_d value becomes important the amout of sand that needs to be deposited may turnsout to be very much larger than the projected amount.

1. Introduction

The region of Lazio is situated on the central Italian coast that adjoins the Tyrrenian Sea and has about 350 km of coastline. About 230 km of the Lazio coast is characterized by sandy beaches, very crowded with tourists during the summer months. Along the Lazio coast, there are many shore protection works, some of which are very old (1910–1920) but

C. L. Goudas et al. (eds.), Soft Shore Protection, 301-310.

the problem of shore protection began to be pressing from the early sixties onwards when several direct and indirect factors contributed to make the situation serious:

- Increasing tourism which made new demands upon the beach area
- Increasing urbanization along the coast with dune destruction and many infrastructures built close to the shoreline
- Decreasing sediment transport by rivers (dams, sand quarries, soil protection, etc.) causing regression of the shoreline

Fig. 1. The Ostia and Anzio coastal region

According to the last studies that the departmental administration has carried out, over 65% of the beaches are in regression (about 150 km) and more than 30 km of these are situated on highly infrastructured coastline. The national government, and since 1980, the departmental administration of Lazio, have been engaged in shore protection with works costing about €100 million in the last twenty years, using breakwaters, groins, submerged dams, seawalls and some nourishment with gravel-sand mixtures. The results have not been very positive and many hard protection schemes have generated new erosion phenomena along the coast.

The new point of view is that the beache crisis is not a series of local problems but a structural effect due to developmental patterns (tourist economy, hinterland soil protection, renewable energy and use of river water by dams, etc.). With this in mind, the administration started a new shore protection program in 1997 based on these points:

- Monitoring of the entire shoreline to estimate the extent of a total program involving structural and maintenance works mainly by unprotected nourishment of beaches

- Check the actual effectiveness of the hard and hard-soft shore protection measures
- Cost-benefit analysis of hard protection, hard-soft protection and soft protection schemes to find the break-even cost of sand in soft protection only schemes;
- Exploitation of marine sand deposits along the Lazio continental shelf
- Review of the actual dredging technology
- Experimental works for unprotected beach nourishment;
- Assessment of the environmental and morphological aspects of the experimental works

2. Monitoring of the entire shoreline

Analysis of several shoreline aerial photos (1990, 1994 and 1996) has made it possible to evaluate the extent of the erosion in global terms. The analysis shows a balance between deposition and erosion, with an overall positive balance of about 300 000 m^3/year along the coast of Lazio but many sectors where erosion is active in varying degrees. Particularly obvious was the erosive crisis on the Tiber delta, where the overall local erosion reaches about 400,000 m^3/year. The erosion intensity has reached 12–20 000 m^3/y/km (Ladispoli, Fiumicino, Ostia) and others in other regions where the erosion intensity is more moderate (3–5,000 m3/y/km at, for example, Tarquinia, Nettuno, and Sabaudia).

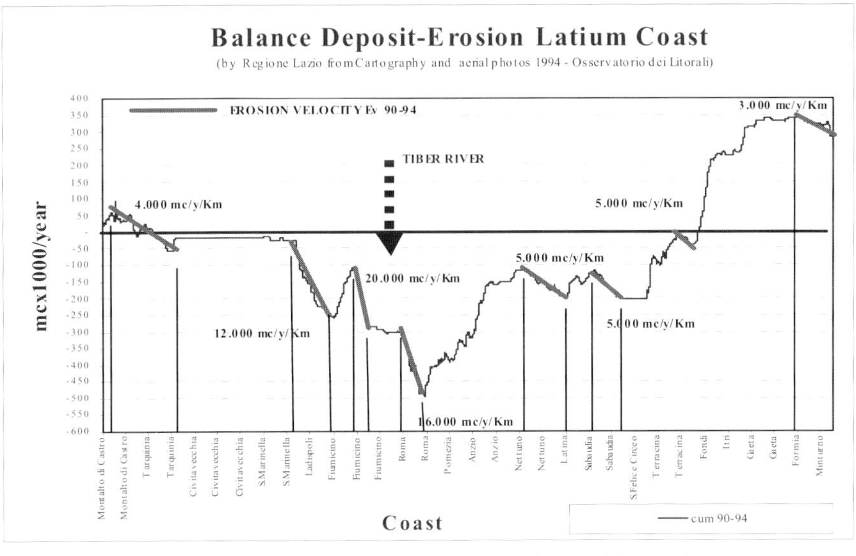

Fig. 2. Balance sedimentation (Deposit-Erosion) of the Latium coast

This kind of analysis has enabled us to consider the erosive problem in detail and also to plan programs for estimating the maintenance costs. The results show a global and average yearly need of about 1 million cubic meters of sand for the maintenance of the entire coast of Lazio which is eroding. Taking into consideration that many millions cubic meters of sand are necessary for the structural needs (high priority beach remodeling), the monitoring results raise the issue of the best way of supplying sand and what defensive systems actually reduce erosion intensity.

3. Cost-benefit analysis

Before starting the Ostia East remodeling works (3,500 m and about 1 million cubic meters of sand), an economic comparison was made between an intervention with hard protection (submerged breakwaters) and an intervention with nourishment only. The comparison investigated by the results of earlier work in a neighbouring littoral (Middle Ostia) in 1990. Although the monitoring does not show any appreciable reduction in erosion intensity, a comparison was made using the following parameters:

- Hard protection (submerged breakwater) can reduce the erosion intensity to 60%
- Hard protection has no maintenance cost
- Unprotected nourishment does not reduce the intensity of erosion.

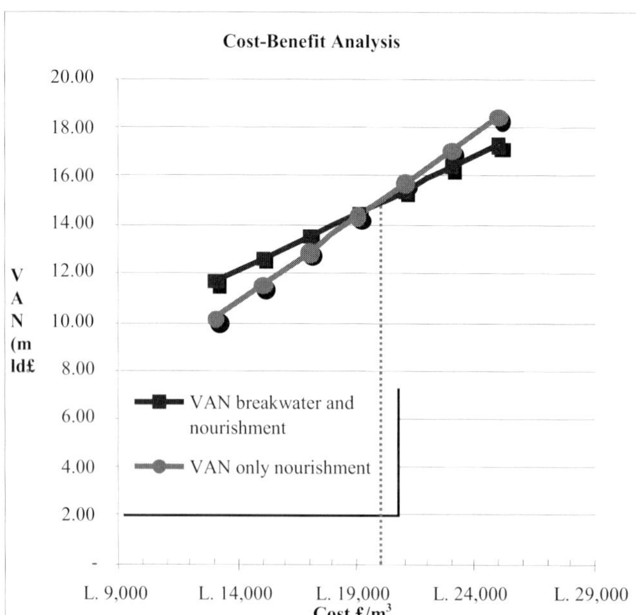

Fig. 3. Cost-benefit comparison between the alternatives VAN breakwater and nourishment and VAN only nourishment

With these conditions, we found that if the sand price is under €10/m^3, nourishment only is more cost effective than protected nourishment because the costs of maintenance are smaller than breakwater amortization costs.

4. Marine sand supply

In this region, it is very difficult to find sand from terrestrial quarries at a price lower than €14–15/m^3. This circumstance, along with several environmental considerations (e.g., lower impact of the works, more affinity between on-site sand and nourishment sand), has pushed the departmental administration towards a new research program designed to find extraction sites at sea (sand at low cost) and an appropriate dredging, replenishment and maintenance technology.

Starting from general research (Mare del Lazio 1977 – University of Rome) into the location of some feasible marine quarries on the Lazio continental shelf are located, the administration and the University of Rome (Prof. G.B. La Monica, Prof. F. Chiocci) decided to collaborate in performing more detailed exploitation. The first results of this research have enabled a plan of the potential sand borrow areas along the coast of Lazio to be drawn up and the available quantity, estimated as being 4 billions m^3 to be evaluated. The distribution of these potential borrow areas centres on six principal sites, located at different depths (from 30 to 100 m) and in different morphological-environmental contexts. On the basis of this collaboration, it has been possible to precisely describe the Anzio borrow site for the Ostia Beach replenishment as being at 45–50 m depth, with an overlying clay layer (thickness of about 1.5–2 m) and with medium-size sand D_{50} = 300–400 μ.

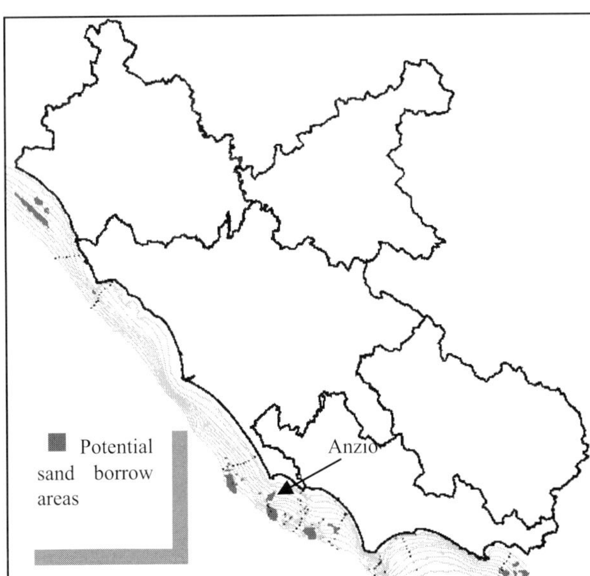

Fig. 4. The potential sand borrow sites on the Latio coast

5. Nourishment works at Ostia 1999

5.1. GENERAL OUTCOMES

The Ostia Beach works, performed in 1999, involved a littoral of about 3 500 m and about 1 million cubic metres of sand. The projected increased beach area was about 155 300 m^2, with 44 m as the average beach progression, considering a sand need/ beach progression ratio (SB) of about 6.2 m^3/m. After one year, the general outcome is:

1. The real increase beach area is about 134 500 m^2 (86% projected beach area) with 38 m as average beach progression.
2. The erosion intensity is unaltered (about 20 000/m^3/y/km).
3. The erosion is not uniform along the littoral.

5.2. NOURISHMENT STABILITY AND SAND SIZE

Sand size is usually considered a fundamental factor for nourishment stability. In the Ostia 1999 experience, more circumscribed limits were used. In fact, because of technical problems at the borrow site, the Ostia nourishment was done with different sand: for advancing the beach between 600 to 1,800 m, the approximate average size was D_{50}= 400 μ, while in the other zones of advance, the average sand size was D_{50}= 270 μ.

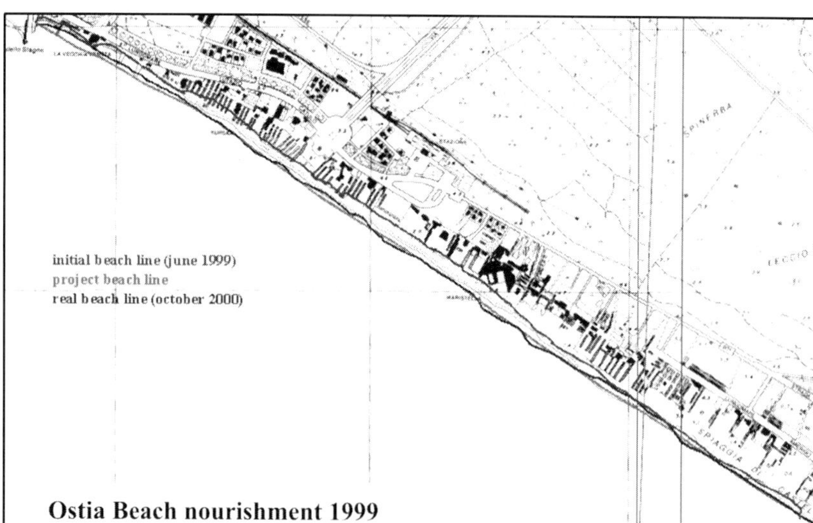

Fig. 5. Ostia Beach showing the shoreline postions in the years 1999 and 2000 and its projected position after nourishment

No relationship could be found between the distribution of sand size along the littoral and the cumulative sand loss distribution, and there was an even greater sand loss where more coarse sand had been placed. Thus, within the sand size range useful for nourishment for tourist purposes (approximately D_{50} = 100 – 1,000), it would seem pointless to try to obtain significantly more stability by using coarser sand. This can be explained by two effects verifiable on the ground:

1. Mapping of the size sand on the submerged beach shows that the bar formation is attributable to the finer sand and thus these natural beach self-defense structures will not occur or will occur with greater difficulty when nourishment takes the form of coarse sand.
2. The longshore transport mechanisms associated with fine and coarse sand are quite different; the fine sand moves toward the longshore component traveling along all the active profile; the coarser sand and gravel move towards the longshore component traveling along a narrow strip, close to the beach scarp where the breaking waves energy is very high.

The problem of the choice of two or more borrow sites depending on the sand size is consequently a matter of comparison of sure parameters such as the higher pumping cost of the coarser sand, and less certain parameters such as the higher supposed stability of the coarser sand.

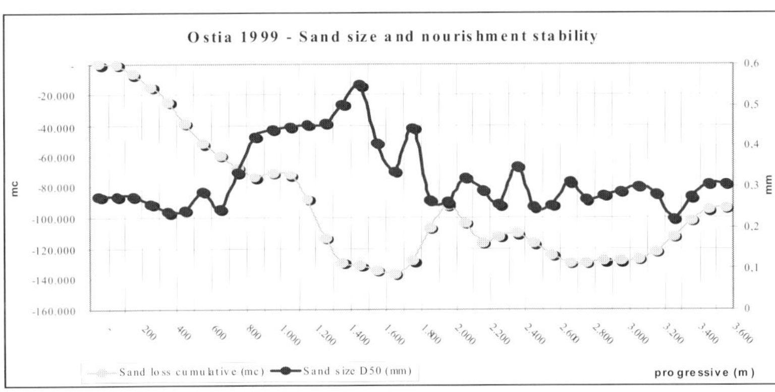

Fig. 6. Ostia 1999 – sand size and nourishment stability

5.3. EQUILIBRIUM PROFILES

In the Ostia nourishment, we found an insufficient projected SB ratio (unit quantity of sand for unit advancement of beach) and the high transverse sand demand of the existing profile was found to be related to unexpected beach recession. The beach profile monitoring clearly showed that the high demand for sand comes from the natural necessity of structuring the bar system. The usual method of computing the quantity of sand necessary to produce a pre-determined beach-line advancement is based on such equations as Dean's or simply on existing profile shift. Neither system takes into account the need for sand of bar construction. Sometimes, the existing

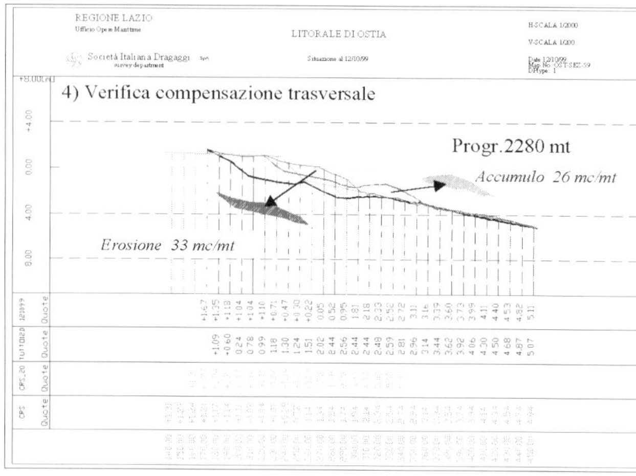

Fig. 7. Verification of the trasverse compensation

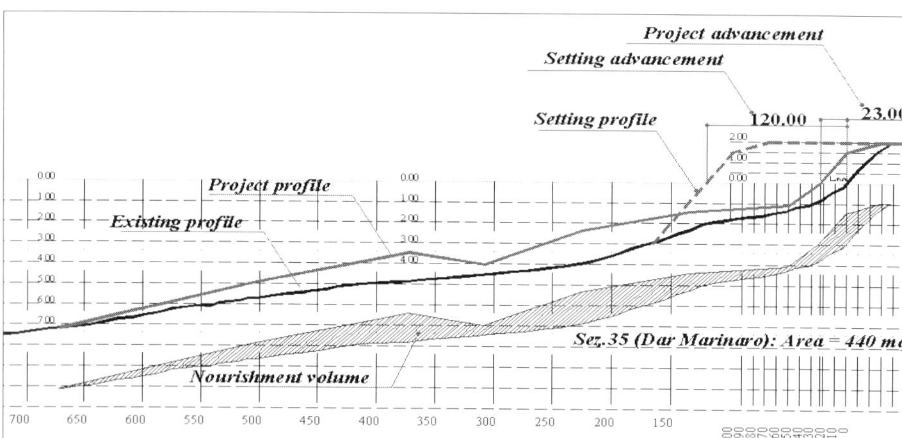

Fig. 8. Diagram of the Ostia Beach profile showing current and projected position of seabed

submerged profile is without sand bars or they are very poor (compared with the nearest beach equilibrium profiles), this sand volume also has to be restored if we want a relatively stable beachline. For this kind of design project, based on the simulation of an existing equilibrium profile, the SB ratio must be much greater than the 6–8 m^3/m of nourishment that the conventional method suggests for these shorelines. In the Anzio Beach project in the most critical zones, we have reached ratio values as high as

SB=20–25 m^3/m for each m coastline advancement (see Figure 9: Area/Project advancement = 440/23 = 19 m^3/m).

Fig. 9. Ostia Beach nourishment project in progress

The project profile must be completed with the engineering needs for the replenishment in mind. This cannot be done according to the projected profile for the submerged portion; it is thus very important to design a 'setting-profile' that can built on to the existing beach profile and contains the same projected volume; this kind of approach is very useful for computing works. The possibility that the actual volume deposited after the SB ratio has been taken into account may be much greater than the projected volume need to be taken into account.

6. Conclusions

After twenty years of experience in shore protection, the departmental administration of Lazio territory is adopting a policy of soft protection systems and in particular a policy of beach remodeling with nourishment methods. The first need is a very large supply of sand that is being addressed by an important research program by the University of Rome aimed at identifying marine borrow sites. The program has already allowed the nourishment of Ostia Beach, where 1 million m^3 of sand was placed on 3.5 km of littoral. This experience has been very positive both in terms of the overall results and in terms of the acquired new planning insights. The problem of optimal sand size, for example, has been reduced to a comparison between sure parameters such as the higher pumping cost of the coarser sand and less certain parameters such as the higher assumed stability of the coarser sand.

To ensure the relative stability of the new beach line, the profile design needs an accurate analysis of the existing profiles and of the quantity of sand required by the bar system. The general purpose is to improve the planning system by objective assessments based on monitoring of the natural sites and the works carried out.

ACKNOWLEDGEMENT

The material presented in this contribution is kept in the archives of the Technical Services of Regione Latio, Roma, Italy and is available on request, subject to approval.

LONGSHORE CURRENT MODIFICATION NEAR THE BOUNDARY BY SEABED GROIN ARRANGEMENTS: A NUMERICAL APPROACH

C. L. GOUDAS,[1] G. A. KATSIARIS,[3] N. KAFOUSSIAS,[1] C. MASSALAS,[2] G. PNEVMATIKOS,[3] M. XENOS[1] AND E. TZIRTZILAKIS[1]

[1] *Department of Mathematics, University of Patras, Patras, Greece.*
[2] *Department of Mathematics, University of Ioannina, Ioannina, Greece.*
[3] *Department of Physics, University of Patras, Patras, Greece.*

Abstract

In this work, a numerical approach is presented for determining the longshore current modification caused by the presence upon the seabed (boundary) of a submerged groin system. This system consists of several low groins with one end-point inshore and the other resting several tens of meters offshore and aligned perpendicularly to the shoreline. Such a system of groins has a direct effect upon the sediment transport properties of the bedload and constitutes a soft shore protection technique against coastal erosion.

Due to the longshore current movement parallel to the shoreline, the mathematical solution of the problem of identifying the flow field in the area between and near this groin arrangement, even without reference to change in the sediment transport properties, is still a complicated and difficult task. Any information regarding the behaviour of this flow field could be very informative and useful for understanding and controlling this problem. Thus, in the mathematical formulation of the problem described in Section 2 of this paper, the vorticity-stream functions are adopted for the steady two-dimensional incompressible flow. To obtain the numerical solution of the problem under consideration, the equations of flow are discretized in a square flow domain by applying a finite difference method using central differences. A combination of Liebmann's method with an under-relaxation and iterative technique is used to obtain the numerical solution, a procedure described in Section 3. Finally, in Section 4 the numerical results relating to the flow field, are obtained for different values of the problem's parameters, that is, the total length L of each groin, the distance M between successive groins and the longshore current velocity $\bar{u}_o$ or the corresponding Reynolds number Re.

The height above the seabed of the groins is assumed several times (about 20) in respect of the boundary layer thickness of flow at the seabed. The obtained results show that the flow field is greatly affected by the presence within the areas between the groins of vortices of alternating polarity. In some cases, depending on the ratio L/M, the results show that the existence of these vortices helps in protecting the bedload of the shore from escaping, a fact that can be interpreted as protection from coastal erosion. It is hence concluded that the technique of seabed groin systems can prove useful in counteracting this environmental problem.

1. Introduction

A system of groins, or any other arrangement that aims to defend a shore from erosion can be granted the status of 'softness' if it is not capable by structure and destination to

C. L. Goudas et al. (eds.), Soft Shore Protection, 311-336.

completely absorb and/or deflect the wave and current momentum directed upon it. Although a better definition of soft arrangements is both desired and awaited, a definition perhaps based upon an allowable maximum for absorption/deflection of the total momentum, we can assumie that a system of groins as described above where each member groin is of mean height $\hbar_g$ less than half of the mean water depth $\hbar_w$, with the means calculated for the entire length L of each groin, qualifies to be a soft protection system. For a sea site of characteristic wave height H_s the mean wave momentum $\bar{\mathbf{I}}_h$ below the upper surface of such groins, and hence to be absorbed/deflected by it, is given, on the basis of linear wave theory, by the expression

$$\bar{\mathbf{I}}_{\hbar_g} = \rho \overline{\int_{-h_w}^{-\hbar_g} u(y)dy}, \tag{1}$$

where y is the water depth measured from the water still line (positive taken upwards), u(y) the water particle velocity at depth y, ρ the water density, and the overbar indicates time average over the wave period. For $\hbar_g = \hbar_w$ we have

$$I_{h_w} = \rho \overline{\int_{-h_w}^{0} u(y)dy} = \frac{1}{8}\rho H_s^2 g/c, \tag{2}$$

with g the gravity constant and c the wave celerity. At the depth $y = -\hbar_g$ the corresponding significant wave height $H_{s_{\hbar_g}}$ is given by the expression

$$H_{s_{\hbar_g}} = -H_s c/\sinh(k\hbar_g), \text{ with } k = 2\pi/cT, \tag{3}$$

and hence the momentum faced by the submerged groin reduces to linear approximation, with the square of its height above seabed.

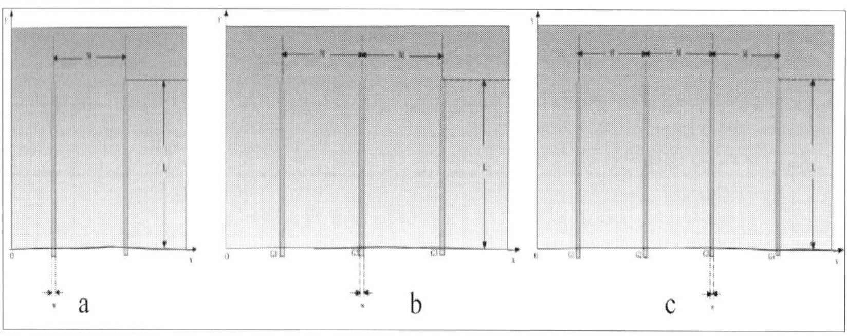

Fig. 1. Configurations of groin systems of two, three and four member groins

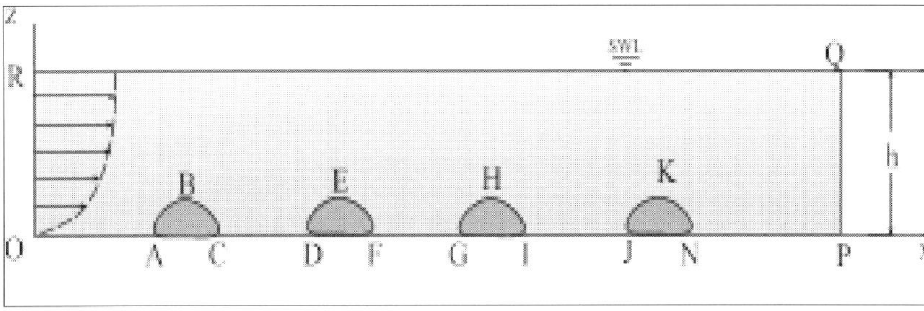

Fig. 2. Configuration of a four-member groin system adopted for the study of vertical water motion. The longshore current velocity profile is also shown

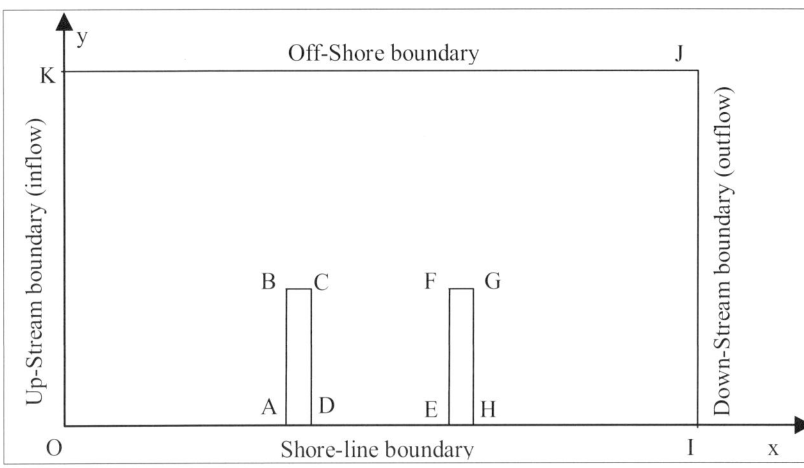

Fig. 3. Flow field area for a two-groin system

A system of submerged groins as described above (see also Figure 1) with groin heights as low as possible, is not capable of acting as a boundary limiting the motion of suspended material. Hence the mathematical model to be used for simulating the groin system impact upon the motion of the sea water will refer to a water layer close to the seabed and below the upper surface of the submerged groins, that is, within the water layer in which bedload transport is actuated by momentum transfer from the moving water to the seabed. The motion of the water near the seabed boundary as modified by the presence of the groin system, irrespective of the water motion of the overlaying layers, can be simulated as a current motion parallel to the seabed with either constant or oscillatory velocity. Analytical and numerical approximation of the width of the boundary layer shaped by both types of motion have shown it to be less than a few centimetres (Goudas & Katsiaris, 1998). The maximum vertical displacement of the parabolic motion performed by sand grains shot from the seabed due to water energy transferred to it, can be several times this width.

Karambas et al (1999) presented a theoretical investigation into coastal processes in the presence of submerged groins used for coastal accretion and erosion control. They applied the WAVE-L model (Karambas, 1999) for wave propagation, wave-induced currents, sediment transport and coastline evolution. They found that the presence of the submerged groins caused the wave heights and angles to be considerably redistributed. Circulation cells bifurcate from the mainstream of the longshore current and are capable of transporting sediment inshore and trapping a proportion of it inshore, contributing to shore accretion. In addition, the effects of undertow, which is an erosion mechanism by transporting the suspended sediment offshore, are reduced while accretion mechanisms are enhanced.

A free nearshore seabed, that is, one without the model groin system may permit the 3D motion of water particles within the small width of the boundary layer to be disregarded. In such a case, study of 2D motion on a plane parallel to the seabed may provide a satisfactory approximation of the longshore current movement and the case studied in this work. However, the presence of the groin system makes the motion genuinely three-dimentional and, in spite of the simplifications already made, a complicated flow problem. Since the model is studied for approximating the water flow below the upper groin surface, it could be assumed that the flow close to the seabed and the groin members could be well approximated by two 2D motions, namely, one on a plane parallel to the seabed and at a distance less than the groin height, and one on a vertical plane parallel to the shoreline. Figures 1 and 2 show the two planes upon which the flow could be studied.

The 2D hydrodynamic problem to be treated on the plane parallel to the seabed, with the appropriate boundary conditions of flow, will be treated according to the method to be outlined below. What seems, however, very essential is the practical interpretation of the results and specifically their use in validation experiments, both *in vitro* and *in vivo*. As will be shown in this work, the pattern of motion on the plane parallel to the seabed, and specifically between successive submerged groins, can be described as a number of closed vortices, ranging from one to several between successive groins. It can also be proven that the flow on the vertical plane does not interfere substantially enough with these vortices so as to destroy their operation. Hence, the numerical treatment shows that the pattern of flow corresponding to the boundary conditions set by a standard longshore steady current is modified, through the development of locally operating vortices, to a flow that is not capable of transporting bedload from the interior areas of the submerged groin system. The same numerical results calculated for several values of the three basic parameters of the groin system, namely, the total length L of each groin, the distance M between successive groins, and the constant longshore current velocity or the Reynolds number Re, and hence a large variety of flow patterns, are obtained.

2. Physical Problem and Mathematical Formulation

We consider the two-dimensional steady laminar flow due to a longshore current with a constant velocity parallel to the shoreline, of a viscous and incompressible fluid (seawater) in the region confined between the shore-line boundary OI, the off-shore boundary JK and the line segments OK and IJ, respectively, as is shown in Figure 3. In this square region of size L^* a system of two can be placed. Three or more seabed groins and a horizontal projection or a vertical cross section of such a system can be seen in Figures 1 and 2, respectively. The dimensions of the seabed groins are 1 m in height above the seafloor, 1 m wide in the shoreline direction and L metres in total length, respectively ($L<<L^*$) and they are placed M metres apart.

Under these assumptions, and for a system of two groins for example, the flow field is confined in the area defined by the line C (polygonal line OABC…JKO, Figure 3) and in the absence of body forces it is described by the continuity and momentum equations which for a Cartesian coordinate system can be written as

$$\frac{\partial \bar{u}}{\partial \bar{x}}+\frac{\partial \bar{v}}{\partial \bar{y}}=0\,,\ \text{(continuity)}, \tag{4}$$

$$\bar{u}\frac{\partial \bar{u}}{\partial \bar{x}}+\bar{v}\frac{\partial \bar{u}}{\partial \bar{y}}=-\frac{1}{\bar{\rho}}\frac{\partial \bar{p}}{\partial \bar{x}}+\bar{\nu}\left(\frac{\partial^2 \bar{u}}{\partial \bar{x}^2}+\frac{\partial^2 \bar{u}}{\partial \bar{y}^2}\right),\ \text{(x-momentum)}, \tag{5}$$

$$\bar{u}\frac{\partial \bar{v}}{\partial \bar{x}}+\bar{v}\frac{\partial \bar{v}}{\partial \bar{y}}=-\frac{1}{\bar{\rho}}\frac{\partial \bar{p}}{\partial \bar{y}}+\bar{\nu}\left(\frac{\partial^2 \bar{v}}{\partial \bar{x}^2}+\frac{\partial^2 \bar{v}}{\partial \bar{y}^2}\right),\ \text{(y-momentum)}. \tag{6}$$

In the above equations $\bar{u}$, $\bar{v}$ are the velocity components of the fluid in $\bar{x}$ and $\bar{y}$ direction, respectively, $\bar{\rho}$ is the density, $\bar{p}$ is the pressure and $\bar{\nu}=\bar{\mu}/\bar{\rho}$ is the kinematic viscosity of the fluid.

Boundary conditions to suit the system (1)-(3) depend on the problem being considered. If a solid surface forms the boundary of the field, it is necessary to set all the velocity components equal to the velocity components of the solid surface. For flow around an immersed body, far afield boundary conditions are also required.

It must be pointed out that specifying the boundary conditions is a difficult task: the governing equations and the boundary conditions have to constitute a well-posed problem (Oliger & Sundstrom, 1978, Gustafsson & Sundstrom, 1978).

For the problem under consideration, the velocity components are specified on all boundaries of the flow domain, taking into account that the following global constraint is satisfied.

$$\int_C \vec{q}\cdot\vec{n}\,ds=0\,, \tag{7}$$

where $\vec{q}=\bar{u}\,\vec{i}+\bar{v}\,\vec{j}$ and $\vec{n}$ is the unit vector normal to the boundary C of the flow. Equation (7) is the global equivalent of the continuity equation (4).

Under the above assumptions, the boundary conditions of the problem can be written as:

$$\left.\begin{array}{ll}\text{Inflow conditions on OK } (\bar{x}=0,\ 0\le\bar{y}\le L^{*}): & \bar{u}=\bar{u}_{o},\ \bar{v}=0, \\ \text{Outflow conditions on IJ } (\bar{x}=L^{*},\ 0\le\bar{y}\le L^{*}): & \bar{u}=\bar{u}_{o},\ \bar{v}=0, \\ \text{Off-shore condition on JK } (\bar{y}=L^{*},\ 0\le\bar{x}\le L^{*}): & \bar{u}=\bar{u}_{o},\ \bar{v}=0, \\ \text{On the line OABC...HI } (0\le\bar{y}\le L,\ 0\le\bar{x}\le L^{*}): & \bar{u}=0,\ \bar{v}=0. \end{array}\right\} \tag{8}$$

For the solution of the problem, the vorticity-stream function formulation is adopted for two-dimensional steady incompressible laminar flow. By introducing the vorticity function $\bar{J}=\bar{J}(\bar{x},\bar{y})$ and the stream function $\bar{\Psi}=\bar{\Psi}(\bar{x},\bar{y})$ by the expressions

$$\bar{J}(\bar{x},\bar{y})=\frac{\partial\bar{v}}{\partial\bar{x}}-\frac{\partial\bar{u}}{\partial\bar{y}}, \tag{9}$$

$$\bar{u}=\frac{\partial\bar{\Psi}}{\partial\bar{y}},\quad \bar{v}=-\frac{\partial\bar{\Psi}}{\partial\bar{x}}, \tag{10}$$

the continuity equation is automatically satisfied whereas the momentum equations (5) and (6) produce, by eliminating pressure $\bar{p}$, the vorticity transport equation

$$\frac{\partial\bar{\Psi}}{\partial\bar{y}}\frac{\partial\bar{J}}{\partial\bar{x}}-\frac{\partial\bar{\Psi}}{\partial\bar{x}}\frac{\partial\bar{J}}{\partial\bar{y}}=\bar{\nu}\left(\frac{\partial^{2}\bar{J}}{\partial\bar{x}^{2}}+\frac{\partial^{2}\bar{J}}{\partial\bar{y}^{2}}\right). \tag{11}$$

Also, substitution of equations (10) into (9) produces the following Poisson equation for the stream function

$$\frac{\partial^{2}\bar{\Psi}}{\partial\bar{x}^{2}}+\frac{\partial^{2}\bar{\Psi}}{\partial\bar{y}^{2}}=-\bar{J}(\bar{x},\bar{y}). \tag{12}$$

In the stream function-vorticity formulation in the steady incompressible flow, described by the system of equations (11), (12), which is of an elliptic type in space, two boundary conditions are required if velocity components are given on the boundary and these boundary conditions specify derivatives of $\bar{\Psi}$ through (10) such as

$$\bar{\Psi}\Big|_{C}=a \quad \text{and/or} \quad \frac{\partial\bar{\Psi}}{\partial\eta}\Big|_{C}=b, \tag{13}$$

where η is the direction normal to the boundary C. It may be noted that no boundary conditions are specified for $\bar{J}$. This is appropriate for solid surfaces since they are sources of vorticity, which is diffused and convected into the flow field (Lighthill,

1963). For flow around an immersed body, far afield boundary conditions such as $\bar{J}=0$ may be substituted for $\left.\frac{\partial\bar{\Psi}}{\partial\eta}\right|_C = b$ if the flow is locally uniform.

However, in obtaining numerical solutions, it is often desirable to have available equivalent boundary conditions for $\bar{J}$, particularly for solid surfaces. In the past, these boundary conditions have often been constructed from the discrete form of (12) to ensure that $\left.\frac{\partial\bar{\Psi}}{\partial\eta}\right|_C = b$ is satisfied. Quartapelle and Valz-Gris (1981) demonstrated that there is no strictly equivalent local boundary condition available for $\bar{J}$. The vorticity is to be computed from the velocity field, but it cannot be specified at the boundaries before the problem is solved.

Consequently, the solution of the problem under consideration is not as simple as it first appears to be, and special techniques are needed to utilize the numerical methods, devised originally for solving Poisson and Laplace equations, to solve the present problem. Such a technique is presented in the next section and at the moment the boundary conditions (8) in terms of stream function $\bar{\Psi}$ can be written as :

$$\left.\begin{array}{ll}
\text{Inflow conditions on OK } (\bar{x}=0,\ 0\le\bar{y}\le L^*): & \bar{\Psi}=\bar{u}_o\bar{y},\quad \dfrac{\partial\bar{\Psi}}{\partial\bar{x}}=0,\\[2ex]
\text{Outflow conditions on IJ } (\bar{x}=L^*,\ 0\le\bar{y}\le L^*): & \bar{\Psi}=\bar{u}_o\bar{y},\quad \dfrac{\partial\bar{\Psi}}{\partial\bar{x}}=0,\\[2ex]
\text{Off-shore condition on JK } (\bar{y}=L^*,\ 0\le\bar{x}\le L^*): & \bar{\Psi}=\bar{u}_oL^*,\quad \dfrac{\partial\bar{\Psi}}{\partial\bar{y}}=\bar{u}_o,\\[2ex]
\text{On OABCDE...GHI } (0\le\bar{y}\le L^*,\ 0\le\bar{x}\le L^*): & \bar{\Psi}=0,\quad \dfrac{\partial\bar{\Psi}}{\partial\bar{y}}=0.
\end{array}\right\} \quad (14)$$

Before proceeding to the numerical solution of the problem under consideration, it is convenient to non-dimensionalize the system of equation (11)-(12) subjected to the boundary conditions (14).

Introducing the non-dimensional variables

$$x=\frac{\bar{x}}{L^*},\qquad y=\frac{\bar{y}}{L^*},\qquad u=\frac{\bar{u}}{\bar{u}_o},\qquad v=\frac{\bar{v}}{\bar{v}_o}, \qquad (15)$$

$$\Psi=\frac{\bar{\Psi}}{L^*\bar{u}_o},\quad J(x,y)=\frac{\bar{J}}{(\bar{u}_o/L^*)},\qquad Re=\frac{\bar{u}_oL^*}{\bar{\nu}}\ \text{(Reynolds number)},$$

the system of equations (11), (12) and the boundary conditions (14) can be written now as :

$$\nabla^2 \Psi = -J(x,y), \tag{16}$$

$$\nabla^2 J = \mathrm{Re}\left\{\frac{\partial \Psi}{\partial y}\frac{\partial J}{\partial x} - \frac{\partial \Psi}{\partial x}\frac{\partial J}{\partial y}\right\}, \tag{17}$$

$$\left.\begin{array}{l}
\text{Inflow conditions on OK } (x=0,\ 0 \le y \le 1):\ \Psi = y,\ \dfrac{\partial \Psi}{\partial x} = 0, \\
\text{Outflow conditions on IJ } (x=1,\ 0 \le y \le 1):\ \Psi = y,\ \dfrac{\partial \Psi}{\partial x} = 0, \\
\text{Off-shore condition on JK } (y=1,\ 0 \le x \le 1):\ \Psi = 1,\ \dfrac{\partial \Psi}{\partial y} = 1, \\
\text{On OABCDE...GHI } (0 \le y \le {}^{l}\!/_{L^*},\ 0 \le x \le 1):\ \overline{\Psi} = 0,\ \dfrac{\partial \Psi}{\partial y} = 0,
\end{array}\right\} \tag{18}$$

where $\nabla^2 = \dfrac{\partial^2}{\partial x^2} + \dfrac{\partial^2}{\partial y^2}$ is the Laplacian operator in two dimensions.

3. Numerical solution

For the numerical solution of the problem under consideration, the square space bounded by the line OAB...JKO is subdivided by vertical and horizontal lines at a constant distance $\Delta x = \Delta y = h$ apart. Grid points that are formed at the intersections of these two sets of perpendicular lines have coordinates (x_i, y_j) where i,j = 1,2,...,n. The values of stream function Ψ and vorticity J evaluated at these grid points are called $\Psi_{i,j}$ and $J_{i,j}$, respectively. Since equation (17) is non-linear, it is necessary to employ an iterative algorithm. At each step of the iteration, equation (17) and (16) is used to update the J and Ψ solutions, respectively. The finite difference method, with central differences for the first and second-order derivatives is applied, with truncation error $O(h^2)$. The system of equations (16) and (17) is discretized and an under-relaxation technique is used for determining the new values of Ψ and J at any interior point (i, j). However, to solve the vorticity transport equation (17), it is necessary to determine the boundary values of the vorticity J, and this is not an easy task. As we have already mentioned, vorticity is to be computed from the velocity field, but it cannot be specified at the boundaries before the problem is solved. To overcome this difficulty, at each iteration step numerical boundary conditions are constructed for the vorticity J as follows (Chow, 1979):

We consider an arbitrary boundary grid point (i, m) as shown in Figure 4.

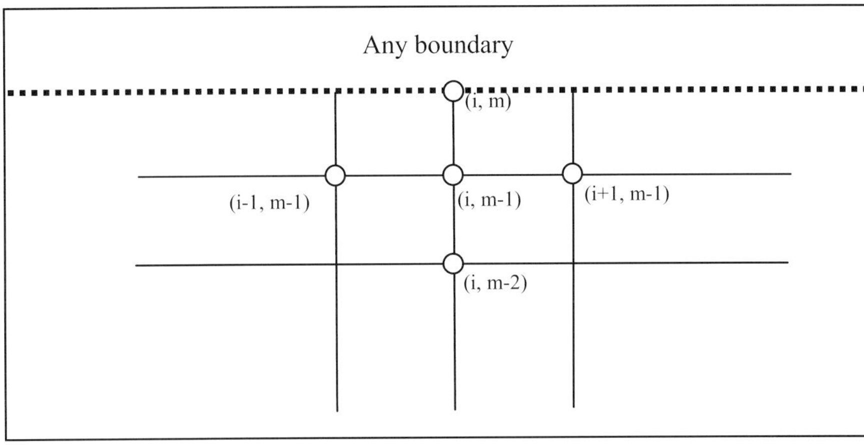

Fig. 4: Grid points for the calculation of $J_{i,m}$

Our purpose is to calculate the vorticity at this point $J_{i,m}$ based on the information of Ψ at the same point and at four neighbouring grid points shown in the above Figure. Using equation (16) for the vorticity J at (i,m), we assume the following form:

$$
\begin{aligned}
J_{i,m} &= -\left(\frac{\partial^2\Psi}{\partial x^2}+\frac{\partial^2\Psi}{\partial y^2}\right)_{i,m} \\
&= a_1\Psi_{i-1,m-1}+a_2\Psi_{i,m-1}+a_3\Psi_{i+1,m-1}+a_4\Psi_{i,m-2}+a_5\left(\frac{\partial\Psi}{\partial y}\right)_{i,m}
\end{aligned} \tag{19}
$$

Substituting from the Taylor's series expansions

$$
\Psi_{i\pm1,m-1}=\Psi_{i,m}\pm h\left(\frac{\partial\Psi}{\partial x}\right)_{i,m}+\frac{h^2}{2}\left(\frac{\partial^2\Psi}{\partial x^2}\right)_{i,m}-h\left(\frac{\partial\Psi}{\partial y}\right)_{i,m}+\frac{h^2}{2}\left(\frac{\partial^2\Psi}{\partial y^2}\right)_{i,m}+O\left(h^3\right)
$$

$$
\Psi_{i,m-n}=\Psi_{i,m}-nh\left(\frac{\partial\Psi}{\partial y}\right)_{i,m}+\frac{1}{2}n^2h^2\left(\frac{\partial^2\Psi}{\partial y^2}\right)_{i,m}+O\left(h^3\right)
$$

and retaining only terms up to $O\left(h^2\right)$, (19) becomes

$$
\begin{aligned}
-\left(\frac{\partial^2\Psi}{\partial x^2}+\frac{\partial^2\Psi}{\partial y^2}\right)_{i,m} &= (a_1+a_2+a_3+a_4)\Psi_{i,m}+(a_3-a_1)h\left(\frac{\partial\Psi}{\partial x}\right)_{i,m} \\
&+\left(\frac{a_5}{h}-a_1-a_2-a_3-2a_4\right)h\left(\frac{\partial\Psi}{\partial y}\right)_{i,m}+(a_1+a_3)\frac{h^2}{2}\left(\frac{\partial^2\Psi}{\partial x^2}\right)_{i,m} \\
&+(a_1+a_2+a_3+4a_4)\frac{h^2}{2}\left(\frac{\partial^2\Psi}{\partial y^2}\right)_{i,m}
\end{aligned}
$$

The constants a's are determined by equating the coefficients of like terms on the two sides of this equation. Substitution of these values into (19) gives

$$
J_{i,m}=\frac{1}{h^2}\left(-\Psi_{i-1,m-1}+\frac{8}{3}\Psi_{i,m-1}-\Psi_{i+1,m-1}-\frac{2}{3}\Psi_{i,m-2}\right)-\frac{2}{3h}\left(\frac{\partial\Psi}{\partial y}\right)_{i,m} \tag{20}
$$

where $(\partial\Psi/\partial y)_{i,m}$ has a known value according to the boundary conditions (18).

The vorticity boundary conditions just derived enable us to solve (17) provided that the right hand side is already known from the previous iteration. However, the determination of Ψ from (16) depends on the distribution of vorticity within the bounded domain. Thus Ψ and J are coupled and an iterative scheme is employed, applying a modification of Liebmann's method with an under-relaxation technique, to obtain the numerical solution following the steps shown in the flow chart given in the next page (Chow, 1979).

4. Results and discussion

In order to determine the flow field in the area between the groins as well as in the area ahead and beyond them, numerical calculations were carried out for different values of the parameters raised for the problem under consideration. These parameters are the total length L of each groin, the distance M between successive groins and hence the ratio L/M, the sea water velocity $\bar{u}_o$, parallel to the shore line, or the Reynolds number Re and the number n of the system of seabed groins.

Figure 5 shows the two-dimensional current flow for n=1, L=20m and Re=2860. It will be observed that a large erosion vortex past the seabed is created with clockwise (CW) polarity.

Figure 6 shows the two-dimensional current flow in the presence of two seabed groins 40 m in length placed 20 m apart. The flow that we are concerned with is that below the upper surface of the groins. Between the groins a permanent counter-clockwise moving vortex is created. No longshore transport can be expected from the seabed area between the two groins. A small accretion-bound vortex at the foot of the upstream side of the first groin and a large erosion-bound vortex past the second downstream groin can be observed. The direction of vortex rotation depends upon the distance between successive groins, the free-stream velocity and the Reynolds number.

FLOW CHART OF THE ALGORITHM

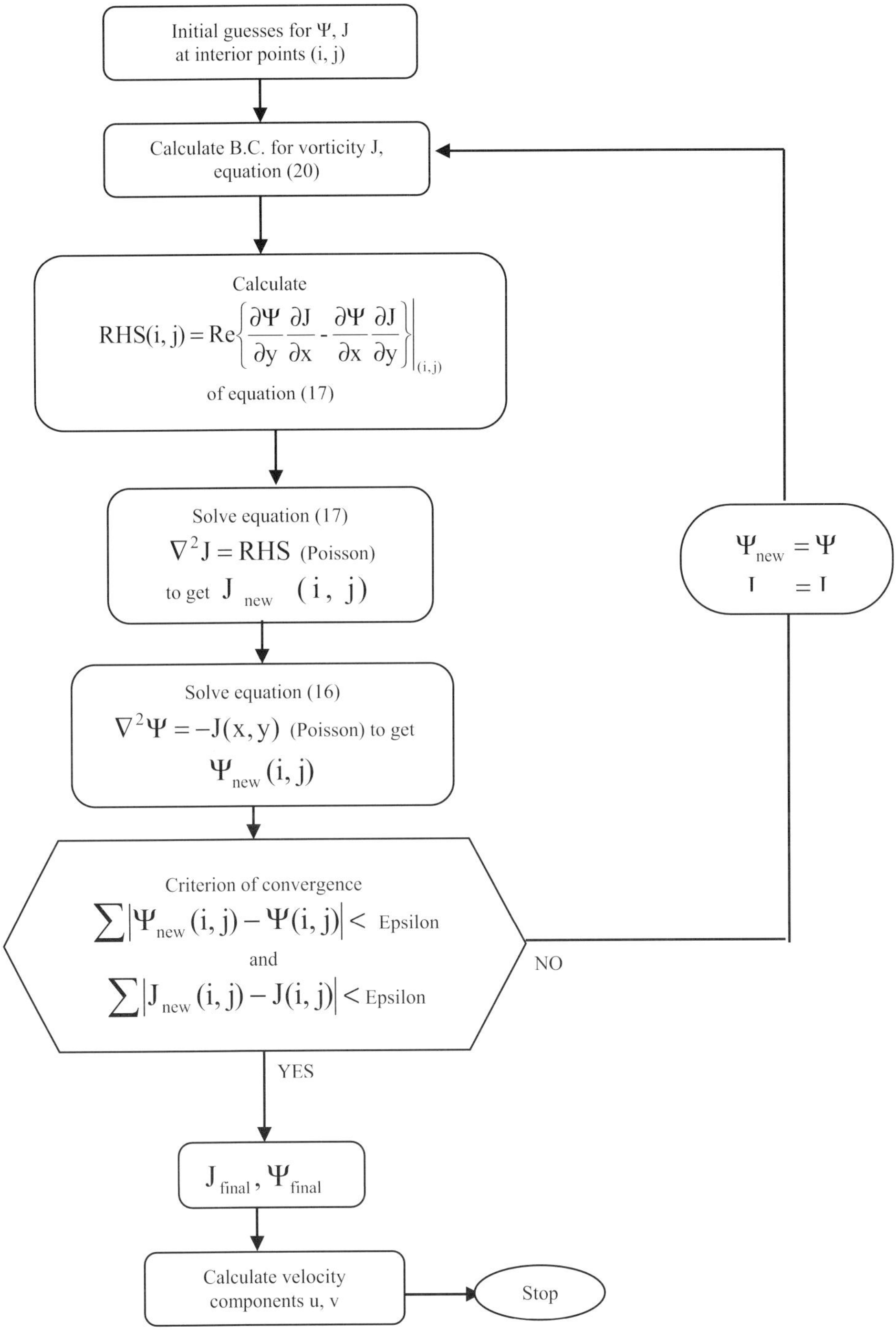

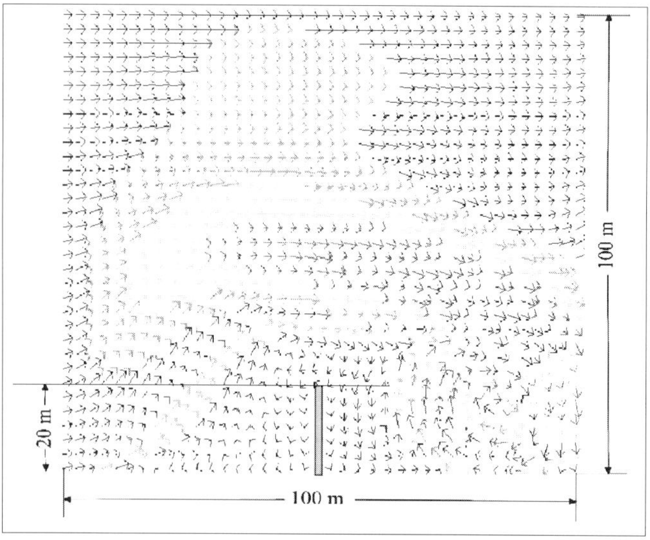

Fig. 5. The two-dimensional current flow in the presence of one seabed groin

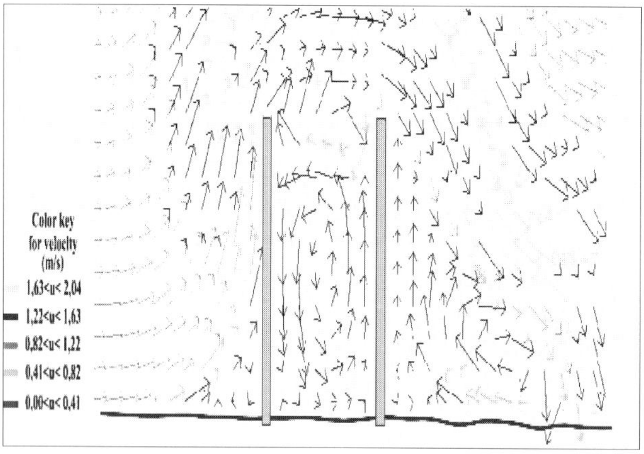

Fig. 6. The two-dimensional current flow in the presence of two seabed groins 40 m in length placed 20 m apart

Two-dimensional current flow, focused between three successive seabed groins 40m in length and placed 20m apart, is shown in Figure 7. The low-current velocity near the shoreline and the opposite vortex polarity are the essential characteristics to note. Figure 8 shows the two-dimensional current flow in the presence of four seabed groins 10 m in length placed 5 m apart. Between each pair of consecutive groins two vortices of comparable size and opposite polarity can be observed. It is again significant that the outer vortices rotate CW and do not prevent escape of bedload. The ratio L/M (L=10 m, M=5 m) is equal to 2.

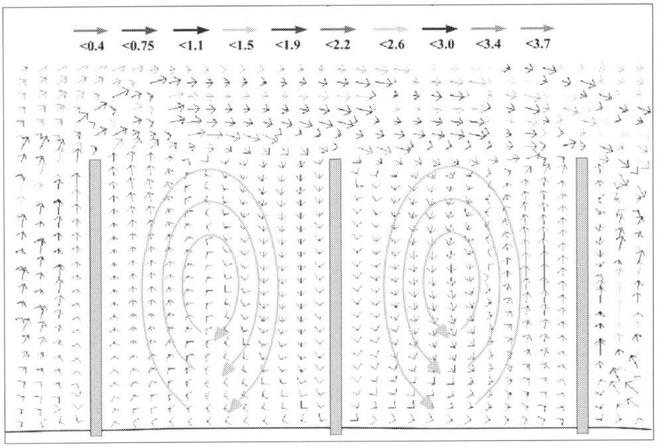

Fig. 7. The two-dimensional current flow focused between three successive seabed groins 40 m in length and placed 20 m apart. The arrows in shades of grey and the numbers below each arrow at the top of this figure and some figure that follow indicate the local current velocity in m/s. In the CD ROM that accompanies this book all figures appear in colour

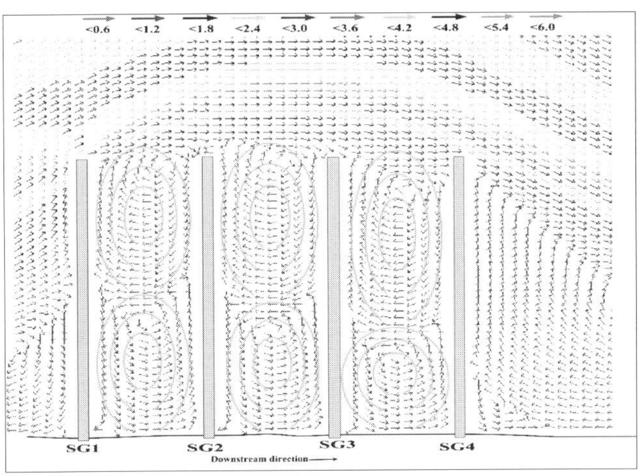

Fig. 8. The two-dimensional current flow in the presence of four seabed groins 10 m in length placed 5 m apart

Two-dimensional current flow in the presence of four seabed groins 50 m in length and placed 24 m apart is shown in Figure 9. Between each pair of consecutive groins two vortices of comparable size and opposite polarity can be observed. It is again significant that the outer vortices rotate CCW and prevent escape of bedload. The CCW vortices between SG2 and SG3 and between SG3 and SG4 are confined within the limits of the system and prevent escape of bedload. The ratio L/M (L=50m, M=24 m) is greater than 2. It could be numerically shown that for even greater values of this ratio, the system offers better protection of a shore from longshore drift.

Two-dimensional current flow in the presence of four seabed groins 50 m in length and placed 28 m apart is shown in Figure 10. Between SG1 and SG2, as well as between SG2 and SG3, two vortices of comparable size and opposite polarity can be observed, with the ones of CW polarity located farther from the shoreline and extending beyond the offshore limit of the groins. Between SG3 and SG4 are three vortices of comparable size and of interchanging polarity, but with the outer vortex of CW polarity and reaching beyond the limit of the two groins. Such a system offers better protection to the shore than the corresponding one with the same in-between distance but shorter groins.

Figure 11 shows the two-dimensional current flow in the presence four seabed groins 50 m in length placed 20 m apart. Between each pair of consecutive groins two or three basic vortices of comparable size and opposite polarity operate. It is again significant that the outer vortices rotate CCW and prevent escape of bedload. The CCW outer vortices are confined within the limits of the system and prevent escape of bedload. The ratio L/M (L=50 m, m=20 m) is equal to 2.5, that is, it is greater than 2, and the system is expected to secure preservation of the bedload of the areas between the seabed groins.

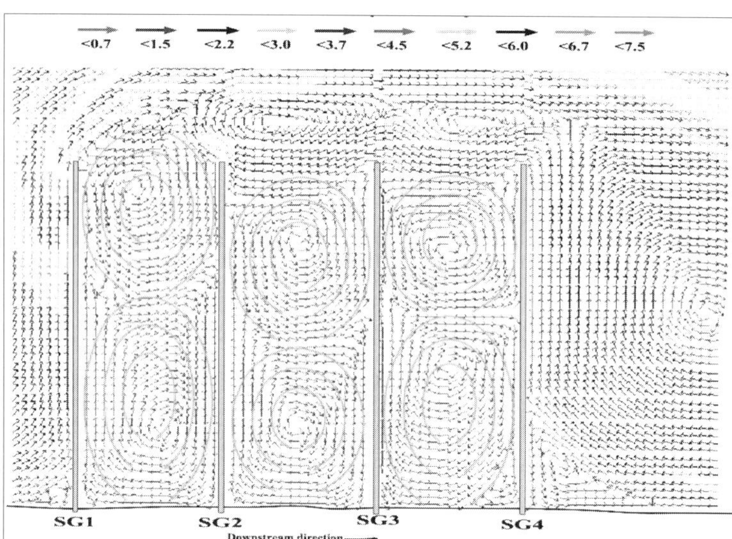

Fig. 9. The two-dimensional current flow in the presence of four seabed groins 50 m in length placed 24 m apart

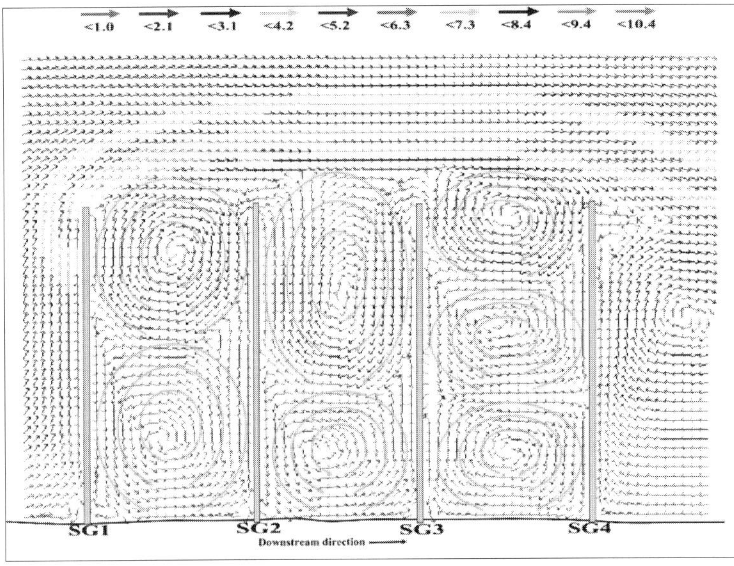

Fig. 10. Two-dimensional current flow in the presence of four seabed groins 50 m in length placed 28 m apart

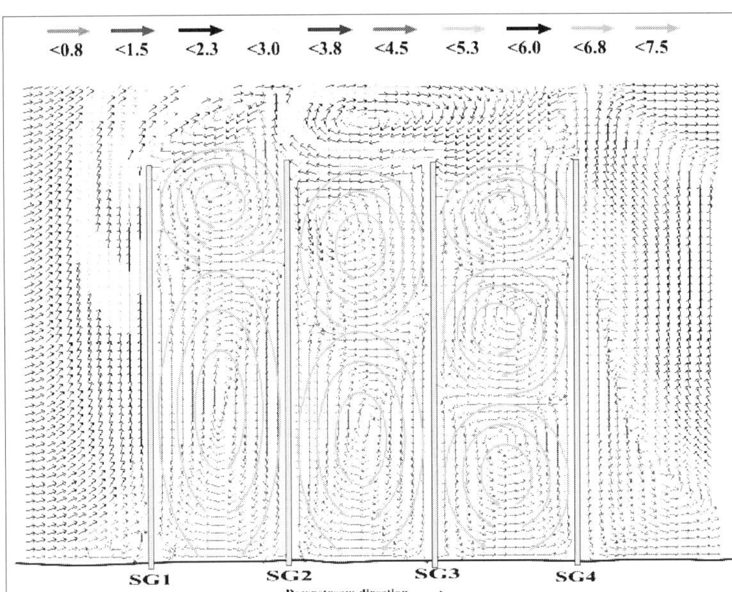

Fig. 11. Two-dimensional current flow in the presence of four seabed groins 50 m in length placed 20 m apart

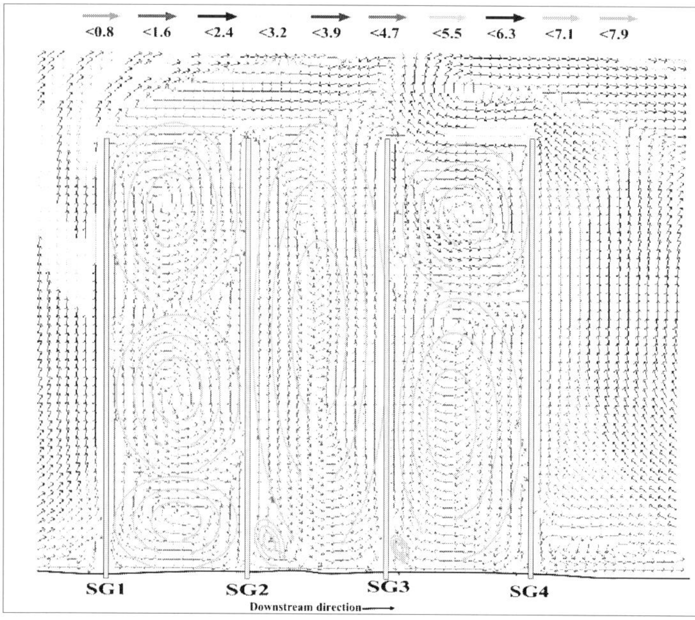

Fig. 12. Two-dimensional current flow in the presence of four seabed groins 50 m in length placed 22 m apart

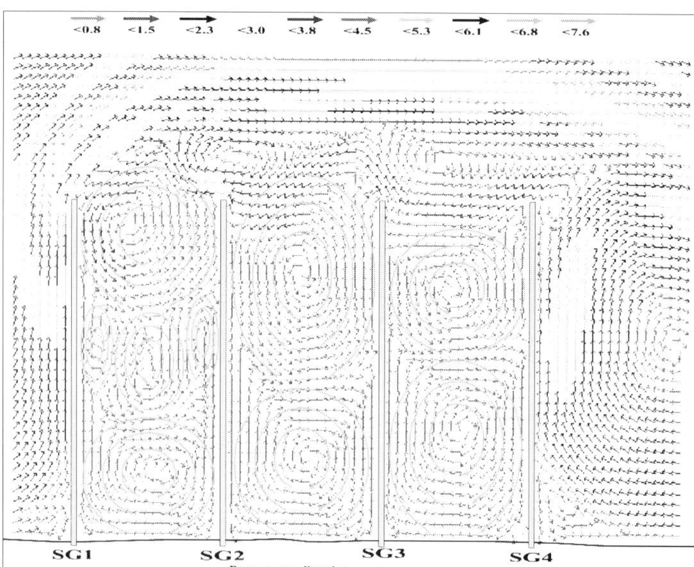

Fig. 13. Two-dimensional current flow in the presence of four seabed groins 50 m in length placed 26 m apart

Two-dimensional current flow in the presence of four seabed groins 50 m in length and placed 22 m apart is shown in Figure 12. Between pairs of consecutive groins two or three vortices of varying size and opposite polarity are in action. It is again significant that the outer vortices rotate CCW and prevent escape of bedload. The CCW outer vortices are confined within the limits of the system and prevent escape of bedload. The ratio L/M (L=50m, M=22m) is equal to 2.27, that is, it is greater than 2.

Figure 13 shows two-dimensional current flow in the presence of four seabed groins 50 m in length placed 26 m apart. Between SG1 and SG2, five vortices of various size and polarity can be observed. It is significant that the outer vortex rotates CCW and prevents escape of bedload. The rest of the vortices play a stabilizing role, and their polarity is unimportant. Between SG2 and SG3, as well as SG3 and SG4, two vortices of comparable size and of opposite polarity are in action with the outer vortices rotating in the CCW direction. Again, no loss of bedload occurs. The ratio L/M (L = 50 m, M = 26 m) is nearly equal to 2. It appears that for values of a ratio greater or equal to 2, the system safeguards a shore from longshore drift.

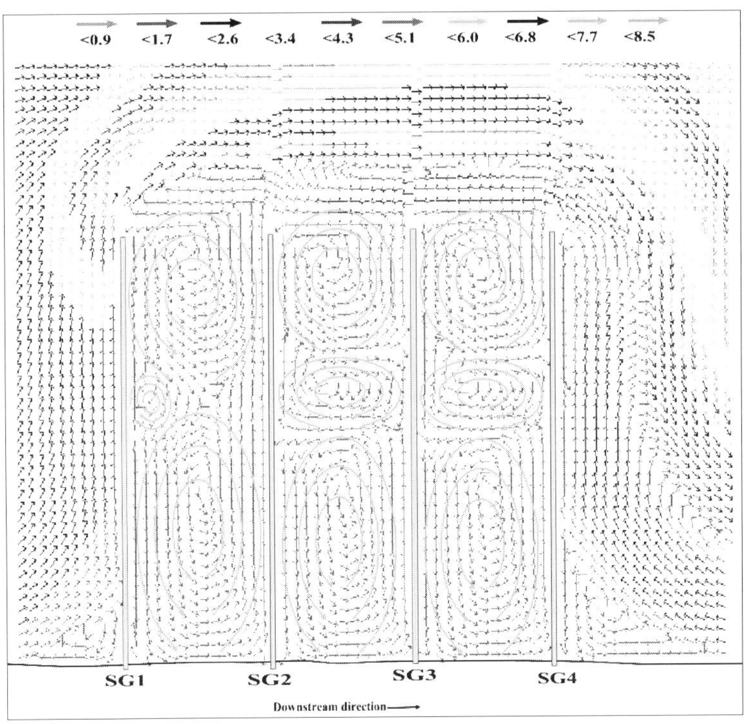

Fig. 14. Two-dimensional current flow in the presence of four seabed groins 60 m in length placed 20 m apart

Two-dimensional current flow, in the presence of four seabed groins 60 m in length placed 20 m apart is shown in Figure 14. Between each consecutive pair of groins two vortices of comparable size, while a third vortex operates between the two large ones. Between SG1 and SG2 the third vortex is small enough and displaced to the left and permits contact of the large ones. Between the other groins, the third vortex clearly separates the two larger ones. It is again significant that the outer vortices rotate CCW and prevent escape of bedload. The CCW outer vortices are confined within the limits of the system and prevent escape of bedload. The ratio L/M (L=60m, M=20m) is equal to 3, that is, it is much greater than 2. Such a system is recommended.

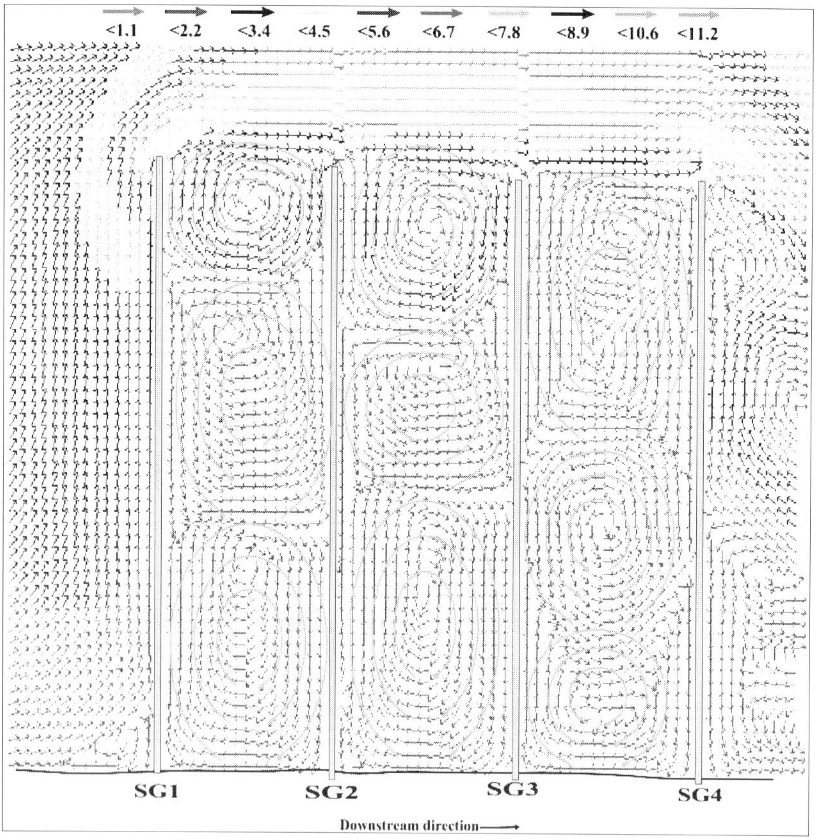

Fig. 15. Two-dimensional current flow in the presence of four seabed groins 60 m in length placed 25 m apart.

Figure 15 shows two-dimensional current flow in the presence of four seabed groins 60 m in length placed 25m apart. Between each pair of consecutive groins operate three vortices of comparable size and opposite polarity. One of the vortices that appeared when M=20m has disappeared. In all cases the vortex in the middle rotates in the CCW direction and hence defines the limits of the protected area. Between SG1 and SG2 the bedload protection covers 83% of the groin length, between SG2 and SG3 75%, and between SG3 and SG4 only 55%. The limited protection offered by the last pair of groins is now greater than in the case of M=20 m.

Two-dimensional current flow in the presence of four seabed groins 60 m in length placed 28 m apart is shown in Figure 16. Between each pair of consecutive groins operate three vortices of varying size and opposite polarity. The CCW vortices that appeared when M=25m are considerably reduced in size between SG2 and SG3 and between SG3 and SG4.

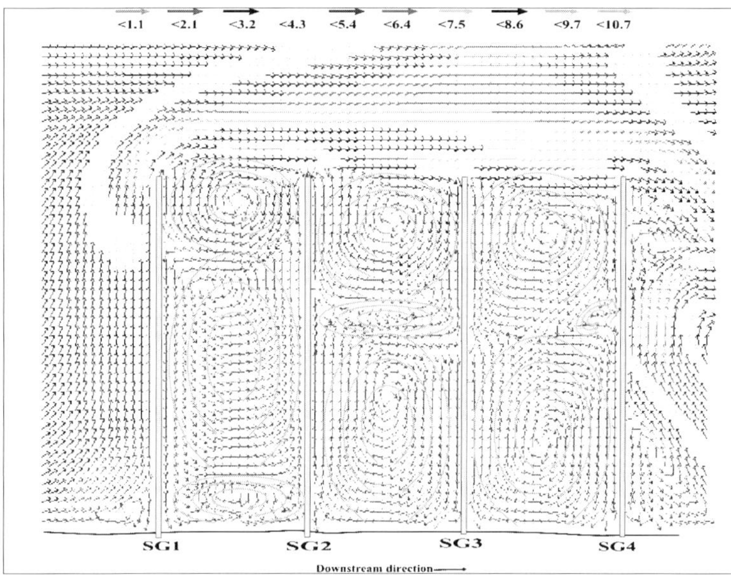

Fig. 16. Two-dimensional current flow in the presence four seabed groins 60 m in length placed 28 m apart

Figure 17 shows two-dimensional current flow in the presence of four seabed groins 40 m in length the first three placed 28 m apart and the fourth 56 m from the third. Between the first and second groin as well as between the second and the third, starting from the upstream end of the system, two vortices of unequal size and of opposite polarity have formed, the small ones of CCW polarity and appearing near the shoreline. The corresponding large vortices, of CW polarity go distinctly beyond the offshore terminal points of the groins and at their limit coalesce with the external free-stream, thus exposing the bedload driven by them to the possibility of escaping the area protected by each respective pair of seabed groins. This possibility is dramatically greater for the area between the third and fourth groin, where two vortices of very unequal size and again of opposite polarity are in operation. In fact, on the upstream side of the foot of SG4 and at the same position of SG1 the same condition prevail, while the large vortex between SG3 and SG4 is clearly facilitating the escape of sand displaced by the vortex flow at its offshore limit.

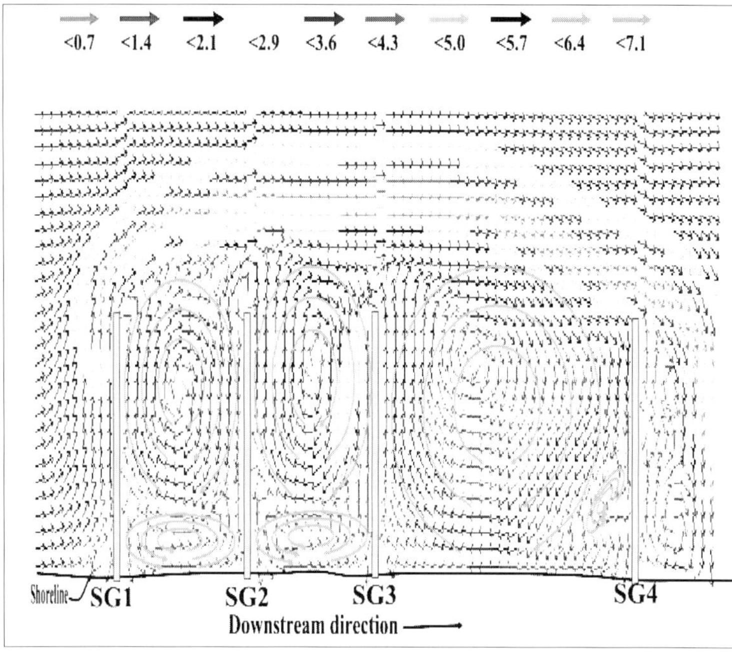

Fig. 17. Two-dimensional current flow in the presence of four seabed groins 40 m in length the first three placed 28 m apart and the fourth 56 m from the third

Two-dimensional current flow in the presence of four seabed groins 40 m in length placed 24 m apart is shown in Figure 18. Between the first two groins, starting from the upstream end of the seabed groin system, two vortices of comparable size and of CW polarity have formed. Between the third and fourth groin, two vortices of comparable size and of opposite polarity have also appeared. Finally, between the second and third groin one large vortex of CW polarity has formed. The high current velocities that develop on account of the greater distance between the seabed groins produce variations in the current regime and instability that can cause escape of seabed materials.

In Figure 19, two-dimensional current flow in the presence of four seabed groins 40 m in length placed 28 m apart, is shown. Large vortices of CW polarity dominate the field between each pair of consecutive groins and reach beyond the offshore terminal line of the system. Smaller vortices of CCW polarity, varying in number but always closer to the shoreline, are present. Such a system offers shore protection but does not secure retention of bedload, since the protruding part of the large vortices permits escape of low grain matter.

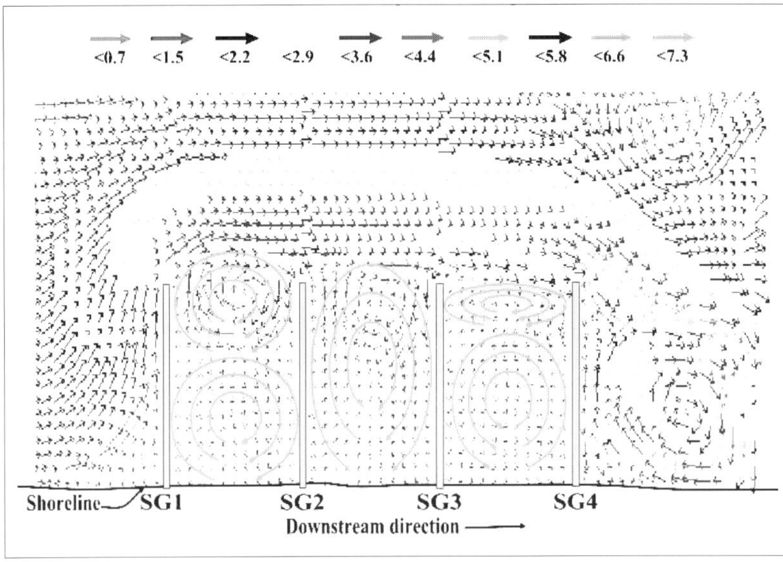

Fig. 18. Two-dimensional current flow in the presence of four seabed groins 40 m in length placed 24 m apart

Finally, Figure 20 presents the impact of one groin of an entire system *in situ*. The deposition of sand on both sides is evident and is the result of the development of vortices within the seabed area between successive groins. Sedimentation can also be observed all along each of the groins on both sides and around their offshore ends. Figure 21 shows the site of the project that includes the groin shown in Figure 20.

Further evidence that validates the numerical model of the flow pattern presented in the previous paragraph is provided by the results of experimental application of the system. Figure 22 shows the sedimentation induced at the Oasis Resort at El Hegaz on the Egyptian shores of the Red Sea.

The contours shown represent the increase in elevation above the seabed level prior to installation of the groin system. The contours are closed, just like the water flow vortices shown in many of the figures in this contribution. The level of the added sediments exceeds 50 cm at some locations and the distribution of added sediment is asymmetrical, cover the entire seabed between successive groins with the maximum amounts near the offshore ends of the groins and on their updrift sides. Since most of the sediment was nourished, it would appear that the groin system in this and all other sites, is able to retain most of the added material. An issue requiring investigation not permitted by the brief *in situ* experimentation, is what the half-life time of nourishment conservation is, as well as the distribution of sediment at half-time.

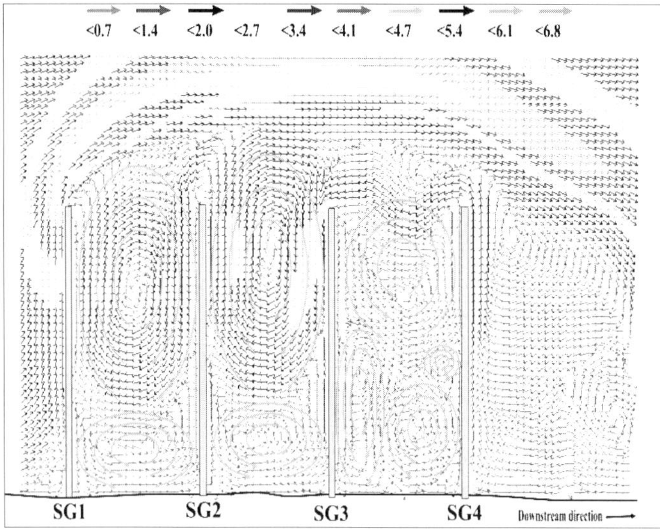

Fig. 19. Two-dimensional current flow in the presence of four seabed groins 40 m in length placed 28 m apart

Fig. 20. Example of one groin of a 13-groin system installed on the beach of Ag. Marina, S. Evoikos Gulf, Greece, photographed shortly after installation. Sedimentation on both sides and all along the body of this low-height groin is evident

Fig. 21. The 13-groin system installed in April 1996 on the beach of Ag. Marina, S. Evoikos Gulf, Greece, photographed in January 1999

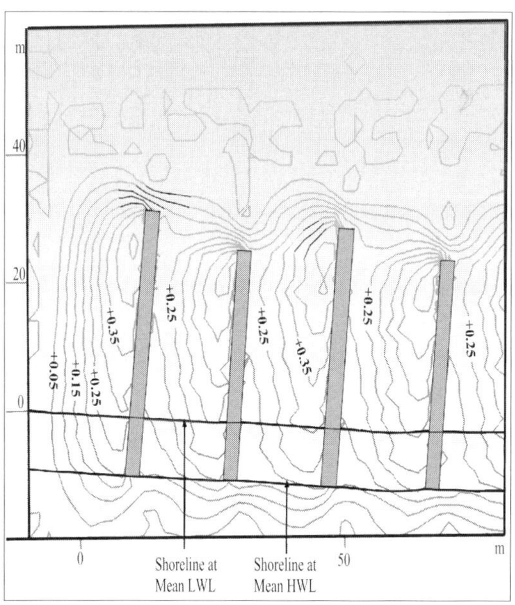

Fig. 22. Monitoring diagram of the 26-groin system installed in 1998 on the Oasis Resort Beach at El Hegaz, Red Sea, Egypt. The contours shown give the sediment height added to the seafloor during the first six months of placement of the system

Fig. 23. The 'hard' defence structure that made the beach along the Red Sea of Oasis Resort at El Hegaz, Egypt, unfriendly to man and had to be replaced by a 'soft' protection system.

Fig. 24. The 'soft' replacement of the system shown in Figure 23 about six months after installation. There is hardly any resemblance between the present and the previous condition of this beach. The 'soft' system, besides obviously being more friendly to people and the environment, was installed at a smaller cost and serve this tourist unit for a long time

The impact of the seabed groin systems installed at various sites in three countries appears everywhere the same and uniform. An example is shown in Figures 23 (beach before installation) and 24 (beach six months after installation).

5. Conclusions

Seabed groin systems whose geometric parameters are adapted to the wave-current climate of the sites they are designed for modify the pattern of water motion near the seabed and eliminates loss of small-sized materials, while at the same time trapping sediment driven alongshore by longshore currents. The computed model and the results of real experiments indicate satisfactory agreement. This programme of research requires continuation for identification of specific essential parameters such as the longevity of the system and the protection offered to the installation site.

Such groin systems cannot be designed using methods based on traditional coastal enginnering.

Although the theoretical-numerical model of the seabed groin systems and the *in situ* experiments provide coherent evidence of being a reliable way of providing 'soft' protection of eroding shores, it is essential that further experimentation be carried out at sites of varying wave-current climates so that the system can be adated to cover a wider range of applications and still provide optimal protection for each and all sites of application.

REFERENCES

Chow, C. Y.: *An Introduction to Computational Fluid Mechanics*, New York: John Wiley & Sons, 1979.

Goudas, C. and Katsiaris, G. A.: 'Direct Utilization of Wave Energy for Achieving Coastal Protection against Erosion-The CPNS and Its Applications', *Proceedings of the Third European Wave Energy Conference, incorporating Waves, Tidal and Marine Currents*, 30 September – 2 October 1998, Patras, Greece. Editor Dr. W. Dursthoff, Franzius Institut, University of Hannover, Germany (to appear).

Gustafsson, B. and Sundstrom, A.: 'Incompletely Parabolic Problems in Fluid-Dynamics', *SIAM J. Appl. Math.*, **35**(2), pp. 343–357, 1978.

Karambas, Th. V. and Koutitas, C.: 'Non-Linear Wave and Sediment Transport Model for Beach Profile Evolution With Application to Soft Shore Projection Methods', (in this volume).

Karambas, Th. V. and Kriezi, E. E.: 'Numerical Simulation of Linear Wave Propagation, Wave-Induced Circulation, Sediment Transport and Beach Evolution', in *Coastal Engineering and Marina Developments*, Wessex Institute of Technology Press, Eds C.A. Brebbia and P. Anagnostopoulos, pp. 253–274, 1999.

Lighthill, M. J.: 'Introduction. Boundary Layer Theory", in *Laminar Boundary Layers*, ed. L. Rosenhead (Oxford University Press, Oxford), pp. 46–109, 1963.

Oliger, J. and Sundstrom, A.: 'Theoretical and Practical Aspects of Some Initial Boundary-Value Problems in Fluid-Dynamics', *SIAM J. Appl. Math.* **35**(3), pp. 419–446, 1978.

Quartapelle, L. and Valz-Gris, F.: 'Projection Conditions on the Vorticity in Viscous Incompressible Flows', *Int. J. Numer. Methods Fluids* **1**(2), pp. 129–144, 1981.

COASTAL EROSION PROBLEMS ALONG THE NORTHERN AEGEAN COASTLINE: THE CASE OF THE NESTOS RIVER DELTA AND THE ADJACENT COASTLINES

G. S. XEIDAKIS AND P. DELIMANI

Department of Civil Engineering, Democritus University of Thrace, 67100 Xanthi, Greece.

Abstract

The coastal zone is becoming increasingly important to human activities and life. In Greece where the majority of the population and its activities are concentrated near the sea, the coastal zone is of paramount importance. This paper reports on research into the geomorphological characteristics of a stretch of around 45 km of the Northern Aegean shoreline, including the delta of Nestos river. Topographic maps and aerial photographs together with field investigation, mapping and measurements of the shoreline changes, were employed. The research showed that most of the coastline is under erosion and retreat, apart from a few stretches such as at Keramoti Cape and Abdera Bay, where sedimentation is taking place. The rate of erosion along the shoreline varies from zero up to 25 m/year and can be attributed to a combination of agents such as river entrenchment, delta deforestation and change of land use, construction of high dams in the river course and the capture of sediments in their basins, compaction and subsidence of the loose delta sediments, and sea level increase. The coasts in this area were classified according to the Valentin classification as low relief, straight, deltaic, submerging and retreating coasts. Some soft protection measures such as nourishment, submerged groins and/or breakwaters, should be employed in the stretches of the coast where severe erosion problems exist.

1. Introduction

Today, the majority of the earth population lives or works near or in the coastal zone. This trend appears likely to continue in the future. The coastal zone will thus become more and more important, with settlements, fishing, swimming, navigation, transport of materials and goods, resorts and so on all centred on it. It is known, however, that human activities are directly affecting the geomorphology of the coastline. For the Greek population this zone is even more important than elsewhere, since the continental area is restricted, mountainous and relatively unproductive. It has been estimated that the total length of the coastline of Greece is around 16 000 km, and about 70% of the population, 80% of industrial activities, 90% of tourism and most of the agricultural land is located near or in the coastal zone. It can be claimed that the survival of the population of the Greek peninsula is directly dependent upon the coastal zone.

In this paper the effects of natural processes and human activities upon the coastal delta of the river Nestos and the adjacent coastlines is examined, together with possible protective measures for mitigating the shoreline retreat. The zone studied extends from Cape Akroneri, in the Kavala prefecture, in the west, eastwards towards Cape Baloustra (Abdera) in the prefecture of Xanthi, on the Northern Aegean (Thracean) sea. The overall length of the stretch discussed here is about 45 km long.

C. L. Goudas et al. (eds.), Soft Shore Protection, 337-348.

2. Geology of the area

The area under consideration belongs to the Rodope geotectonic zone and forms part of the Tertiary basin of Prinos-Nestos. The coastal zone under examination is situated along the delta of the Nestos river and consists mainly of delta sediments (gravels, sands, silts and clays). The thickness of these sediments is estimated as being between 2 500 m and 6 000 m (Psilovikos et al, 1988). The broader area of Rodope has participated in a number of tectonic activities, the last one being the Alpine orogeny. Today the area is considered tectonically stable. During the Quaternary (that is, the period that began 1.5 million years ago) the area is estimated to have been subjected to a tectonic subsidence of 1 m–100 m (Figure 1). Geomorphologically, the coastal zone is flat, consisting of loose sediments (gravels, sands, silts, clays, etc.), susceptible to erosion and rapid retreat of the coastline, which sometimes severely affects the immediately adjacent roads and structures.

3. Wind climate

The region of the Northern Aegean and Thrace area is characterised by Mediterranean-type climate with dry summers and mild, rainy winters.

3.1. WINDS.

The winds in the Northern Aegean Sea are quite strong, with a mean speed of 34 – 40 knots (17.2 – 20.7 m/sec) creating waves in the open sea of up to 7 m. Near the coast, which has quite shallow waters, the sea waves are lower but still very active, creating wide surf and swash zones. According to the Kavala meteorological station, which is located west of the area under examination, there is a prevalence of SE winds with a speed of 2–4 Beaufort at a frequency of 22.16%, whereas on Thasos, the prevailing winds come from the NE direction and have a frequency of 21.50% and speed of 1–3 Beaufort (Figure 2).

In Kavala Bay, the maximum wave height is around 5 m, whereas on the west coast of the island of Thasos, the wave height is 2–2.5 m and in Keramoti Bay, with south-eastern winds, the wave height is 2–3 m (*Greek Waters Pilot*, 1989).

3.2. LONGSHORE CURRENTS

Due to the prevailing SW winds during the winter and NW winds during the summer and autumn and the E-W trend of the shoreline in the Northern Aegean Sea, there is a prevailing longshore sea current from the east to the west. This current is modified by the local morphology of the shoreline. In the area under study, the longshore currents from Cape Akroneri (to the west) up to the bay of Keramoti (to the east) tend to be westerly. However, when strong SE winds blow in the area, the longshore current adopts a reverse direction, flowing to the east (Fig. 3A). From the Nestos river mouth up to Cape Keramoti there is a westwards longshore current which becomes quite strong when SW winds blow in the area. The sea current in the Thasopoula–Keramoti straits usually moves in a westerly direction at a speed of up to 1 m/sec (15 knots) (*Greek Waters Pilot*, 1989). When this current moves further west, out of the straits, it slows down and deposits coarse sand along Cape Keramoti (Delimani, 2000). From Cape Keramoti the longshore current turns to the north and goes to the opposite coast near the village of Piges and Cape Akroneri carrying less sediment, resulting in a strong coast eroding effect (Figure 3, photos 1 and 2). Another eastward longshore current starts from the mouth of the Nestos and flows to the east coast as far as Cape Baloustra (Abdera), where it joins the westwards sea current offshore (Figure 3B).

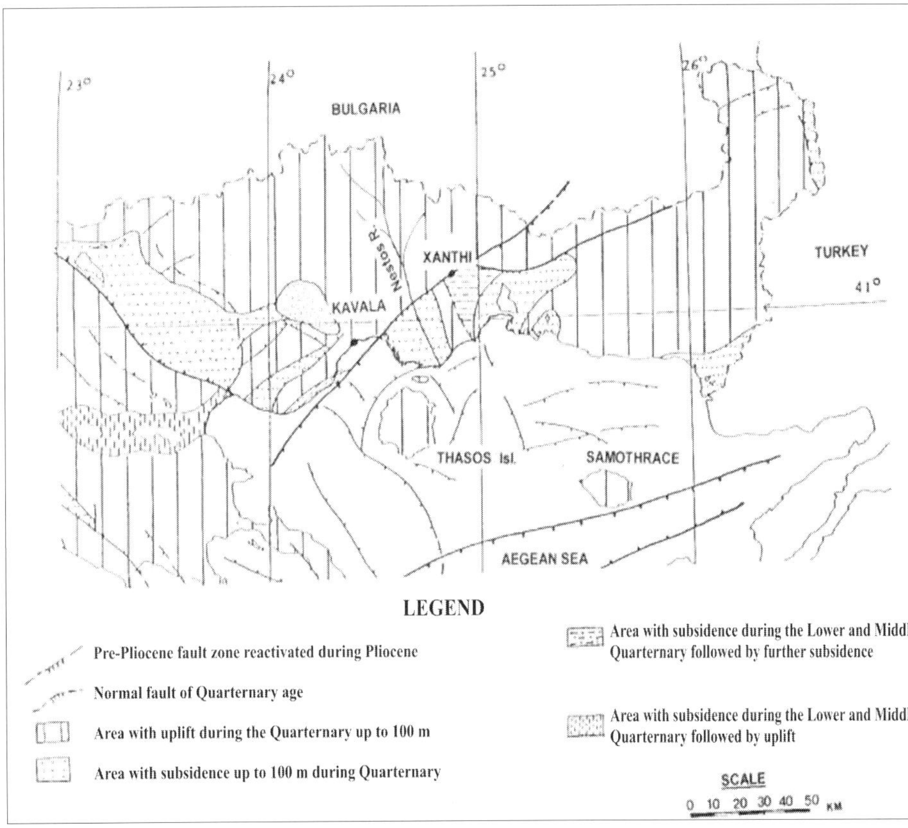

Fig. 1. Tectonic map of Thrace (Mpornovas, 1973, IGME)

4. Classification of the coasts

The coasts under study have been classified according to various classification systems. Here the Valentin classification is used. According to this classification, the coasts under examination are classified as low relief, almost straight coasts. They are also of a retreating, submerging, delta type with river sedimentation. Along the south and west side of Cape Keramoti, the coast is advancing and prograding with both fluvial and sea sedimentation.

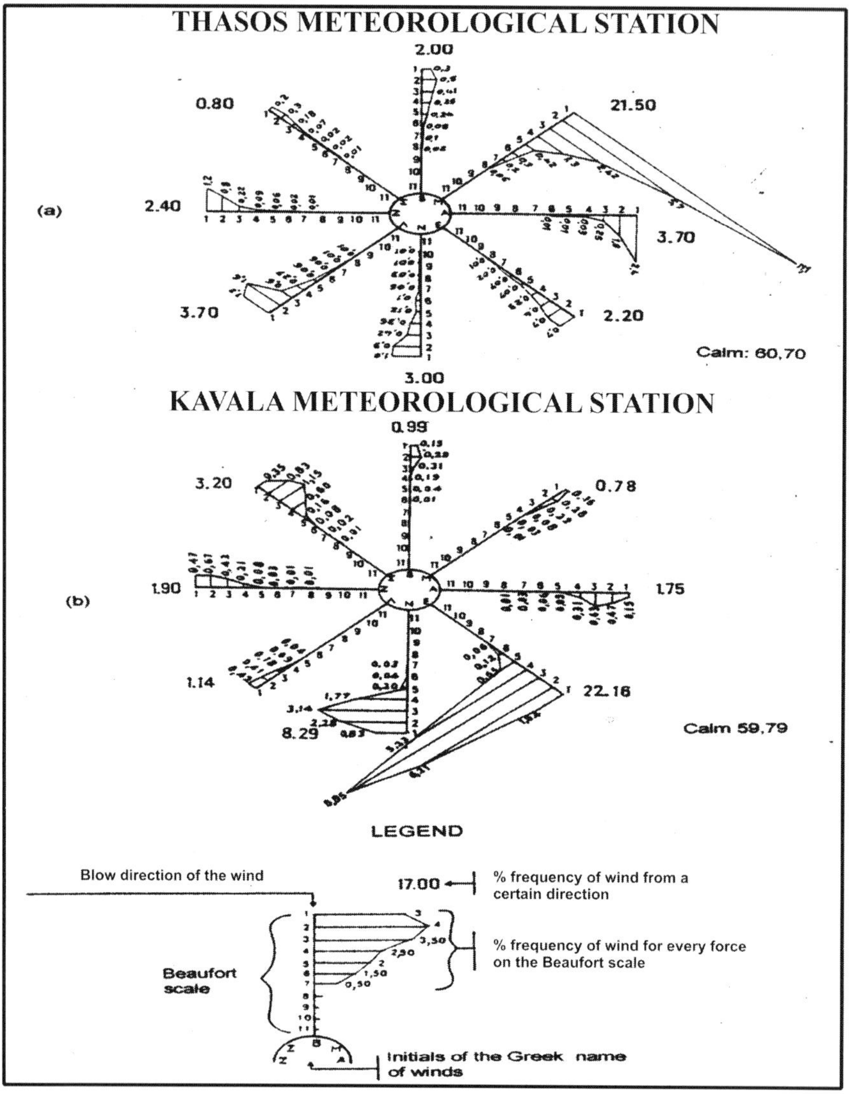

Fig. 2. Mean annual percentage frequency of wind direction and force (according to the Beaufort scale) from observations during the years 1951–1978 (a) on Thasos Island and (b) at the Kavala meteorological station (National Meteorological Service of Greece, Athens)

The coast stretch from the Keramoti peninsula up to the mouth of the river Nestos is of a low relief, deltaic, eroding, submerging and open plain type. The same classification holds for the coast east of the river Nestos apart from the coast inside a few embayments, for example, near the Mangana village fishpond. The section of the coast from the Laspias torrent up to Abdera Bay is advancing due to sedimentation. Cape Baloustra itself consists of hard metamorphic rocks of the Rodope range (it is a horst) and the coastline is rocky and with pocket beaches, retreating (eroding) and submerging. In this area is the submerged (under the sea) port of ancient Abdera, built about 2500 BC.

5. Methods of study

For studying the shoreline changes over time, the position of the coastline at various time periods was examined on the basis of topographic maps and aerial photographs.

The topographic maps used were:

1. The topographic sheets 1:50 000 drawn up by the Geographic Service of the Greek Army, Chrysoupolis sheet, issue 1966, and Abdera sheet, issue 1970.
2. The topographic sheets 1:50 000 drawn up by the Hydrographical Service of the Greek Navy, issue 1966.
3. The topographic sheets 1:50 000 drawn up by the Hydrographical Service of the Greek Navy, issue 1939.
4. Aerial photographs at various scales and taken at various times.
5. The office studies were supplemented by field measurements and mapping.

6. Results and discussion

From the aforementioned studies it became evident that:

- Most of the shoreline under consideration has serious erosion and retreat problems with some exceptions (a) on the Keramoti peninsula and east of the Laspias torrent, where the coast is advancing.
- The rate of shoreline retreat is not constant along the whole length of the coast but ranges from a few centimetres per year up to 25 m per year.

The retreat along almost all of the Nestos Delta coast, which is a Holocene delta, as well as the adjacent shorelines, can be attributed to:

- Sea effects (waves, currents, winds, etc.)
- Terrestrial agent effects (dam construction, irrigation systems, regulation of river courses, etc.
- Human interference (construction of ports, breakwaters, groins, fishing ponds, etc.)
- Relative increase of the sea level, which is considered to have been between 13–15 cm in the last 100 years (Titus et al., 1995, Maroukian, et al, 1999)

- Low consistency of the sediments in the coastal zone (loose sandy silts with some clay) as well as the low elevation and the flatness of the delta area
- Low permeability of the surface deposits due to higher clay content from seasonal flooding
- Relatively high water table, 0.5–1 m below the ground surface
- Consolidation and settlement of the delta deposits
- Entrenchment of the course of the River Nestos itself and the construction of an irrigation system in most of the delta area during early 1950s, deprived the delta plain of the ponding waters and sedimentation inland and the longshore sediments current. This intervention increased the velocity of the water flow along the river course and reinforced the spread of the sediments further to the sea. Some of the sediments arriving at the sea were moved offshore by the sea currents and waves.
- Construction of two big dams (one 170 m and another 95 m high) along the course of the river Nestos during early 1990s was another reason for the reduction in the load transported by the river. These dams act as catchment areas for water floods and river flow throughout the year, resulting in even lower amounts of sediments transported to the sea.
- Deforestation of the delta plain and the use of the land for agricultural purposes has upset the equilibrium of sedimentation along the coastal zone.

In the following section specific stretches of the shoreline with a higher rate of erosion or deposition are examined in more details.

6.1 THE COASTLINE AT CAPE AKRONERI AND THE VILLAGE OF PIGES, KAVALA

At Cape Akroneri, Kavala (Figure 3) severe sea erosion and coastline retreat is occuring. The first lighthouse, built in 1929, has been moved inland due to coastal erosion: in 1935 by 120 m, in 1937 by 86 m, in 1953 by 150 m, in 1956 by 350 m and in 1964 by 180 m, a total 886 m during the period 1929 to 1964 (35 years). This represents a mean rate of retreat of 25.3 m/year (historical data from the archives of the Lighthouses Service of the Greek Navy, 2000, Athens).

About 1 km east of Cape Akroneri, the shoreline retreat between September 1999 and April 2000 was found to be 2 m. 3 km further to the east, the rate of retreat, measured in a dune, was found to be about 2 m/year. In the eastern part of this coastal stretch (on the coast near the village of Piges), the coastline retreat between 1939 and 1969 was estimated from the topographic maps to be between 300 m and 400 m. That means a shoreline retreat of 10–13 m/year (photos 1 and 2).

There are many signs of the enormous shoreline retreat in this area. For example, the coastline near a topographic post retreated by 2.5 m in a year. The coastal zone marshes are covered by about 2–3 m/year of sea sand, brought by winds and sea waves. The roots of the swamp shrubs and bushes can still be seen in the sea some 30–40 m away from the coastline. Two freshwater wells dug near the coastline are now under the sea. One of them, dug in 1986 50 m inland, is now (1999) 40 m offshore.

The severe erosion of this part of the coastline has been attributed to the diffraction of the sea waves at Cape Keramoti and to the deposition of sediments by the longshore currents along the Keramoti Peninsula and to strong winds blowing in the area from the NE (21.50%) and the SE (22.16%) (Figure 3, photos 1and 2).

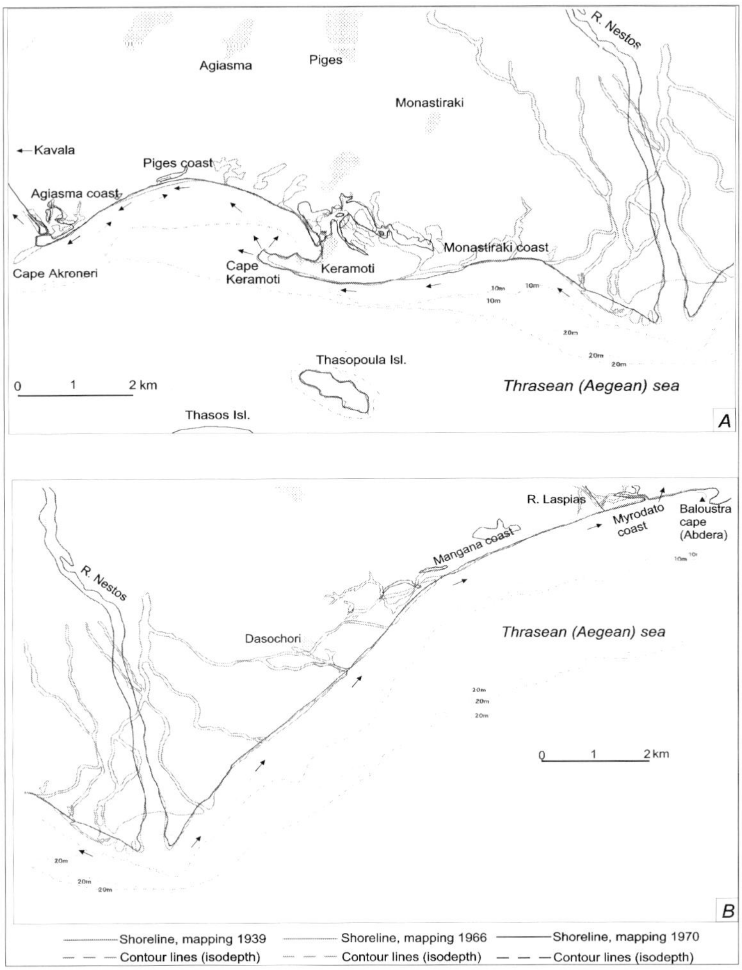

Fig. 3. Maps of shoreline changes between 1939 and 1970

6.2 DEPOSITION ALONG THE KERAMOTI PENINSULA

On the W and SW side of the Keramoti Peninsula substantial deposition of coarse to medium sand, mainly of marine origin has occurred, according to data of the Hydrographical Service of the Greek Navy, the total progradation of Cape Keramoti westwards (seawards) between 1949 and 1982 was in the vicinity of 150 m, that is, about 4.5 m/year. The lighthouse on the cape has been moved once towards the shoreline by 70 m. The pedestal of the old lighthouse, built in 1950, is now (2000) about 950 m from the west shoreline, while the pedestal of the modern lighthouse, constructed in 1982 50 m away from the shoreline, is now around 350 m from the same shoreline and is about to be again moved nearer to the shoreline (Figure 3A).

Photo 1. Cape Akroneri, Kavala, 1998. The modern lighthouse (torch) constructed in 1964 standing about 150 m inland of the west shoreline. The pedestal of the first lighthouse, built in 1929, has been about 900 m offshore today. The rate of shoreline retreat has been reduced during the last decades due to the construction of protective rock breakwaters for protecting the sand bar separating the fish-rearing lagoon from the sea

Photo 2. The coastline near Piges, Kavala, July 1999. The soil heap for the foundation of the topographic post has been reached by the sea and has started to erode. It had been built some 50 m inland in the 1950s

The sand deposition along the Keramoti Peninsula has been attributed to the reduction in speed of the east to west longshore current due to widening of the sea further west and to the diffraction of both the current and the waves along the cape to the N–NE, which enter the bay of Keramoti and also travel towards the opposite coast where the village Piges is situated (Figure 3A).

6.3 EROSION ALONG THE MONASTIRAKI COAST

Further east of the Keramoti Peninsula, has occurred along the Monastiraki coast severe erosion during the last four years. The coastal dunes are retreating quickly and the beach sand is moving inland and covering the shrubs and bushes of the coastal marshes. Patches of beach up to 100 cm high have been formed here and there. During the last three years (1997–2000) a rural road along and adjacent to the coastline has been completely destroyed by sea waves (Figure 3A).

The erosion along this stretch of the coast must have started during the last two to three decades, since it is not evident in older topographic maps (1930, 1970). It should be mentioned that the mouth of the river Nestos was situated in this area until 1953, when it was diverted about 5 km eastwards. The increase in the erosion rate of the last decades could be attributed to the diversion and the entrenchment of the river Nestos (Figure 3).

Photo 3. Monastiraki coast, Kavala, April 2000. The western part of the destroyed road. View from west to east

6.4 THE PRESENT MOUTH OF RIVER NESTOS

The present Nestos delta dates mainly from the Holocene age. It prograded to the sea at a quite high rate up to the 1950s. After the diversion of the river's course in 1953 about 5 km to the east, the sedimentation and progradation of the delta itself was restricted mainly to the front of the new river mouth and the surrounding area. In the mid sixties, the rate of the river's delta progadation was reduced even more due to the construction of a small dam about 25 km inland from the river's mouth. The dam was constructed for diverting the river's water within two canals to the west towards the plain of Kavala and to the east towards the plain of Xanthi, for the purpose of land-irrigation (Figure 1 and 3). This deprived the river's mouth of sediments and waters. The situation became worse when two big dams 170 m and 95 m in height were built further inland, in the mountainous area about 100 km from the river's mouth.

Photo 4. Monastiraki coast, Kavala, July 1999. Partly destroyed coastal road running along the coast about 15 m inland from the shoreline in the winter 1999. This site is about 200 m eastwards of the site in Figure 8. The road was repaired in the summer of 1999, and destroyed completely again in the winter of 2000. The eastern part of the road viewed from east to west is shown

Today, both sides of the delta coasts are retreating, but the retrogression of the west coasts is more severe than that occurring along the east coast due to the prevailing NW and SW winds. It should be mentioned that the river's course has a more or less N-S trend, whereas the shoreline has a E-W trend (Figure 3). As mentioned before, two longshore currents start from the river's mouth, one travelling along the east coast reaching up to the Cape of Baloustra (Abdera) and another along the west coast passing

Cape Akroneri. The westward currents appear to be stronger than the eastward, judging from the geomorphologic forms shaped in the coastal area. (Figure 3, photo 2).

6.5 THE COAST NEAR THE VILLAGE OF DASOCHORI, EAST OF THE RIVER'S MOUTH

This coast stretches about 5 km east of the river's mouth. It is a sandy beach with intense erosion and a high retrogression rate. In 1970s there was a resort there with a nice swimming shore and a sandy beach more than 20 m wide. For the need of the visitors fresh water well was opened some 50 m inland from the shoreline. Today the whole sandy beach has disappeared, the sea has gone inland and the well is located around 50 m out to sea. This would indicate that the rate of erosion is 3.3 m/year.

Further east the shoreline has not shown signs of back or forth movements in modern times. It appears to be in equilibrium. Only some 10–11 km east of the Nestos is some erosion (Figure 3B).

6.6 MYRODATO COAST

Some sediment from the Nestos River is being deposited about 12 km to the east along the coastline near the village of Myrodato as far as Cape Baloustra and the small port of Abdera, which has filled with sediments and needs dredging from time to time. Cape Baloustra, which is of hard metamorphic rock, acts as a barrier, preventing movement of the sediment further east (Figure 3B).

7. Conclusions

On the basis of the above, the following conclusions could be drawn:

Most of the shoreline from Cape Akroneri, in the Kavala prefecture, up to Cape Baloustra (Abdera), in Xanthi, including the Nestos delta, is under erosion and retreat. Only at Cape Keramoti and in some small stretches east of the mouth of the Nestos is the shoreline appearing to advance or to be in equilibrium.

The rate of the shoreline retreat varies from a few centimetres a year up to 25 m a year, depending upon the local geomorphology, human activities in the area and the prevailing winds.

The erosion of the coastline can be attributed to a combination of factors such as the Nestos entrenchment, the low consistency of the delta sediments (soft sediments), the high groundwater level in the delta zone, capture of the sediments by dams upstream, subsidence of the delta materials and increase in the sea level in the area during the last 15 000 years.

There is a quite strong east to west longshore current caused by the waves and the prevailing winds. This current, together with the sea waves and the winds, shape the geoforms in the coastal zone of the area.

The coasts in the region are classified according to Valentin as low relief, almost straight, deltaic, submerging and retreating.

Some soft protection measures such as nourishment, submerged groins and/or breakwaters should be undertaken in the stretches of the coast with severe erosion problems.

REFERENCES

Greek Waters Pilot, 1987, Greek Navy, Athens, Greece.

Delimani, P.: Geological changes to the shoreline in the Thrace area and their influence on land use in the coastal zone. Ph.D. thesis, Civil Engineering Department, University of Thrace, Xanthi, 2000, Greece (in Greek).

Maroukian H.: The evolution of the delta plains and other low-lying coasts of continental Greece in the 21st century, *Hellenic Congress on Management and Improvement of the Coastal Zone,* 22–25 November 1999, Laboratory of Port Works, Technical University of Athens, Athens, Greece.

Map 1:50 000, Hellenic Army Geographic Service (GYS) edition, July 1970, Abdera sheet.

Map 1:50 000, Hellenic Army Geographic Service (GYS) edition, July 1969, Chrysoupoli sheet.

Map 1:50 000, Thasopoula to Cape Fanari, published by the Hellenic Navy Hydrographic Service, 1966.

Map 1:50.000, Cape Vrasidas to Thasopoula and Thasos, published by the Hellenic Navy Hydrographic Service, 1966.

Mar 1:50.000, Greek Coast of Thrace: Thasos strait to Vistonikos kolpos (bay) (Porto Lagos), published by the Hellenic Navy Hydrographic Service, 1939.

Psilovikos, A. Vavliakis, E. and Laggalis Th.: Natural and anthropogenic processes in the recent development of the Delta of the Nestos River. *Bulletin of the Greek Geological Society*, 1988, Vol.XX pp. 313–324, Athens, Greece (in Greek).

Titus, G.J. and Narayanan V.K.: The probability of sea level rise, in CLIMATE CHANGE IMPACTS: SEA LEVEL, report published by the US Environmental Protection Agency, (EPA 230-R-95-008), 186 pp., 1995, Washington D.C., USA.

BEACH NOURISHMENT: A SHORT COURSE

ROBERT G. DEAN

Department of Coastal and Oceanographic Engineering, University of Florida, Gainesville, FL 32611-6580, USA.

Abstract

Beach nourishment, the placement of large quantities of sand in the nearshore, causes both physical and environmental changes. The physical changes and processes which are the main focus of this chapter include induced cross-shore and longshore sediment transport components, the equilibration time scales of which are important to the project performance. Both simple and detailed approaches are available to develop and evaluate alternate designs and to predict their performance characteristics. The simple approaches provide the interrelationships of the relevant parameters and reasonable estimates of project performance. Because the detailed approaches, which require the use of numerical models, are currently in an evolutionary state and some are proprietary, this chapter emphasizes the more simple approaches.

The cross-shore and longshore project performance characteristics are treated separately. The cross-shore processes and performance are best represented by available equilibrium profile methodology. The equilibrium beach width resulting from placement of a particular sand size and volume density is shown to be very sensitive to the size of the nourishment sediment relative to that of the native sediment. The simple approach to planform evolution is through the Pelnard Considère differential equation and associated solutions. Many results of design interest can be established, including a simple relationship showing that the longevity of a beach nourishment project constructed on a long straight beach is proportional to the square of the project length and inversely proportional to the wave height to the 2.5 power. A brief review is provided of some of the general characteristics of numerical models which are available to carry out detailed calculations of beach nourishment project performance.

1. Introduction

Beach nourishment comprises the placement of large quantities of good-quality sediment to advance the shoreline seaward. 'Good-quality' means a grain size that is similar to or larger than that of the existing (native) beach sediment. In most cases, beach nourishment is placed where a background (historical) erosion has caused shoreline recession to a degree that remedial action is deemed warranted. Of the three alternatives to coping with a beach erosion problem – retreat, armouring and beach nourishment – beach nourishment is the only alternative which has the capability of maintaining the shoreline position and natural character of an eroding area. Beach nourishment is reasonably costly and usually can be justified only in areas where a significant upland economic base exists; examples include Miami Beach, Florida and Ocean City, Maryland. However, with the increasing commercial, residential, recreational, and environmental values and investments, as well as other interests along the shoreline, beach nourishment has become much more prevalent in the last decade and is expected to be applied more often in the foreseeable future.

C. L. Goudas et al. (eds.), Soft Shore Protection, 349-394.

Beach nourishment commenced in the United States in the early 1900s and most early projects were usually in reasonably contained beach segments. Other projects have included those where the nourishment was a by-product of dredging for other purposes such as excavation for a harbor. The longevity of a nourishment project can be increased by structures such as off-shore (detached) breakwaters, terminal structures, and groins. Although structures have the potential of increasing the longevity of a beach nourishment project, in some locations the effects of structures on the adjacent shoreline are not as predictable as might be required in order to justify their incorporation into the project. Throughout this chapter we will see that sand quality (size) is a key measure of the potential performance of a project both from a physical performance measure and also as an environmental measure.

This chapter describes the parameters usually required to justify a beach nourishment project, the basic mechanics of beach nourishment placement, and the subsequent evolution of the project. Examples of major projects will be presented together with a discussion of their performances. The major performance measures of project longevity and dry beach width will be examined in some detail and the role of structures will be considered. Relatively simple design and prediction methodologies will be examined and their appropriate uses discussed.

2. General discussion

2.1 BEACH NOURISHMENT JUSTIFICATION

Usually there a combination of parameters will determine whether a nourishment project can be economically justified. These parameters include the background erosion rate, the value of the upland assets to be protected and the value of the beach as a recreational and/or environmental resource. In rare cases, nourishment may be carried out to protect an historic site or simply to maintain a natural beach system.

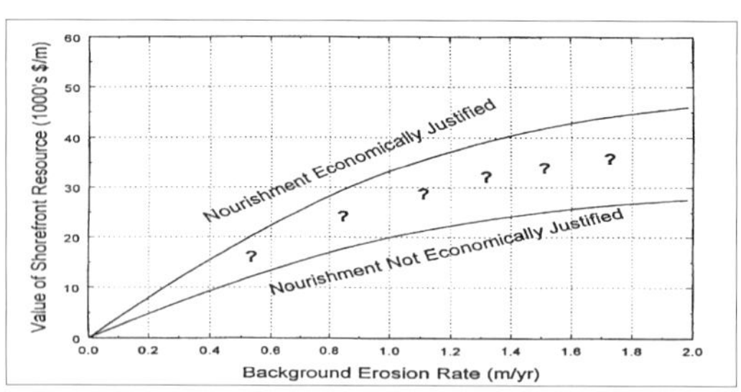

Fig. 1. Qualitative illustration of interrelationship between shorefront value and background erosion rate for nourishment justification

Figure 1 illustrates, for the more normal situation requiring economic justification of the project, the qualitative relationship between the background erosion rate and the value per unit beach front length of the resource. In this figure the value of the resource includes the upland structures, and the recreational and environmental value of the beaches. As will be evident later, a significant parameter not evident in Figure 1 is the length of the asset to be protected. Beach nourishment is most effective when carried out over substantial longshore distances.

2.2 BEACH NOURISHMENT PLACEMENT

In general, nourishment can occur through placement of the material on the beach or as an underwater (nearshore) mound (see Figure 2). The advantages of placement on the beach are that the beneficial effects can be seen and used for recreational purposes immediately; if the material is placed as an underwater mound the full benefits are not as noticeable nor as early. However, in some cases, if the quality of the material is considered marginal, mound placement may allow the use of this material whereas the material would be deemed unsuitable for direct placement on the beach. Materials placed as an underwater mound have been carried out on a number of beach nourishment projects with varying degrees of success. The possible benefits of placement as an underwater mound include the better quality (coarser) portion of this material migrating to shore and gradually nourishing the beach or, if the mound is stable, providing wave energy reduction such that the beach landward of the mound will experience some stabilization against erosion.

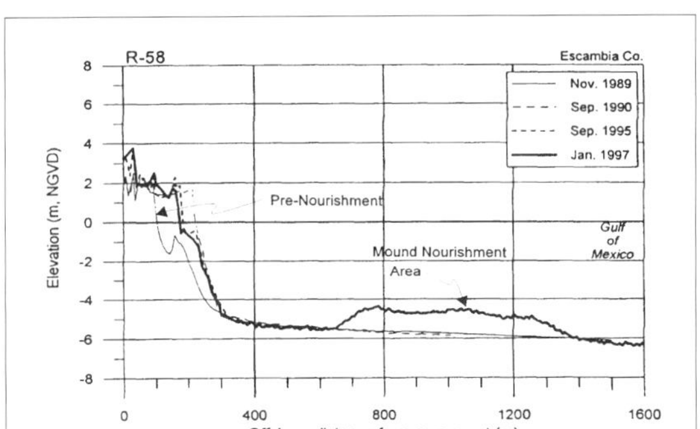

Figure 2. Example of beach and offshore mound nourishment (Browder and Dean, 1997)

Beach nourishment material is usually placed such that the beach is extended seaward and the initial underwater slope is substantially steeper than would be associated with an equilibrium beach profile. It is reasonable to establish as an upper limit of the designed beach nourishment project the natural berm height which represents the upper limit of wave uprush. Although this elevation depends on the tidal range and wave height, in many locations this elevation is in the order of 2 to 3 meters above mean sea level. One approach is to design the placed berm elevation slightly lower than is anticipated under natural conditions and to allow the natural uprush and overwash processes to increase the berm height to its natural elevation. Upon the placement of beach nourishment, the profile will commence to equilibrate and additionally the planform, which now represents a perturbation in the system, will commence to spread laterally to the adjacent beaches (Figure 3). The time scales associated with these two processes are important project performance measures. Valuable references on beach nourishment have been prepared by Rijkswaterstaat (1987) of the Netherlands and the National Research Council (1995).

2.3 IMPORTANT PERFORMANCE MEASURES

Two types of performance measures will be discussed; physical performance and environmental performance measures. The two relevant physical performance measures are primarily the project longevity and the dry beach width. The project longevity is a measure of the time between which successive nourishments are required to be carried out to maintain some desired minimum project conditions. Because the waves are the mobilizing factor which cause the spreading out of the nourishment, it will be shown that the project longevity is sensitive to the wave energy level in which the project is placed and that the alongshore length of the project also figures centrally in project longevity.

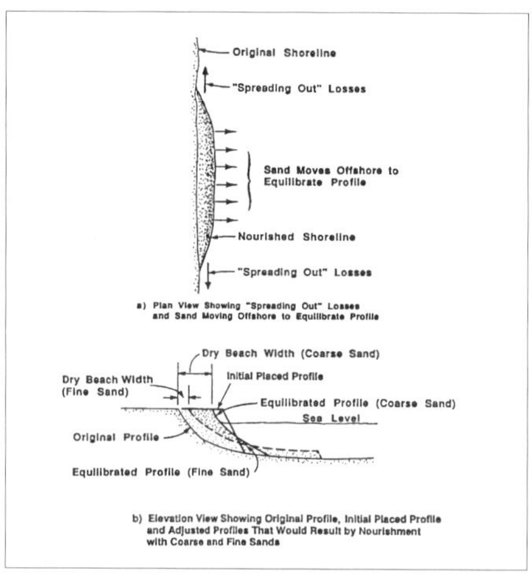

Figure 3. Sand transport 'losses' and beach profiles associated with a beach nourishment project

The dry beach width and the additional dry beach planform areas are also relevant measures of project performance. The nourishment grain size relative to the native is a significant determinant in these two measures. Specifically, because coarse sands are naturally associated with steeper slopes than fine sands, if the nourishment sand is coarser than the native, the additional dry beach width after equilibration will be substantially greater than if the nourishment sands were finer than the native. Thus, the dry beach width and the associated additional planform area are sensitive to nourishment sediment size.

The primary environmental considerations are the nourishment sand characteristics relative to the native. The nourishment sand is usually more poorly sorted than the native sand and may thus contain a greater proportion of fine material which can cause two adverse environmental effects. The first occurs during placement when silt and clay sized particles remain suspended and can reduce light penetration, interfere with the respiration of fish, and benthic flora and fauna, including corals. The second possibility is through a partial weak cementation of sand grains together, resulting in a beach that is less suitable for sea turtle nesting. The latter can usually be improved by tilling the beach surface with normal agricultural equipment, (Trindell et al, 1998). Although beach nourishment with large quantities of compatible sands can smother some of the resident biota, it has been found that recolonization usually occurs within one or two years. This relatively rapid recovery is interpreted as a result of the capability of the biota which reside in the nearshore area to cope with and recolonize after large naturally occurring changes. Beach nourishment can also be of substantial environmental benefit through the creation of wide beaches suitable for sea turtle and bird nesting.

2.4 ROLE OF STABILIZING STRUCTURES

As noted previously, structures can be used to increase the longevity of the project either internally within a project or as terminal structures. In some cases, for example on barrier islands, projects are bounded on the two ends by fairly long jetties which serve as terminal structures. If the structure lengths exceed the active width of the beach, there is no possibility for good-quality sand to be carried outside the project limits and the sand can only be redistributed between the two terminal structures. In such a case, if the rate of longshore sediment transport is large, it may be worthwhile to consider including intermediate structures within the project, for example as a field of groins. Periodic backpassing of sand from accreted downdrift beaches to eroded updrift beaches is an effective technical solution, but may not be politically acceptable.

2.5 EXAMPLES OF MAJOR PROJECTS

Many major beach nourishment projects have been constructed and in place for a sufficient duration to quantify their performances. Three such projects located in the United States are discussed below.

2.5.1 Miami Beach, Florida

The Miami Beach nourishment project was constructed in five phases over a five-year period from 1976 to 1981, and thus has been in place for approximately 20 years at the time of writing this chapter, and has been monitored sufficiently to establish its performance. Most of this project is located on Miami Beach and is bounded by a large curved jetty at the north end of the island (at Baker's Haulover Cut) and a second large jetty at the southern end of the project at Government Cut (Figure 4). This portion of the project extends over a 16 km length and comprised the placement of approximately seven million cubic meters of sand over the five-year construction period. The placement of this material resulted in a widening of the beach by approximately 100 m. The source of the sand was five borrow areas located offshore and consisted primarily of calcarious material, whereas approximately 35% of the native material was quartz sand. The Miami Beach area is interesting because the narrowed beach width, which resulted in the need for nourishment, was substantially a result of encroachment of the upland hotels on the beach. These hotels were initially constructed with a small but reasonable set-back from the waterline. As tourism developed, the hotels responded to a need for a deck and swimming pool by expanding their facilities seaward and encompassing much of the remaining dry beach width within the expanded hotel structure. This seaward encroachment occurred, of course, prior to the time that the State of Florida had instituted adequate restrictions regarding such development. The hurricane of September 1926 caused considerable erosion of the Miami Beach area and resulted in construction of many long groins in an attempt to stabilize the beach. Under the action of the relatively small net longshore sediment transport (estimated at approximately 10 000 m^3/year to the south) in conjunction with the excavation of and construction of jetties at Bakers Haulover Cut (Inlet) at the north

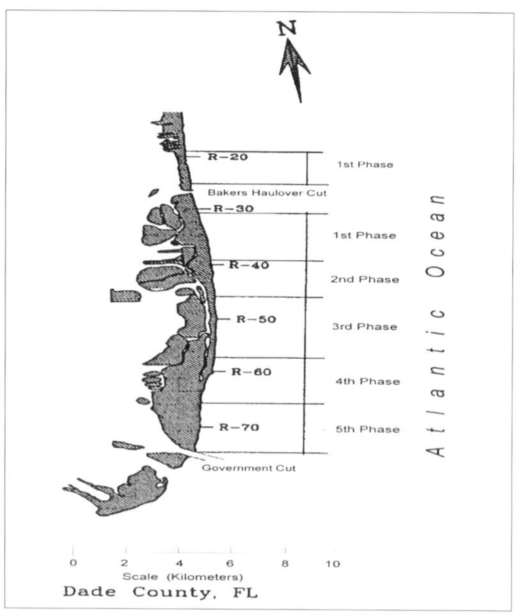

Figure 4. The five phases of the miami beach nourishment project

end of Miami Beach in 1925, the beach gradually narrowed such that there was little beach left in many locations at normal tide and no beach at high tide. The Miami Beach tourist industry declined consistent with the condition of the beach. This decline and the recognition of the relevance of the beach to the economy of the area resulted in support for and implementation of the beach nourishment project.

The borrow areas for this project were located between shore-parallel coral reefs located approximately 5 km offshore. There was a considerable amount of fine material in at least one of these borrow areas resulting in the placement of substantial quantities of silts and clays (greater than 10%). Borrow area quality requirements in Florida now preclude the use of nourishment materials with greater than approximately 4 to 5% silts and clays based on concerns of reduced light penetration, smothering of live reefs and clogging of fish gills.

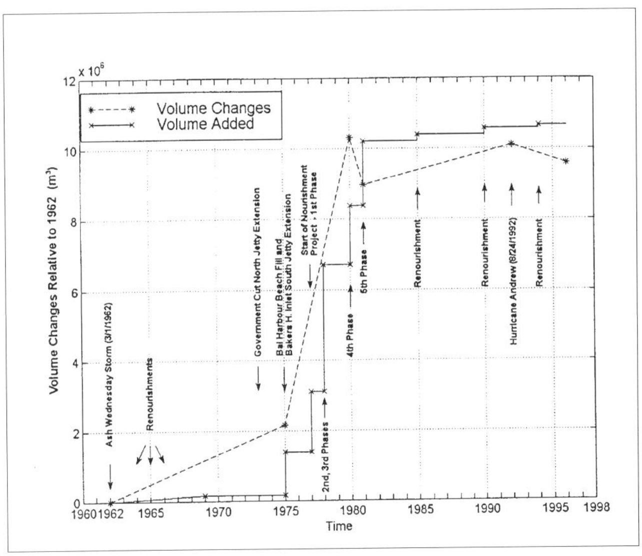

Figure 5. Total volume changes and volumes added relative to 1962 between Bakers Haulover Cut and Government Cut (based on shoreline changes).

Surveys have shown that the total volume remaining within the Miami Beach area is within 9.5% of the total amount placed (see Figure 5), a consequence of the substantial terminal structures at the north and south ends of the project making it difficult for sand to be transported from the project area. However, two erosional hot-spots have occurred: areas that erode more rapidly than anticipated and/or more rapidly than the adjacent beaches. The occurrence of these erosional hot-spots has led to the requirement for additional limited renourishments totaling approximately 400 000 m^3, about 6% of the original amount placed.

The Miami Beach project is credited with revitalizing the economy of Miami Beach. Today, when airplanes arrive at the Miami International Airport from South America they are greeted with the recorded message 'Welcome to Miami and its beaches'. An excellent summary of the Miami Beach nourishment project is provided by Wiegel (1992).

2.5.2 Ocean City, Maryland

The Ocean City, Maryland project was constructed in two phases, with the first and second phase constructions occurring in 1988 and 1990–1991, respectively, and comprising a total of 5.0 million cubic meters. Soon after the completion of the Ocean City, Maryland beach nourishment project, the so-called 'Halloween Storm of 1991' caused significant modifications to the project especially in terms of reduction in beach width. The project was immediately hailed as a great success by its proponents and a failure by its detractors. The proponents noted that the damage prevented by the beach nourishment project during this one storm was substantially greater than the cost of the project itself. Thus the benefits of the project had exceeded the project cost within the first year of the project life. The critics of the project termed the reduction in beach width a 'loss of sand'. However, surveys documented that the reduction in beach width was due dominantly to profile adjustment; most of the placed sand remained in the project limits. The Halloween Storm of 1991 was clearly a rare storm with a number of vessels lost at sea. A valuable series of articles documenting the construction and performance of the Ocean City, Maryland project appears in the January 1993 issue of *Shore and Beach* (Kraus; Grosskopf & Stauble; Grosskopf & Behnke; Jensen &Garcia; Stauble & Grosskopf; and Kraus & Wise).

2.5.3 Santa Monica Bay, California

Santa Monica Bay is located in the vicinity of Los Angeles, California and is some 64 km in length. Near the southern end of the Bay, the head of a submarine canyon approaches close to shore. The natural beach was narrow, with widths ranging from 15 to 45 m (Leidersdorf et al, 1993). As a result of harbor excavation projects, sand was placed on the beach and through compartmenting the beach by a number of structures, the beach widths today are in the order of 60 to 150 m. (An excellent paper describing the history of this beach is Leidersdorf et al, 1993). With the subsequent construction of King Harbor in 1939, an additional volume of material was placed on the beach; however, the construction of King Harbor also caused erosion to the south of the harbor. It was concluded that the King Harbor breakwater blocked the southerly directed longshore sediment transport and that the submarine canyon was draining the sand from the now shorter compartment to the south of King Harbor. A revetment and south breakwater to King Harbor were constructed updrift of the submarine canyon, thereby preventing additional losses to the canyon, which functioned as a sand sink. As a result of the placement of a total of 22.2 million m^3 of sand, the present beaches of Santa Monica Bay are much wider than they were naturally. Similar comments apply qualitatively to Huntington Beach, California and some other California beaches.

3. Simplified methodology for representing project evolution

3.1 GENERAL

Figure 3 shows that with the placement of a beach nourishment project, two components of sediment transport are induced; cross-shore and longshore sediment transport. The longshore sediment transport is due to the planform anomaly associated with the project and this transport will be termed here 'spreading out losses'. The time scales associated with these two components of transport figure centrally in the decision to finance a particular project and methodologies to develop estimates of the project longevity will be presented in this chapter. Although the cross-shore and longshore processes occur simultaneously, for a project of reasonable length the cross-shore equilibration time scales are considerably shorter than the longshore equilibration time scales, thus providing a reasonable basis for separate consideration of these two components of evolution.

3.2 CROSS-SHORE CONSIDERATIONS

3.2.1 Profile closure depth

Construction of a beach nourishment project usually occurs at a slope that is steeper than equilibrium; this equilibrium slope is related to the grain size and water depth. A consequence of the steeper slope is a beach width that is initially greater than that associated with an equilibrated profile. A necessary parameter for the calculation of the equilibrated dry beach width is the depth to which the waves will mobilize the placed sand. Greater depths of equilibration will result in smaller additional dry beach widths for a given placed sand volume per unit beach length and vice versa. This depth of equilibration is referred to as the 'closure depth' or 'limiting depth of motion' and for our purposes here will be denoted as h_*, which is referenced to mean low water datum. The closure depth concept represents an idealization of the actual processes; however, it is very useful in design and prediction of the performance of beach nourishment projects.

Hallermeier (1977, 1978, 1980, 1981) describes the results of an examination of profiles from laboratory and field environments and characterizes the relevant depths in terms of an inner limit and an outer limit of motion. The inner limit is considered to be the depth corresponding to seasonal changes of the profile whereas the changes at the outer limit occur less frequently. Hallermeier recommended that the inner limit be used for engineering purposes and this limit is usually considered to represent the closure depth for engineering applications. Hallermeier recommended the following for the annual closure depth

$$h_* = 2.28H_e - 68.5\left\{\frac{H_e^2}{gT_e^2}\right\}, \qquad (1)$$

in which H_e is the significant wave height which is exceeded 12 hours per year, T_e is the associated wave period, and g is acceleration of gravity. It is noteworthy that the sediment size does not appear in this or other expressions for h_* which are examined by Hallermeier (1981).

The Hallermeier equation was modified by Birkemeier (1985) using more accurate measurements of field profiles, resulting in

$$h_* = 1.75H_e - 57.9\left\{\frac{H_e^2}{gT_e^2}\right\}, \tag{2}$$

with a reasonable estimate provided by

$$h_* = 1.57H_e, \tag{3}$$

Finally Nichols, et al. (1996) using accurate profile data from Duck, North Carolina, generalized these results by considering a multi-year approach. The original Hallermeier-recommended relationship [Eq. (1)] for the closure depth, h_*, was found to provide a conservative estimate for multi-year considerations; however, it overpredicted h_* increasingly with longer periods of time considered. For multi-year considerations H_e and T_e are now to be determined on the basis of a 12-hr exceedance for the multi-years considered. In summary, the closure depth should be regarded as a concept providing a useful approximation for engineering applications, but is not an absolute limit of sediment motion. It is also a simplified measure of processes which are poorly understood.

3.2.2 Equilibrium beach profiles and the role of sediment size

Two principal measures of the success of a beach nourishment project are the resulting equilibrated dry beach width and total planform areas, both within the project area and including the adjacent beaches. For our immediate purposes, we will limit our discussions to the equilibrium dry-beach width. The most simple available equilibrium beach profile (EBP) methodology provides a useful approach to evaluate the additional dry beach width associated with the placement of a particular volume of material of a specific grain size. Although beach nourishment material may consist of a fairly wide range of sizes, in most of our discussions here we will characterize the material in terms of the median grain size which is the size separating the weight of the sample equally, that is, one-half of the sample by weight is coarser and one-half of the sample by weight is finer than the median grain size. Equilibrium beach profile methodology has established (Bruun, 1954; Dean, 1977, 1991) that natural beach profiles can be represented approximately by the following equation

$$h(y) = Ay^{3/2}, \tag{4}$$

in which $h(y)$ is the depth at a distance, y, from the mean water line and A is the so-called 'profile scale parameter' with dimensions of length to the one-third power. This form of the equation has two disadvantages and one major advantage. The two disadvantages are the infinite slope at the shoreline and the monotonic nature of the profile; that is, this form cannot represent the presence of bars. The major advantage of this form of the equilibrium beach profile is its simplicity and its ready applicability to many significant problems of coastal engineering interest.

The profile scale parameter, A, has been investigated for more than a thousand beach profiles, and the recommended relationship developed by Moore (1982) and modified by Dean (1987) is shown in Figure 6, where it is seen that the profile scale parameter increases with median sediment size. Table 1 presents a tabulation of the profile scale parameter A versus sand sizes for the normal beach material size range for sand size increments of 0.01 mm. For example, the A value for a sand size of 0.27 mm is 0.119 $m^{1/3}$. Because the profile scale parameter A increases with sediment size D, equilibrated profiles composed of sediments which are coarser and finer than the native will be steeper and more mildly sloped than the native, respectively.

Table 1. Recommended A values ($m^{1/3}$).

D(mm)	0.00	0.01	0.02	0.03	0.04	0.05	0.06	0.07	0.08	0.09
0.1	0.063	0.0672	0.0714	0.0756	0.0798	0.084	0.0872	0.0904	0.0936	0.0968
0.2	0.100	0.103	0.106	0.109	0.112	0.115	0.117	0.119	0.121	0.123
0.3	0.125	0.127	0.129	0.131	0.133	0.135	0.137	0.139	0.141	0.143
0.4	0.145	0.1466	0.1482	0.1498	0.1514	0.153	0.1546	0.1562	0.1578	0.1594
0.5	0.161	0.1622	0.1634	0.1646	0.1658	0.167	0.1682	0.1694	0.1706	0.1718
0.6	0.173	0.1742	0.1754	0.1766	0.1778	0.179	0.1802	0.1814	0.1826	0.1838
0.7	0.185	0.1859	0.1868	0.1877	0.1886	0.1895	0.1904	0.1913	0.1922	0.1931
0.8	0.194	0.1948	0.1956	0.1964	0.1972	0.198	0.1988	0.1996	0.2004	0.2012
0.9	0.202	0.2028	0.2036	0.2044	0.2052	0.206	0.2068	0.2076	0.2084	0.2092
1.0	0.210	0.2108	0.2116	0.2124	0.2132	0.2140	0.2148	0.2156	0.2164	0.2172

Applying EBP methodology as represented by Eq. (4), it can be shown that the three generic types of equilibrated nourished profiles depicted in Figure 7 are possible: intersecting profiles, non-intersecting profiles, and submerged profiles (Dean, 1991). Nourishment (fill) and native profile scale parameters are denoted as A_F and A_N, respectively. Intersecting profiles require a nourishment sediment that is coarser than the native. Non-intersecting profiles occur for nourishment sediments that are of the same size as the native and can also occur for sediments that are coarser or finer than the native. The third type, submerged profiles, requires that the nourishment sediment be finer than the native. Figure 8 presents an example in which the same volume of different sized sediments is placed as nourishment. In all four examples presented in Figure 8, the native sand size and the nourishment volume are the same. In the upper panel (Figure 8a), the native sand size is 0.2 mm and the nourishment sand size is 0.375 mm. It is seen that through the placement of 340 m^3 per meter of beach length, intersecting profiles occur with an equilibrated dry beach width of 92 m. In Figure 8b,

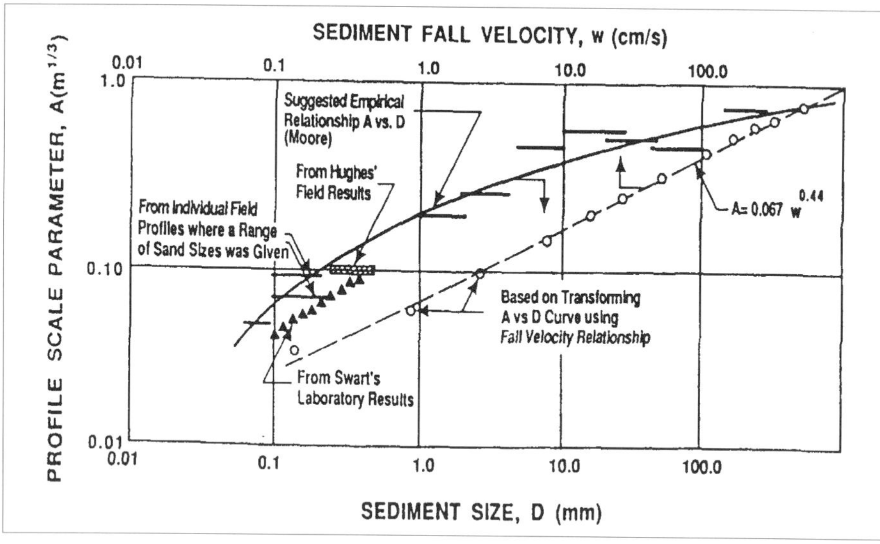

Figure 6. Variation of profile scale parameter A with sediment size and fall velocity (Dean, 1987, modified from Moore, 1982)

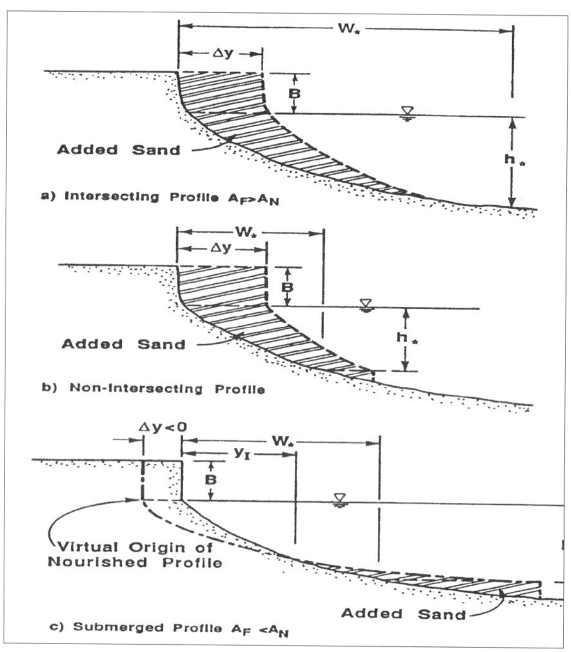

Figure 7. Three generic types of nourished profiles (Dean, 1991)

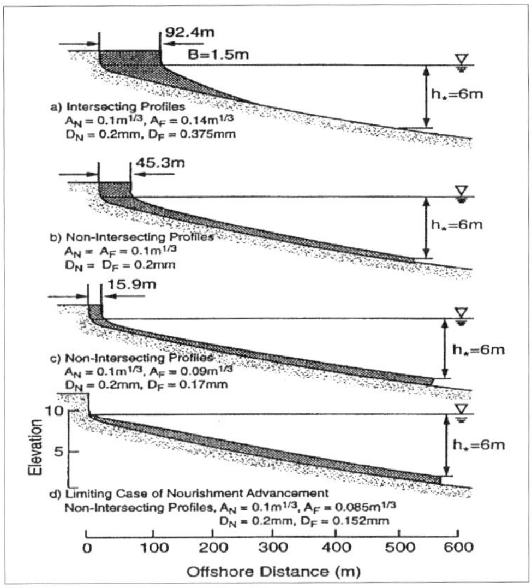

Figure 8. Effect of nourishment scale parameter A_p, on width of resulting dry beach. Four examples of decreasing A_p with same added volume per unit beach length. Berm height, *B,* and closure depth, h_* are 1.5 m and 6 m, respectively (Dean, 1991)

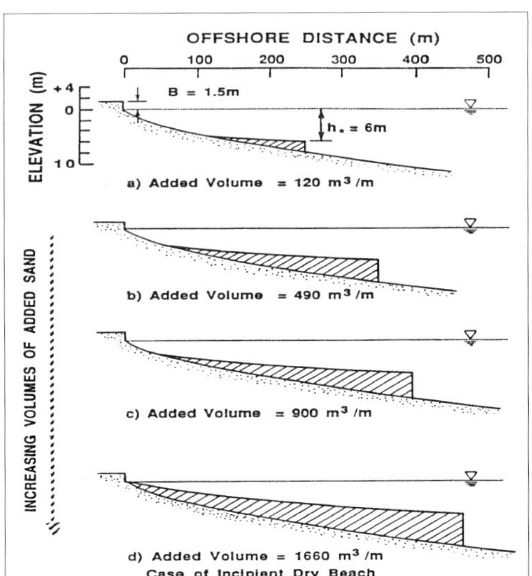

Figure 9. Effect of increasing volume of sand added on resulting beach profile. $A_F = 0.1m^{1/3}$, $A_N = 0.2\ m^{1/3}$, $h_* = 6.0$ m, $B = 1.5$ m (Dean, 1991)

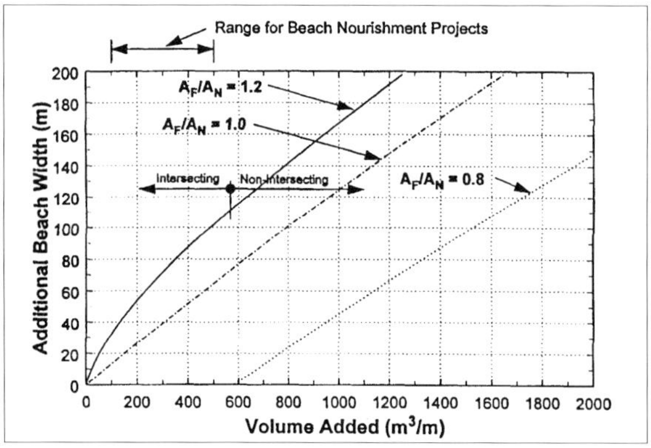

Figure 10. Effect of increasing volume of sand added on resulting beach profile. A_F = 0.1 $m^{1/3}$, A = 0.2 m$^{1/3}$, h_*= 6.0 m, B = 1.5 m (Dean, 1991)

a nourishment sand size of 0.2 mm (the same as the native) is considered, non-intersecting profiles occur, and the equilibrated dry beach width is 45 m, approximately one-half that shown in Figure 8a. In Figures 8c and 8d, nourishment sand sizes of 0.17 mm and 0.152 mm, respectively, are considered; and it is seen that the equilibrated dry beach widths decrease to 16 m and 0 m, respectively. The example in Figure 8d represents an incipient dry beach in which all of the placed sand is configured as an underwater 'submerged' profile, the seaward end of which is located at the so-called depth of closure, h_*. Figure 9 shows a similar example of placement of different volumes of the same size sand, which is finer than the native. Progressing from the upper to the lower panel, greater and greater volumes of nourishment sand are placed. It is seen that with a volume of 1 660 m^3 per meter, an incipient dry beach occurs. In all four cases shown in Figure 9, the seaward depth of the nourishment profile is 6 m. The significance of sand size can be illustrated further, as shown in Figure 10, where the additional dry beach width is plotted as the ordinate and the volume of nourishment sand added is plotted as the abscissa. In this plot, nourishment with sands of three different sizes is represented with a native sand size of 0.2 mm for all three cases. For a nourishment sand that is 0.275 mm, some 38% larger than the native sand, it is seen that placement of 400 m^3 per meter increases the dry beach width over that which would be obtained with sand of the same size as the native by approximately 69%. It is also noted that for the coarser material, the additional dry beach width increases rapidly for small volumes added and then the relationship becomes approximately parallel to the case in which the nourishment and native sand sizes are the same. The volume at which the two relationships become parallel corresponds to the transition from intersecting profiles to non-intersecting profiles. It is at this condition that the depth of intersection of the nourished profile on the native profile becomes h_*. For a nourishment sand that is significantly finer than the native {A_F =0.14 mm), approximately 600 m^3/m is required before an additional dry beach width occurs. This threshold volume is required due to the small equilibrium slopes associated with the fine nourishment sand and the need for the profile to extend from the closure depth landward to the mean water line. As indicated in Figure 10, the volume densities of most nourishment projects range from 100 m^3/m to 500 m^3/m.

It can be shown from equilibrium beach profile methodology that the non-dimensional beach width $\Delta y'$ can be related to the following three non-dimensional quantities: non-dimensional volume ‘′, non-dimensional sediment scale parameter A_F/A_N, and non-dimensional berm height, $B' = B/h_*$, i.e.,

$$\Delta y' = f(\forall', A_F/A_N, B/h_*), \tag{5}$$

in which $\Delta y' = \Delta y/W_*$ and ‘′=″/(W_*B). W_* is the distance from the mean water line to the depth of closure, h_*, on the *native* profile, and ‘ is the nourishment volume per unit length. For nourishment sands which are the same as the native, the shoreline advancement, Δy, can be expressed approximately for any profile form as

$$\Delta y = \frac{V}{h_* + B}, \tag{6}$$

or

$$\Delta y = \frac{\forall'}{1 + \frac{1}{B'}}, \tag{7}$$

Returning to idealized profiles described by Eq. (4), there are three specific equations of the form of Eq. (5), one for each of the nourished profile types shown in Figure 7. These are:

a. Intersecting profiles:

$$\forall' = \Delta y' + \frac{3}{5B'}(\Delta y')^{5/3} \frac{1}{\left[1 - \left(\frac{A_N}{A_F}\right)^{3/2}\right]^{2/3}}, \tag{8}$$

b. Non-intersecting profiles:

$$\forall' = \Delta y' + \frac{3}{5B'}\left\{\left[\Delta y' + \left(\frac{A_N}{A_F}\right)^{3/2}\right]^{5/3} - \left(\frac{A_N}{A_F}\right)^{3/2}\right\}, \tag{9}$$

b. Submerged profiles:

$$\forall' = \frac{3}{5B'}\left\{\left[\left(\frac{A_N}{A_F}\right)^{3/2} + \Delta y'\right]^{5/3} + \frac{(\Delta y')^{5/3}}{\left[\left(\frac{A_N}{A_F}\right)^{3/2} - 1\right]^{2/3}} - \left(\frac{A_N}{A_F}\right)^{3/2}\right\}, \tag{10}$$

where in the last equation $\Delta y'$ is negative and represents the 'virtual shoreline' position relative to the original shoreline (Figure 7c). The critical volume, $(\forall')_{c1}$ associated with intersecting/non-intersecting profiles is

$$(\forall')_{c1} = \left(1 + \frac{3}{5B'}\right)\left[1 - \left(\frac{A_N}{A_F}\right)^{3/2}\right], \tag{11}$$

and applies only for $A_F/A_N > 1$. In Eq. (11), if the volume $(\forall')_{c1}$ is exceeded, the profiles are non-intersecting. The critical volume $(\forall')_{c2}$ which will just yield a finite shoreline for non-intersecting profiles is

$$(\forall')_{c2} = \frac{3}{5B'}\left(\frac{A_N}{A_F}\right)^{3/2}\left(\frac{A_N}{A_F} - 1\right), \tag{12}$$

and, of course, only applies for $A_F/A_N < 1$.

Because the non-dimensional equilibrated dry beach width, $\Delta y'$, is a function of three variables (Eq. 5), it is not possible to present a general solution on a single graph; however, solutions for values of $B' = B/h_* = 0.25$ and 0.5 are presented in Figures 11 and 12, respectively. The general character of these solutions can be understood by discussing Figure 11, which is applicable for a value *of* $B' = 0.25$. The horizontal axis is the ratio of fill to native sediment scale parameters, A_F/A_N, and the vertical axis is the non-dimensional additional dry beach width, $\Delta y'$. The isolines on this figure represent $\forall'$, non-dimensional volume of sand added per unit beach length. The form of the non-dimensional volume can be interpreted as the number of layers of sand of the berm height, B, which are spread out to W_* which is the beach width out to h_* on the *native profile*. For example, $\forall' = 0.1$ is equivalent to a volume of l/10th of a berm height spread out over a distance W_*, on the native profile.

3.2.3 *Profile evolution*

The evolution of a beach profile placed steeper than its natural slope is of interest because there will be a wider beach for some indefinite length of time as the profile approaches equilibrium. It is necessary to inform the general public in advance of the magnitude and the probable range of time scales of the evolution. The available monitoring results indicate that the time scale required for an equilibration of 50% to the native profile is in the order of 3 to 7 years.

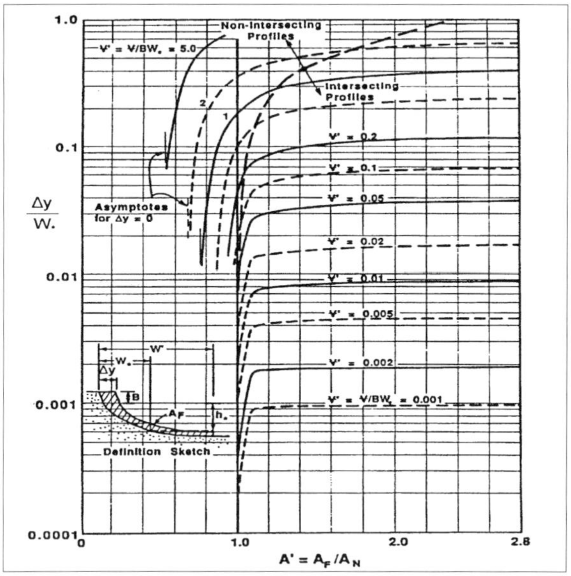

Figure 11. Variation of non-dimensional shoreline advancement *Δy/W*∗, with *A′* and $\mathcal{V}'$. Results shown for *B′(=B/h∗) = 0.25* (Dean, 1991)

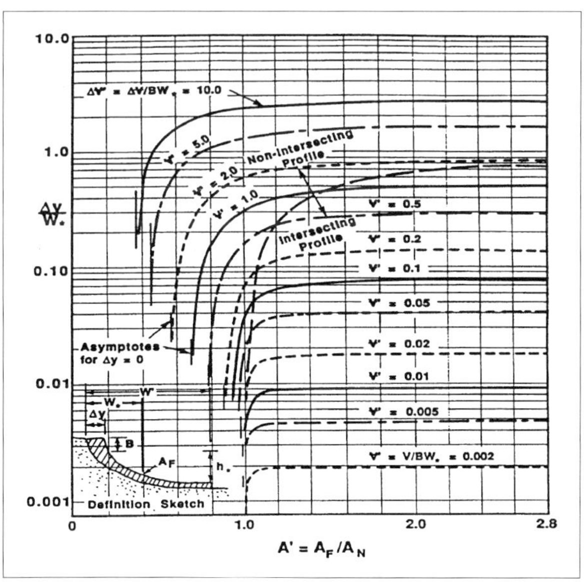

Figure 12. Variation of non-dimensional shoreline advancement *Δy/W*∗, with *A'* and $\mathcal{V}'$. Results shown for *B' (=B/h∗) = 0.25* (Dean, 1991)

There are no known numerical or analytical models that can reasonably predict the time scales of equilibration. There are many numerical models that have been developed to represent cross-shore sediment transport; however, most of these are more applicable for the case of erosion associated with a condition that is substantially out of equilibrium, such as an elevated water level during a storm and the accompanying large wave heights. To appropriately represent the evolution of an over-steepened profile it is necessary that the slope play a role in the *direction* of cross-shore transport. This characteristic is present in some of the available profile evolution models, for example in the EDUNE model by Kriebel (1982) and Kriebel and Dean (1985), but not present in the SBEACH model by Larson and Kraus (1989), in which the direction of sediment transport is controlled entirely by the wave and sediment fall velocity characteristics. In the application of a model to represent a nourished (over-steepened) profile evolution toward equilibrium, it would be necessary to represent the variability in wave height and period because the larger waves of shorter period are predominantly responsible for the seaward transport.

3.2.4 Underwater mounds

The placement of material as an underwater mound is of interest because of the lower costs of placement in this manner and also the possibility of using material that would not otherwise be acceptable for direct placement on the beach for nourishment. Figure 2 has presented one example of nourishment on the beach and in an underwater mound. One of the major questions that arises in association with placement of an underwater mound is whether the mound will be stable or unstable, and if it is unstable, whether the mound will move landward or seaward. There are two effects by which the beach system can benefit from the presence of an underwater mound. The first is if the mound is unstable and will gradually move shoreward and thus act as a feeder of sand to the beach. The second possibility is if the mound is stable and causes sufficient wave energy dissipation such that the waves are smaller when they reach the shore and the mound, thus functions as an offshore breakwater, thereby contributing to the stability of the beach. During storm conditions, storm tides will increase the depths over the mound, thereby decreasing its effectiveness as a wave absorber. On an intuitive basis, if a profile is initially in approximate equilibrium, it would seem that the placement of an underwater mound on that profile would result in either a mound that is stable or one that moves only landward. The rationale for this is that after placement of the underwater mound, the profile would have an excess of sediment relative to equilibrium and this excess can only be reduced through *landward* sediment transport.

Andrassy (1991) reported on the monitoring of an underwater mound placed off Imperial Beach, California. The mound was placed in approximately 6 m of water depth and was documented to migrate landward over a period of approximately one year. Figure 13 presents a series of profiles from approximately the center of this project. It is seen that the placed sediment appears to have been transported shoreward as a dampened wave of translation. A second well-documented underwater mound project is that off the west coast of Denmark near Torsminde (Laustrup et al, 1996). This project has been monitored exceptionally well, including some 17 surveys over a period of 2 years. There are no known documented incidences of mounds being transported offshore after placement (Ahrens & Hands, 1998).

Hands and colleagues (1980, 1988, 1991, 1994) and Ahrens and Hands (1998) have conducted analyses to develop methods addressing the question of stability or instability of underwater mounds. As noted previously, the methodology of Hallermeier (1978) includes two limits (the inshore and offshore limit) of depths for sediment motion. Hands found that if the mound crest was inside the depth of inner motion as determined by Hallermeier's predictions, the mound always moved landward and if the crest of the mound was at a depth greater than the outer limit, the mound was always stable. For mound crest depths that were between the inner and outer depths as predicted by Hallermeier's methodology, the mound could be either stable or unstable (Hands & Allison, 1991). More recently, Ahrens and Hands (1998) have investigated the appropriate wave-induced water particle velocities for correlation with movement of the mound. These velocities were based on the non-linear numerical stream function wave theory (Dean, 1965; Dalrymple, 1974). Quite surprisingly, Ahrens and Hands found that the best correlation with movement of an underwater mound was with the *trough* velocities as predicted by the stream function wave theory.

Otay (1994) has reported on the monitoring of a substantial underwater mound placed in the Gulf of Mexico (Figure 2) and has applied the earlier methodology of Hands and Allison (1991). This methodology predicted that the mound should move; however, detailed monitoring results conducted 7 years after placement have found that although the irregular surface of the mound tended to be smoothed out, there was no discernable movement of the centroid of the underwater mound.

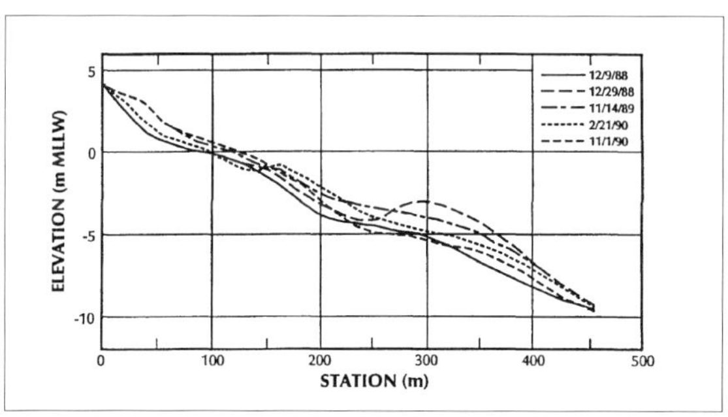

Figure 13. Berm evolution at Imperial Beach, California (Andrassy, 1991)

McClellan and Kraus (1991) have applied a criterion for offshore bar formation to the movement of an underwater mound. A thorough study by Larson and Kraus (1992) of bar movement at the Corps' Field Research Facility at Duck, North Carolina has further developed criteria for bar movement. These criteria would admit both onshore and offshore mound movement. It is not clear that these results are fully applicable because usually the underwater mound lies well outside the limits of the normal surf zone and the motion of an offshore bar, for which the methodology was developed, is associated with the mechanics of breaking waves. The latter study by Larson and Kraus notes the greater likelihood of onshore movement of placed mounds. Offshore placed mounds may be subject to breaking waves during extreme storms.

Douglass (1995) has developed a method for predicting the migrational characteristics of a placed underwater mound. The method employs Stokes' second order wave theory and Bagnold's method for sand transport (Bailard & Inman, 1981) and results in a convective-diffusion equation, thereby predicting both the onshore speed of movement and dispersion of the mound, both of which were found to depend strongly on water depth.

3.3 PLANFORM EVOLUTION

The following sections present the available methodologies for conducting simple predictions of evolution for a number of different initial planforms and conditions. Later, the use of numerical models will be reviewed. Basically, the use of numerical models allows a greater variety of initial planforms, a more detailed wave climate, more flexibility in the internal and boundary conditions, response of the beach to structures, and the capability to include complex details of the wave/sediment interaction.

3.3.1 Methodology/tools for planform evolution

In the simplest models of planform evolution, it is necessary to assume that the profile response to an increase or decrease of volume is a seaward or landward displacement, respectively, of the profile without change of form, Eq. (6). Thus the profile is displaced landward or seaward in accordance with gradients in the longshore sediment transport. These models are termed 'one line models' as they track only one contour, usually the mean sea level contour. The simplest analytical basis for dealing with planform evolution combines the linearized longshore sediment transport equation with the continuity equation, which results in the Pelnard Considere (1956) diffusion equation

$$\frac{\partial y}{\partial t} = G\frac{\partial^2 y}{\partial x^2}, \tag{13}$$

where y is the shoreline displacement from a reference baseline, t is time, and x is the longshore coordinate. This equation is recognized as the familiar heat conduction equation and the quantity G in the above equation is referred to as the 'longshore diffusivity' and can be defined in terms of the breaking wave conditions (denoted by subscript 'b') as

$$G = \frac{KH_b^{5/2}\sqrt{\dfrac{g}{\kappa}}}{8\,(s-1)(1-p)(h_* + B)}, \tag{14}$$

or in terms of deep-water conditions (denoted by subscript 'O') as

$$G = \frac{KH_o^{2.4} C_{G_o} g^{0.4}}{8\,(s-1)(1-p)C_* \kappa^{0.4} (h_* + B)}, \tag{15}$$

in which s is the ratio of sediment density to that of the water in which it is immersed, p is the in-place porosity, C_G is the group velocity, C is the wave celerity, κ is the ratio of breaking wave height to breaking water depth (»0.78), and the subscript '$*$' denotes conditions at the depth of closure. The coefficient K in the above equations is dimensionless and is the so-called sediment transport coefficient and is often taken as 0.77 as determined by Komar and Inman (1970). Based on limited field data, Dean (1987) developed an inverse relationship between the sediment transport coefficient K, and sediment size D. Komar had earlier (1977) considered such a relationship and concluded that over the limited sand size ranges present on most coastlines, a relationship between K and D could not be determined. Komar (1988) reviewed the data employed by Dean, pointed out uncertainties and again concluded that no relationship existed over the range of sand sizes for which data were available.

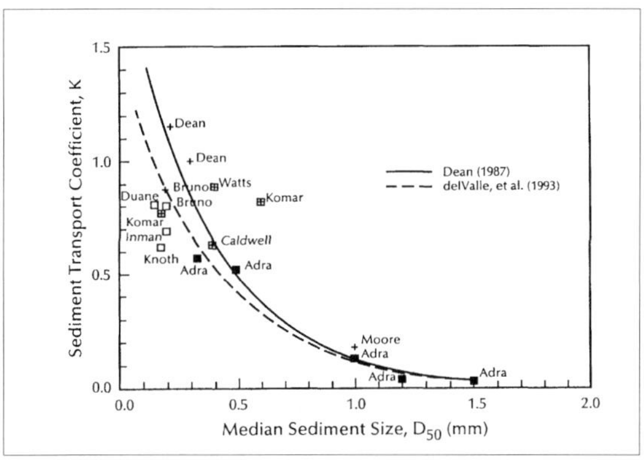

Figure 14. Relationship of sediment transport coefficient with median sediment size, D_{50} (delValle et al, 1993)

DelValle et al (1993) reported on a field investigation of the erosion of the Adra River Delta (Spain) in which transport coefficients were determined for several grain sizes along the delta. Their results are presented in Figure 14 using adjustments in the data recommended by Komar (1988). The solid and dashed curves are the relationships originally recommended by Dean (1987) and that by delValle, et al (1993) including the adjusted original and additional data. In summary, although there is considerable uncertainty in the K vs D relationship, it appears that K is inversely related to D. In view of the significance of K to the performance of beach nourishment projects, it is recommended that the dashed line in Figure 14 be employed.

Returning to the Pelnard Considère equation (Eq. (13)), its two main advantages are: (1) linearity (allowing superposition of various solutions), and (2) frequent occurrence in classical physics, resulting in a number of solutions that can be adopted directly to represent various cases of general relevance to beach nourishment. We will review some of those cases in the following sections.

3.3.2 Examples of Pelnard Considère solutions

The simplest and perhaps most useful solution is that of an initially rectangular beach nourishment project on a long straight beach with no background erosion considered. The resulting evolution is described by the following well-known solution

$$y(x,t) = \frac{Y}{2}\left\{ erf\left[\frac{l}{4\sqrt{Gt}}\left(\frac{2x}{l}+1\right)\right] - erf\left[\frac{l}{4\sqrt{Gt}}\left(\frac{2x}{l}-1\right)\right]\right\}, \tag{16}$$

in which '*erf*' is the so-called error function, *Y* is the initial beach width, and l is the length of the initial fill. It can be shown by integrating Eq. (16) over the beach segment in which the rectangular nourishment was placed, that the proportion of material remaining, *M(t),* at any time, *t*, is given by

$$M(t) = \frac{\sqrt{4Gt}}{l\sqrt{\pi}}\left(e^{-(l/4\sqrt{Gt})^2} - 1\right) + erf(l/\sqrt{4Gt}), \tag{17}$$

Figure 15 is a plot of Eq. (17) and it is seen that initially sediment is lost fairly rapidly from the project limits and later the losses occur much more slowly. Referring to either Eqs. (16) or (17), if the group of terms $(Gt)^{1/2}/\ell$ is the same for two projects, the projects will evolve in a geometrically similar form. Thus the longevity of a project depends only on this parameter. Referring to Figure 15, we can develop a simple and useful equation for the time at which X% of the material placed will remain within the placement area. Thus, this time is a reasonable measure of the longevity of the project. As an example, selecting a value of 50%, that is, the half-life of the project, we see by referring to Figure 15 that the associated parameter, $(Gt)^{1/2}/\ell$, is equal to 0.46 which results in the following equation

$$t_{50\%} = \frac{(0.46)^2 l^2}{G}. \tag{18}$$

Selecting reasonable values for most of the parameters, it can be shown that the half-life is given by

$$t_{50\%} = K'' \frac{\ell^2}{H_b^{5/2}}, \tag{19}$$

where *K'* is equal to 0.179 years – $m^{5/2}/km^2$ for $t_{50\%}$ in years, H_b in meters, and ℓ in kilometers. Thus, if an initially rectangular beach nourishment project is placed in an area where the effective wave height is 1 meter and the project length, ℓ, is 1 km, the half-life will be 0.18 years, slightly in excess of 2 months. If the project length is doubled, the half-life is quadrupled because of the square of the project length appearing in Eq. (19). For a project length of 10 km and a wave height of 1 m, the projected half-life is nearly 18 years! With this as background, we can now return to Figure 15 and see why the losses from the project area occur rapidly at first and then later much more slowly. As losses occur from the project area, the sand from these losses is deposited adjacent to the project area, and in effect, the project is effectively longer than it was initially.

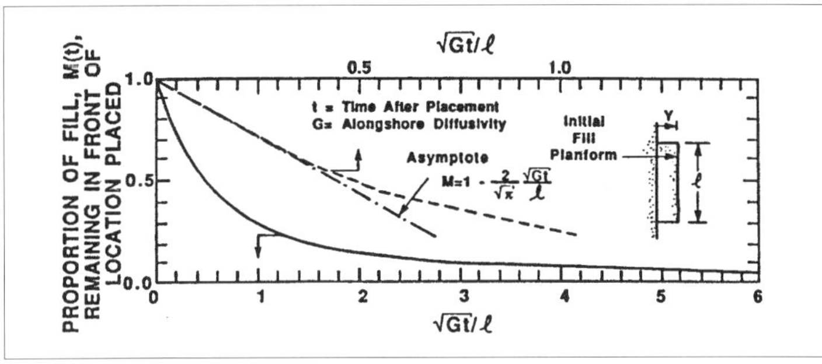

Figure 15. Proportion of material remaining in region placed vs. the parameter $\sqrt{Gt} / \ell$

The recognition that short projects lose sand more rapidly than long projects can be applied immediately to the case of repeated nourishments, a problem first considered by Dette et al (1994). It can be shown through reasonably simple numerical modelling that for a project which is nourished and then allowed to evolve until a threshold proportion of the initial nourishment remains within the project area and the project area is then renourished back to its original nourished volume, the required time intervals between renourishments *increases* with renourishment number. The explanation for this follows from the above discussion. After the first renourishment, there are two nourishments in place that are losing sand (recall the linearity of the Pelnard Considère equation and resulting superimposability of solutions). The first nourishment, now functioning as a longer project, is losing sand more slowly than after the initial nourishment whereas the second nourishment (the first *renourishment*) loses sand at the same proportional rate as the initial nourishment. Thus, the composite losses are slower than they were for the initial nourishment. With continued renourishments at any particular time, the rate at which sand is lost from the nourishment area is a composite of the initial nourishment and all of the preceding renourishments. The longer a project has been in place, the more slowly the composite losses from the project area. However, in the presence of background erosion, the renourishment intervals initially increase with renourishment number, and then decrease for larger renourishment numbers. The interpretation for this, though rational, is somewhat complicated. Figure 16 presents an example of multiple nourishments for a zero background erosion rate.

A second specific example of a Pelnard Considère solution for beach nourishment demonstrates the significance of boundary conditions such as might occur on a barrier island. Consider an initially rectangular planform on a barrier island in which the nourishment can extend to the limits of the barrier island at an inlet or the nourishment can be set back from the inlet. In this case, approximating the shoreline displacement as zero at the inlet is the most logical boundary condition. The results for an initial nourishment extending along an entire barrier island compared with a shoreline of infinite length are presented in Figure 17. The advantages of nourishment on a long beach are clearly evident since the effect of an inlet is to cause the sand to be drained off much more rapidly. Figure 18 shows the proportion remaining for different lengths of the nourishment relative to the length of the barrier island.

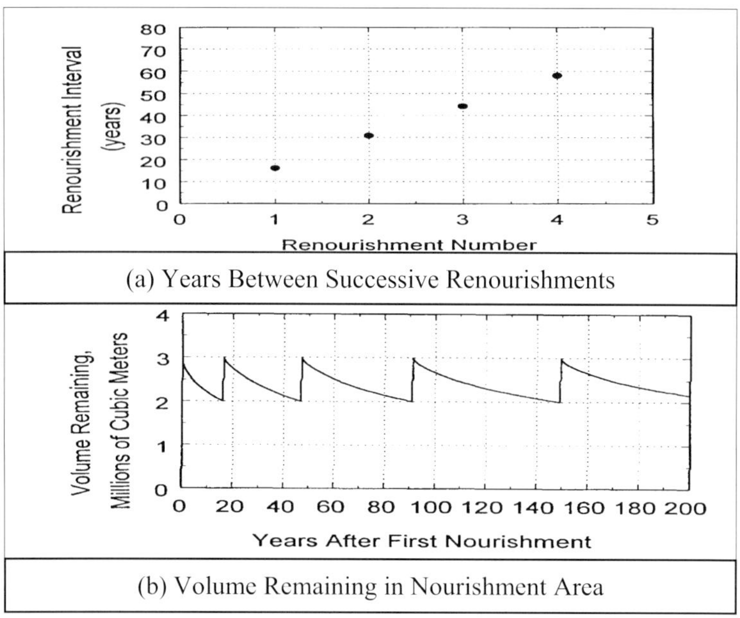

Figure 16. Renourishment characteristics. No background erosion. H_b = 0.5 m. Project length = 6 km, $h_* + B = 9\ m$, V_{max} = 3,000,000 m3, V_{min}= 2,000,000 m^3 (Dean, 1995)

Larson et al (1997) have presented a number of solutions to the Pelnard Considère equation which are of interest to coastal engineering practice. Walton (1994) has presented an analytic solution to the case of an initially trapezoidal beach planform. Figure 19 presents the solutions to four initial planforms of coastal engineering interest including Walton's trapezoidal planform. Recognizing that trapezoidal planforms lose sediment more slowly than rectangular planforms, Hanson and Kraus (1993), following an earlier study by the U.S. Army Corps of Engineers (1982), have investigated the possible benefits of commencing with an initially trapezoidal planform. It can be shown that the apparent benefits of employing an initially trapezoidal planform rather than a rectangular planform are a function of the manner in which the benefits are calculated. For example, if we consider a particular length of shoreline, ℓ, and place an amount of fill in a rectangular planform from $-\ell/2$ to $+\ell/2$ and compare the evolution with that of an initially trapezoidal planform with the same volume of material and with the *mid-point* of the sloping sides at $+\ell/2$ and $-\ell/2$, it can be shown that at any future time the initially rectangular planform will maintain a greater proportion of material within the longshore limits of the initial trapezoidal planform than the initially trapezoidal planform. This is consistent with intuition since the initially trapezoidal planform is similar to the evolved rectangular planform at some later time.

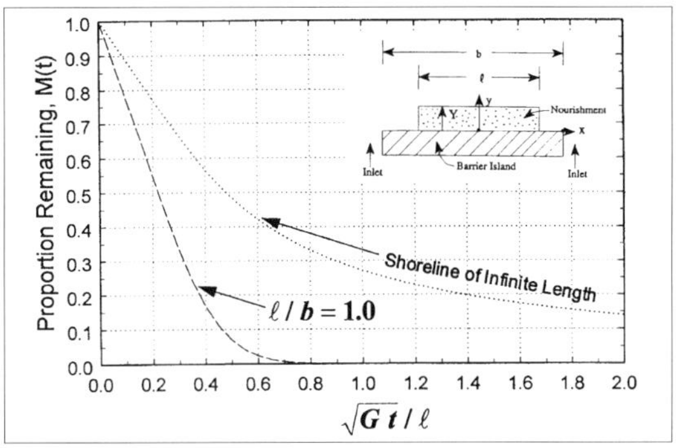

Figure 17. Proportion remaining, M(t), for nourishment along full length of barrier island compared to nourishment on an infinitely long beach, ' ℓ ' = nourishment project length (Dean, 1997)

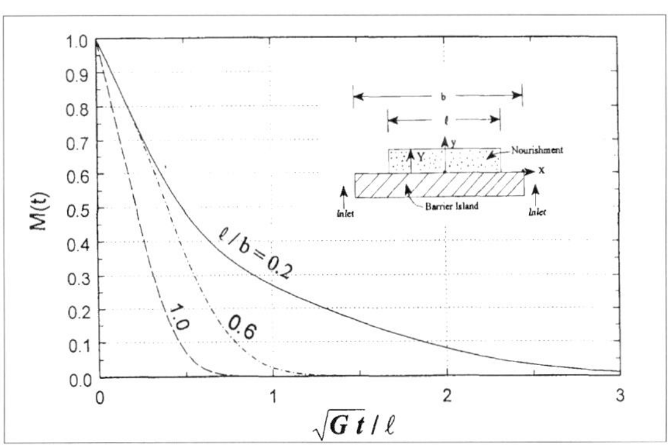

Fig. 18. Proportion remaining , M(t), for nourishment ends set back from ends of a barrier island (Dean, 1997).

There are a number of interesting and useful generic results that can be developed from the Pelnard Considère linearization and associated solutions. One result is that a beach nourishment project on a long straight beach is relatively insensitive to wave direction.

The reason is that if a project is located on a long straight beach, the effect of an oblique wave direction is to superimpose a uniform transport with the spreading out effect due to the planform anomaly. This follows from the superimposability of the Pelnard Considère equation. In this case, one solution is simply the background transport and the second is the result of the perturbation planform causing a symmetric spreading out of the initial planform. Note that Eq. (16) represents a symmetric evolution and wave direction does not occur in the longshore diffusivity, *G,* which is the only parameter in which wave characteristics occur. (Actually, there is a very weak dependency on wave direction in *G;* however, this has been neglected.).

Using the Pelnard Considère formulation, the evolution of a beach nourishment project can be shown to be independent of storm sequencing. As an example, considering a large storm and a small storm, the resulting evolution due to both storms does not depend on which storm occurs first. This can be considered in greater detail by noting that in Eq. (16) or (17) the relevant parameter is equal to the product of the longshore diffusivity *G* and time *t.* If the longshore diffusivity varies with time, usually predominantly due to varying wave heights, then the relevant quantity can be shown to be

$$\int_o^t G(t)dt \ .$$

The effect of wave refraction around the nourished planform is to reduce the losses from a beach nourishment project. Intuitively it is evident that if a beach nourishment project is placed on a long, straight beach and the waves break landward of the perturbed contours, then the contours will cause some refraction around the perturbation such that the breaking wave obliquity is reduced. The result is a reduced transport out of the placed planform. To a first approximation, this result can be represented simply by multiplying the longshore diffusivity in Eqs. (14) and (15) by the ratio of C_b/C_* where C_b represents the wave celerity at breaking and C_*, represents the wave celerity at the depth, h_*. In some cases this effect can be substantial, increasing the predicted project longevity by 20% to 40%.

3.3.3 Representation of background erosion

In almost all cases where implementation of beach nourishment is being considered, a background erosion rate exists. As used here, background erosion refers to the continuing erosion trend and is based on historical shoreline positions. Although the shoreline could be advancing, this is not usually the case in an area to be nourished. This background erosion rate can be relatively uniform within the longshore segment of interest or can vary substantially in the longshore direction. In the interest of realistic representation of the natural conditions and project performance, background erosion must be represented in beach nourishment design and prediction methodologies. Dean and Yoo (1992) have recommended that the background erosion be included in the prediction process simply by considering the background erosion to be associated with a gradient in longshore sediment transport and adding this background transport to the transport associated with the nourishment project in the following form

$$Q_T(x) = Q_B(x) + Q_P(x) , \qquad (20)$$

where Q_T , Q_B and Q_P represent the total, background, and 'project' transport, respectively.

If there are no structures within the shoreline segment considered, it can be shown that it is only the transport gradients $\partial Q_B/\partial x$ that are important and not the background transport itself. However, if internal or terminal structures are present within the project area, it is necessary to quantify the background transport, Q_B , which requires a more complete knowledge of the system. The background transport $Q_B(x)$ can be obtained from the pre-project (background) shoreline change rates, $\partial y_B/\partial t$, as

$$Q_B(x) = Q_B(x_o) - (h_* + B)\int_{x_o}^{x} \frac{\partial y_B}{\partial t}dx \quad (21)$$

where $Q_B(x_o)$ is the background transport rate at some reference location, x_0. As noted, for a long straight unstructured beach, the project evolution is independent *of* $Q_B(x_o)$.

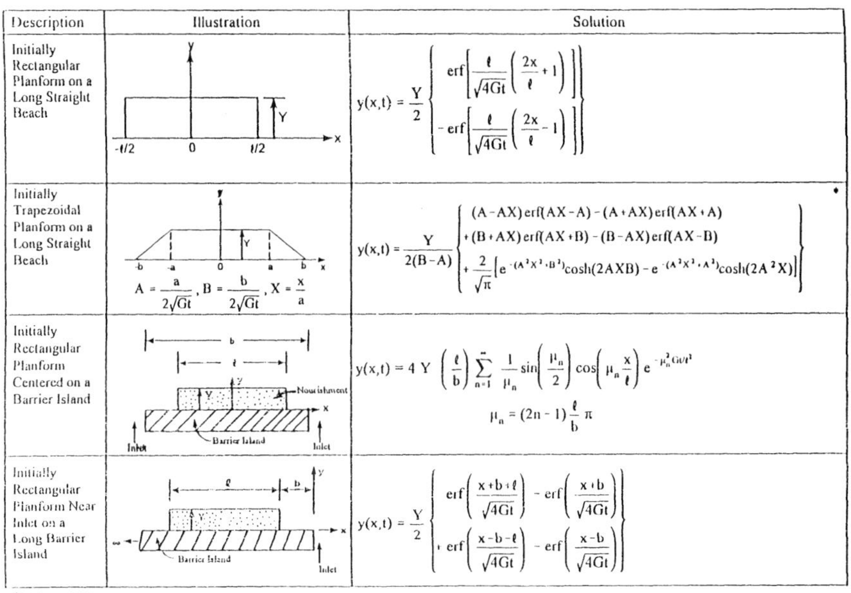

Figure 19. Solutions for four initial platform geometries and boundary conditions.

A second approach to including the effects of the background transport is to attempt to calibrate a numerical model such that the predicted shoreline changes over some previous period are consistent with the historical erosion rates and distributions. Unless the background erosion is a result of a recent large-scale perturbation introduced into the system such as a jetty or groin, this author's position is that in most cases, the causes of erosion are so subtle and possibly variable that it is not possible to calibrate a numerical model such that the background erosion is correctly related to the cause(s).

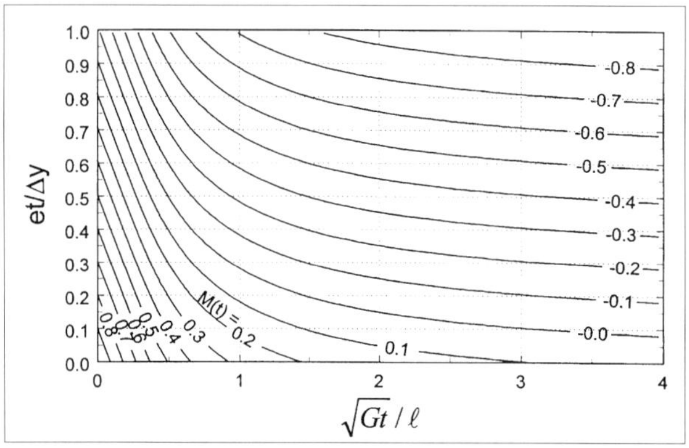

Figure 20. Isolines of *M(t)* vs spreading and background erosion parameters, e = erosion rate, Δy_o = initial equilibrium shoreline displacement, G = longshore diffusivity and t = time. Long straight beach (Dean, 1997)

Figure 20 presents isolines of the proportion of sand remaining in the nourishment area *M(t)* versus $\sqrt{Gt}/\ell$. and non-dimensional uniform background erosion rate, $et/\Delta y_o$, where e is the uniform background erosion rate. For $et/\Delta y_o = 0$ (along the abscissa), the values *of M(t)* are the same as plotted in Figure 15. It is noted that for non-dimensional erosion rates greater than 0.2 the portion of material remaining in the project area, *M(t),* becomes negative after a non-dimensional time (horizontal axis) of approximately 1.5.

3.3.4 Nourishment planform evolution in seawalled areas

This generic problem represents conditions in which a nourishment project is placed in an area where there is no other sediment available to be transported. This might be approximated in a case where a seawall has been placed and there is a rock bottom seaward of the seawall. This problem has been evaluated analytically, numerically, and in the laboratory by Dean and Yoo (1994) and by Yoo (1993). It was found that the wave direction for this problem is a key variable. Since for oblique waves, there is no 'feed' of sediment from the updrift direction as there is for projects on a sandy beach, the beach nourishment is cannibalized on the updrift end of the nourishment and transported to the downdrift direction where it is deposited, thus resulting in a downdrift migration of the planform centroid. For *normal wave incidence* it can be shown from idealized considerations that the project simply spreads out at the same rate as if it were located on a long, straight beach of compatible sand. For an oblique wave approach, the centroid of the project migrates in the downdrift direction, as has been mentioned earlier, and the project becomes longer and the rate of centroid migration increases with time. See Dean and Yoo (1994) for additional details.

3.3.5 Erosional hot and cold spots

Erosional hot spots (EHS) are defined as areas which erode more rapidly than anticipated and/or more rapidly than the adjacent shoreline segments. EHS can occur on natural or nourished beaches and they can be stationary or migrate. EHS on nourished beaches are of concern as they are usually interpreted as indicative of general poor project performance and may require equipment mobilization and renourishment earlier than would have been required otherwise. It is not unusual for an area of accretion to occur adjacent to an EHS. Such locations may be referred to as 'erosional cold spots' (ECS).

Bridges (1995) has identified eight possible causes of erosional hot spots. Some of these causes are related to the manner in which the nourishment was carried out and thus can be remedied, whereas the effects of other causes are unavoidable. However, the anticipation of the EHSs may allow minimization of their consequences, possibly by the placement of additional sand during nourishment or the selective use of structures.

Figure 21-Erosional 'hot' and 'cold' spots on Grand Isle, Louisiana (Combe & Soileau, 1987)

The causes of EHS in beach nourishment that can be avoided include those produced by wave transformation over bathymetric anomalies associated with the project. Combe and Soileau (1987) have documented two salients (ECS) that occurred landward of two borrow areas off Grand Isle, Louisiana (see Figure 21). Adjacent to these ECS were narrow beaches, considered to be EHSs. One possible interpretation of the effects in this area, advanced by Gravens and Rosati (1994), is that wave refraction over the borrow areas and adjacent bathymetry resulted in low waves landward of the borrow areas with a concentration of energy adjacent to the areas of low energy. The resulting currents due to gradients in wave set-up, from the regions of high to low waves caused in sediment transport to and deposition in the salient locations. A second explanation is that the offshore borrow areas reflect and dissipate wave energy, thus acting much like a detached breakwater causing the waves on either side to diffract with the curved crests causing transport into the lee of the dredged holes. Horikawa et al (1977) have conducted computer simulations and laboratory studies of the influence of dredged holes on the shoreline. The shoreline was found to advance in the lee of a dredged hole. The interpretation was that 'quiet water' was created in the lee of the dredged hole, resulting in sand accumulation which is consistent with both of the mechanisms described above. Wave transformation by offshore holes have been investigated by McDougal et al (1996) and the holes were shown to reflect and diffract wave energy.

The previous example has considered the effects of the borrow area on the equilibrium planform. The configuration of the placed material can also modify the equilibrium shoreline. Consider the case in which sand placement occurs in an irregular manner to depths greater than those rapidly configured by the waves. Dean and Yoo (1993) have shown that the orientation of the shoreline, $\Delta\beta_2$, is related to that of the seaward nourished contour $\Delta\beta_1$ by

$$\Delta\beta_2 = \Delta\beta_1 \left[1 - \left(\frac{C_2}{C_1} \right) \right], \tag{22}$$

where C_1 and C_2 are the wave celerities at the outer depths of the nourishment and at the closure depth, respectively and $\Delta\beta_1$ and $\Delta\beta_2$ are the angular deviations of the offshore contours and shoreline, respectively, from uniform alignment. Dean and Yoo refer to this as a 'residual bathymetry' effect.

A second and probably more frequent cause of an EHS is uneven alignment of the prenourished shoreline due to hardened shoreline segments. As an example, consider an eroding shoreline along which a valuable building is located. A shore-parallel seawall of 60 m length has been constructed to protect the building; however, the adjacent shoreline continues to erode to a position which is 20 m landward of the seawall. At this time, there is additional sufficient economic development of the adjacent upland that nourishment is justified, and it is decided to add a *uniform* beach width of 40 m along the entire coast. The 40 m advancement adjacent to the seawalled section represents a 20 m projection relative to the adjacent nourished shoreline. This projection of 20 m in the cross-shore direction by 60 m in the longshore direction is equivalent to a beach nourishment project of the same dimensions. Previous discussion has shown the substantial effect of nourishment length on project longevity. The beach fronting the seawall would erode rapidly from a 40 m width to a 20 m width, to establish uniform alignment with the adjacent beaches. The sand eroded adjacent to the seawall would spread out to the adjacent beaches with the interpreted result of an EHS fronting the building and an ECS on each side of the EHS. In such cases, a groin field or detached breakwater could be considered for stabilization of the protruding shoreline.

Other causes of EHS may include: (1) Not providing sufficient fill adjacent to seawalls at which there is initially no beach present. A 'threshold' amount of sand must be added to these profiles in order to fill the profile up to the mean water line. Rule of thumb application will underestimate the volume required to achieve a desired equilibrated beach width (Figure 10). (2) Profiles oversteepened by a groin field which, through trapping sand from the longshore sediment transport, has maintained the shoreline at a more or less fixed position. However, the seaward portion of the profile has continued to erode. As in (1) above, if the deepened offshore portion of the profile is not accounted for in establishing the required nourishment volumes, the equilibrated profile will be narrower than predicted by the usual rules of thumb. (3) The use of sand that is finer than the native: as described earlier in this chapter, sand size is a central parameter controlling the width of the equilibrated profile. If a borrow area includes regions of sand which are finer than the average and if the contractor is not constrained to those portions where the coarser sand is located, because the costs of pumping fine sand are less than coarse sand, it is to the contractor's advantage to pump the finer sand the greater distances.

3.3.6 Effects of sediment size on dry beach area evolution

The additional planform area resulting from a beach nourishment project is the dominant measure applied by the lay person in evaluating and judging project success and thus is an important performance measure. Preceding sections have considered that the beach is nourished with compatible sands; however, in the earlier discussions of equilibrium beach profiles, we have seen that the equilibrium width of the nourished profile depends substantially on the size of the nourishment sediment relative to that of the native profile. Thus it can be appreciated that if the same volume of nourishment material is used, the initial equilibrated planform area will be substantially larger for a coarser sand than for a smaller nourishment sand size. However, a surprising result is that if the nourishment sand is coarser than the native, the equilibrated dry beach total plan area *increases* with time whereas if the nourishment sand is smaller than the native sand, the equilibrated dry beach planform area *decreases* with time. This can be shown readily by referring to Figure 10, which presents the relationship between the additional dry beach width versus volume per unit length of beach for three different sand sizes. It is seen that for the coarser sand, the ratio of dry beach width to volume added is greatest for the smaller volume densities. Thus, as the beach nourishment project spreads out on a long beach, there will be smaller and smaller values of volume density, resulting in the ratio of additional dry beach width to volume density increasing with time. In the limit for infinite time it can be shown that for the case of nourishment sand characterized by a single size which is coarser than the native, the dry beach width Δy is related to the volume density, V, by $\Delta y(x)=V(x)/B$. Thus, the total plan area, at infinite time, $PA(t=\infty)$ is

$$PA(t=\infty)=\int_{-\infty}^{\infty}\frac{V(x)}{B}dx=\frac{V_{TOT}}{B}, \tag{23}$$

where V_{TOT}- is the total volume added. Following the same argument, the total plan area for a nourished beach composed of sand which is compatible with (of the same size) as the native will result in a plan area that does not change with time and is given by [see Eq. (6)]

$$PA(t) = \int_{-\infty}^{\infty} \frac{V(x)}{h_* + B}\, dx = \frac{V_{TOT}}{\left(h_* + B\right)}, \tag{4}$$

Considering now the situation where a beach is nourished with sand which is finer than the native and referring again to Figure 10, it can be seen that for the smaller volume densities, the additional dry beach width approaches zero at the threshold volume. Thus, for this case the total plan area would evolve to a limit of zero (no additional dry beach plan area) at a time when the greatest alongshore volume density equals the threshold volume density. These two limits for nourishment sands that are finer and coarser than the native are somewhat unrealistic since they predict limiting plan areas which are independent of the amount by which the nourishment and native sand sizes differ. The only condition required for these limits to occur is that the sizes be coarser or finer than the native. Figure 22 presents an example illustrating these results in non-dimensional form for three different sand sizes.

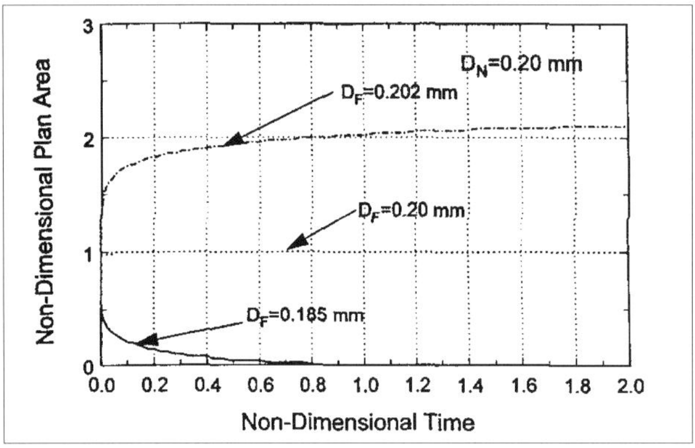

Figure 22. Non-dimensional plan area vs. non-dimensional time for various sizes of fill material. Single nourishment size sand (Dean, 1998)

A more realistic representation is one in which the nourishment sands are considered to be composed of two sizes, as shown in Figure 23. In this case, a recognition of the sorting of the nourishment sediments is considered by representing the nourishment sediments as composed of two fractions of different proportions. The method requires that one of the fractions be of the same grain size as the native and the second can be either coarser or finer than the native. It is assumed, consistent with findings in nature, that the coarser of the fractions is located in the nearshore region. In Figure 23, because the average of the nourishment material is coarser than the native, the sands of the same size as the native are located in the seaward portion of the profile. Characterization of the sand size distribution into these two sizes requires a procedure which will not be discussed here. The profiles for each of the fractions follow Eq. (4). The results of this representation for the nourishment sediments are more realistic since now the limiting plan area (t ® ¥) is dependent on the proportion and size of the fraction of sand which differs from the native. In this case, the evolution of the beach nourishment planform

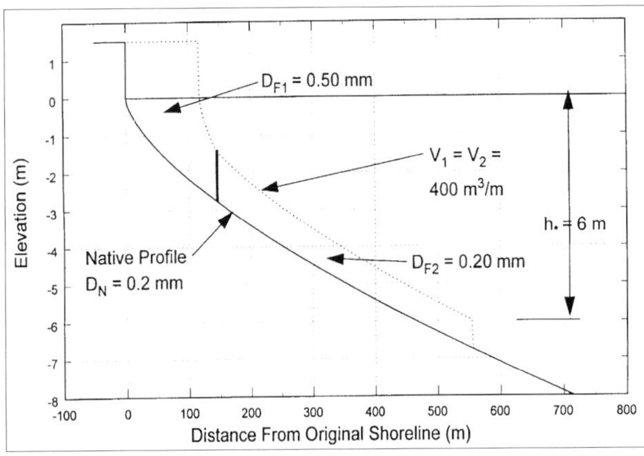

Figure 23. Example of $V_1 = V_2$ with $D_N = 0.20$ mm. $D_{F1} = 0.50$ mm, $D_{F2} = 0.20$ mm, h_*=6 m, B = 1.5 m (Dean, 1998).

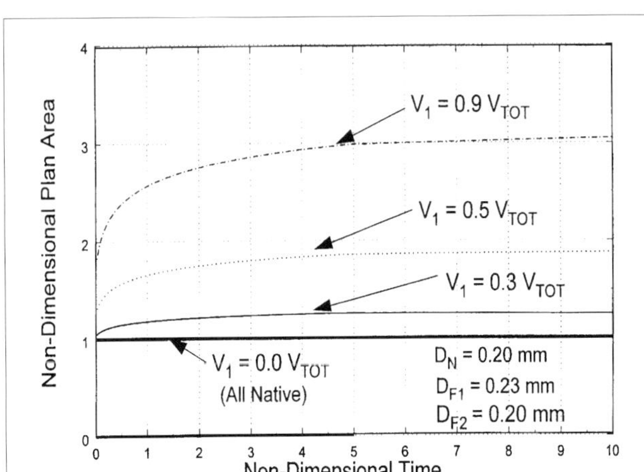

Figure 24. Example calculation for nourishment sand coarser than the native. *Effect of proportions of coarser sand.* Two nourishment sizes considered (Dean, 1998)

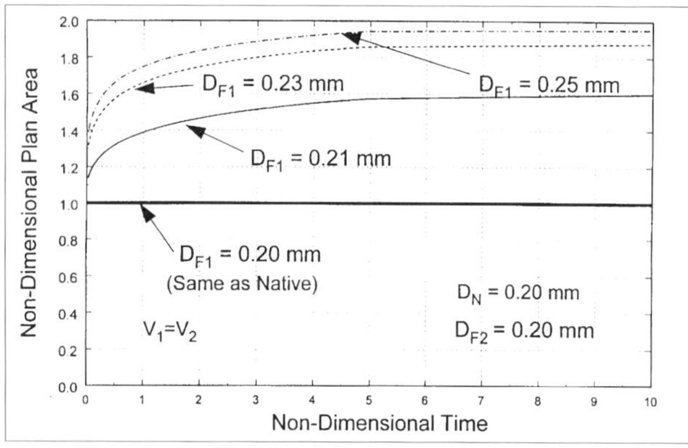

Figure 25. Example calculation for nourishment sand coarser than the native. *Effect of sand size*. Two nourishment sizes considered (Dean, 1998)

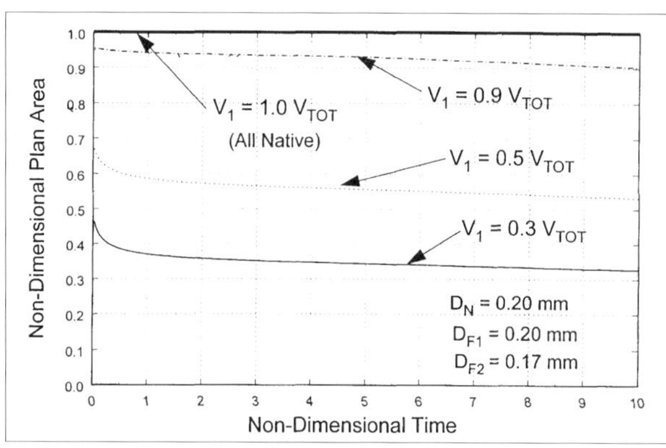

Figure 26. Example calculation for nourishment sand finer than the native. *Effect of proportions of finer sand*. Two nourishment sizes considered (Dean, 1998)

can be determined using a somewhat more complicated numerical model. Several examples will be presented to illustrate this. Figure 24 shows the results for a case in which the native sand is 0.20 mm and the average of the nourishment sand is coarser than the native such that the finer size component is the same size as the native (0.20 mm) and the second component is coarser, in this case 0.23 mm. The proportion of the coarser (0.23 mm) sand ranges from 0.0 to 0.9 of the total volume placed. Figure 25 illustrates the planform evolution for cases in which the average of the nourishment sand is coarser than the native with equal proportions of nourishment sand equal to and coarser than the native. In this example the *size* of the coarser fraction is varied. Again, the sensitivity of both initial and asymptotic plan areas on sand size are substantial. Figures 26 and 27 are the counterparts of Figures 24 and 25, respectively, for the case in which the averages of the nourishment sands are finer than the native.

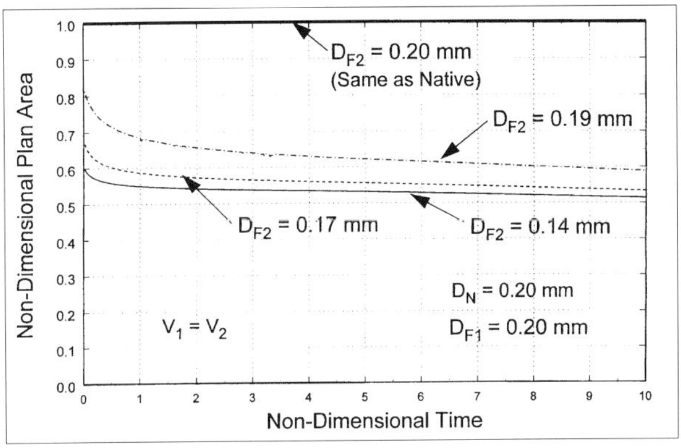

Figure 27. Example calculation for nourishment sand finer than the native. *Effect of sand size*. Two nourishment sizes considered (Dean, 1998).

4. Detailed methodology for representing project evolution

The preceding sections have reviewed simplified methods of predicting the cross-shore and planform characteristics of beach nourishment projects. These methods provide a good basis for understanding the general characteristics of project evolution, an approximate estimate of the time scales and an appropriate framework for preliminary design. However, these simple methods are limited in the detail of forcing that can be represented and in the initial and boundary and internal conditions that can be simulated. The present section describes numerical models which allow greater flexibility and realism to be incorporated into the predictions. The best basis for developing an overall understanding of a particular design is a *combination* of simple and more detailed models.

4.1 MODELS OF PROFILE EVOLUTION

Cross-shore sediment transport processes are so complex that practically all models of profile evolution require numerical solution. Two types of profile evolution exist which can be termed ‘open loop’ and ‘closed loop’. All such models employ a transport equation and the conservation of sand equation.

4.1.1 Open loop profile evolution models

Open loop profile evolution models require detailed or idealized representation of the hydrodynamics and sediment transport processes. As an example, the suspended sediment distribution and mean velocity may be a component, or the dominant transport mechanism. Some models include bed load transport as a principal component. Most, if not all, of these open loop models, if allowed to run indefinitely under the action of steady forcing, will become unstable unless some artificial 'smoothing' of the results is introduced. Examples of open loop models are given in: Bailard and Inman (1981), Dally and Dean (1984), Watanabe (1988), Southgate and Nairn (1993) and Nairn and Southgate (1993). Roelvink and Hedegaard (1993) have presented a review of several open-loop models.

4.1.2 Closed loop profile evolution models

Closed loop profile models are characterized by the specification of a particular 'target profile', to which the computed profile will converge if the forcing mechanisms are held constant indefinitely, thus ensuring computational stability. Eq. (4) has been used as the target profile in at least three of these models. Because Eq. (4) is consistent with uniform wave energy dissipation per unit water volume, D_*, within the surf zone, (where the subscript '$_*$' denotes the value for equilibrium conditions), the seaward transport, q, (in length squared per unit time) is related to the actual (non-equilibrium) wave energy dissipation per unit volume D by

$$q = K'(D - D_*)^m, \tag{25}$$

where K' and m are constants and in two of the closed loop models, $m = 1$. Examples of closed loop profiles are given in: EDUNE (Kriebel & Dean, 1985), and SBEACH (Larson & Kraus 1989). In EDUNE, the cross-shore positions of the contours are specified and it is thus not efficient to represent non-monotonic profiles. SBEACH represents the offshore distance as a series of grids which readily allow representation of offshore bars. The closed loop models are much more effective and reliable for the beach and dune *erosion* phase than for the *recovery* phase, which occurs on a longer time scale and can be described by the sediment moving onshore in 'pulses'. Closed loop models are considered much more suitable for engineering applications than the open loop models, which are more in the research and development arena.

4.2 MODELS OF PLANFORM EVOLUTION

The simplest model for planform evolution is the 'one-line' model, so called because the cross-shore position *y(x,t)* of one contour line is represented. A basic assumption in the one-line model is that the profile is displaced landward or seaward in response to gradients in the longshore transport Q as represented by Eq. (26). All one-line models incorporate a transport equation and an equation representing the conservation of sand. The solution to these equations may be explicit or implicit. One equation for sediment transport is the so-called CERC equation

$$Q = K \frac{(EC_G)_b}{2\rho g(s-1)(1-p)} \sin 2(b - \alpha_b), \tag{26}$$

in which E is the wave energy density, β and α are defined in Figure 28, and the subscript 'b' denotes that the subscripted variables are to be evaluated at wave breaking. The continuity equation is

$$(h_* + B)\frac{\partial y}{\partial t} + \frac{\partial Q}{\partial x} = 0, \tag{27}$$

written in (explicit) finite difference form as

$$y_i^{n+1} = y_i^n - \frac{1}{(h_* + \beta)}\frac{\Delta t}{\Delta x}(Q_{i+1}^n - Q_i^n), \tag{28}$$

in which the superscript represents the time level, and the subscript the cell for which the computations are being carried out. Note that the '*i*' index on the transport references the grid cell onto which positive transport occurs. The grid system on which these equations are solved is shown in Figure 28, where it is seen that the *Q*- values are defined on the grid lines and the *y* values at the center of the cells. The explicit method solves the transport and conservation equations sequentially whereas the implicit method solves these two equations simultaneously. A second difference is that there is a limiting time step, Δt, associated with the explicit method of solution

$$\Delta t_{max} < \frac{\Delta x^2}{2G}, \tag{29}$$

where *G* is defined by Eq. (14) or (15). The solution of the explicit method is described first.

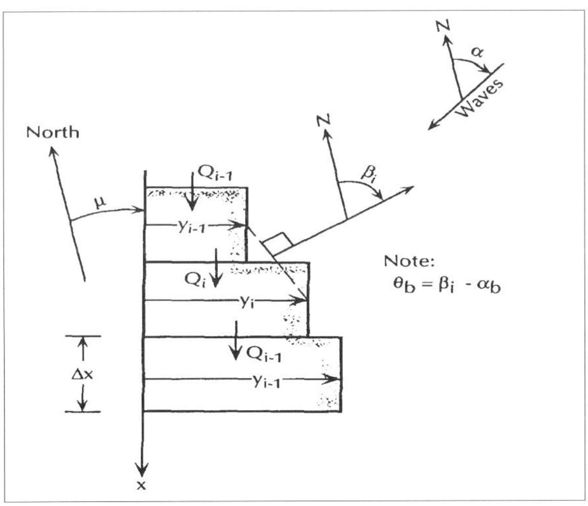

Figure 28. Definition sketch for one-line numerical model

In the explicit method, the shoreline displacements are held fixed and the transport values, Q_i^n, computed. For this purpose the β_i values are

$$\beta_1^n = \mu + \frac{\pi}{2} - tan^{-1}\left(\frac{y_i^n - y_{i-1}^n}{\Delta x}\right). \tag{30}$$

With the new values of $Q^n{}_i$, the y_i values are then updated from time level n to $(n+1)$ by Eq. (28). The process is then repeated until the desired time is reached. One useful characteristic of the explicit method is that it is relatively straightforward to incorporate boundary and internal conditions on shoreline displacements or discharges. An example would be a terminal structure at one or both ends of a computational region which blocks the sediment transport completely, and this would simply require that the transport at these locations be assigned the value of zero. A second boundary condition could be at a groin at the downdrift end of a computational domain at which the shoreline is at the groin terminus. In this case, the shoreline displacement at this point would be specified. Structures such as groins can alternately require representation as an internal boundary condition on y or Q depending on wave direction and shoreline position relative to the structure. Also, because a limitation of the one-line model is the previously noted displacement of the profile without change of form, it may be appropriate to prescribe a gradual transition between full by-passing and no by-passing, depending on the shoreline position relative to the groin terminus. Such a scheme would better mimic nature. GENESIS is an example of a one-line model that incorporates substantial flexibility in terms of boundary and internal conditions (Hanson, 1989; Hanson & Kraus, 1989). Implicit models of shoreline change are usually represented as a tri-diagonal matrix which can be solved by standard numerical procedures.

4.3 COMBINED (THREE-DIMENSIONAL) MODELS

Combined models include consideration of both the cross-shore and longshore sediment transport components simultaneously and incorporate interactions of these two components. This interaction can be especially significant in the vicinity of structures whereby the profile steepens and flattens on the updrift and downdrift sides of the structure, respectively, inducing seaward and landward cross-shore transport. As discussed for the case of cross-shore sediment transport, the cross-shore and longshore transport components of the model can be at the engineering or research level. The engineering level would incorporate a model of the CERC type [Eq. (26)] for longshore sediment transport together with a function that distributes the total transport across the active zone. Examples of this cross-shore distribution function have been presented by Komar (1977), and a comprehensive review has been published by Bodge (1989). For an engineering level model, the cross-shore element would usually be of the ‘target’ type such that the profile would converge to a prescribed form. Solution of the engineering type models can be by an explicit procedure in which the cross-shore and longshore transport components are solved sequentially, by an implicit procedure in which the two components are determined simultaneously, or a combination implicit/explicit procedure, which may be an improvement on the other two methods. An example of the combined engineering models is that of Bakker (1968) which is a ‘two-line’ model, that is, two depth contours are represented in the process. Perlin and Dean (1985) presented an ‘n-line’ model in which the solution was fully implicit, and an arbitrary number (n) of contour lines was represented. Dean (1996) presented the results of applying an n-line model that was solved by a combined implicit-explicit method. A disadvantage of monitoring contours in which the shore-normal coordinates to several contours are represented is that it is difficult to simulate the presence of bars, which cause the profile to be non-monotonic such that there are several distances from the shoreline to the same depth contour. For situations in which bars are present, a grid system in which the depth is the independent variable, and x *and* y are the independent variables is more effective.

The research level models are generally of the type that establish the suspended sediment transport based on calculated fields of hydrodynamics and suspended sediment. The contribution of bedload is calculated separately. The cross-shore and longshore sediment transports are thus components of a transport vector. It should be noted that several of these models are used for engineering problems and some are only available commercially and require a purchase and/or a licensing agreement.

5. Summary

A review has been presented of the available methodology for the design and performance prediction of beach nourishment projects. Both 'simple' and 'detailed' methodologies have been examined. The simple methodologies consider the cross-shore and longshore processes separately, represent the forcing (waves and tides) as constant or by representative values and are useful for preliminary design and for demonstrating the interrelationships between the various parameters. The detailed methodologies employ numerical models and are available at various levels of complexity and detail. They offer advantages of accommodating the time-varying input of the forcing (waves and tides) and greater flexibility with boundary conditions and internal conditions such as a groin field. For the simple treatment of the cross-shore processes, the focus is on the equilibrium beach width resulting from placement of a specified amount of sand per unit beach length. Using equilibrium beach profile methodology, it is shown that three generic types of nourished profiles can result: intersecting, non-intersecting and submerged. The equilibrium beach width varies substantially with the size of the nourishment sediment relative to that of the native. The planform evolution of nourishment projects is investigated through the Pelnard Considère equation and the associated solutions. For the case of nourishment on a long straight beach, the longevity of a project varies with the square of the project length and inversely as the 2.5 power of the wave height. Other results and/or methodologies that are demonstrated at the simple level include background erosion, the increase of renourishment intervals for a project on a long straight beach with renourishment number where the background erosion is small and where a minimum threshold of sediment volume within the nourishment area is specified, a simple method to account for refraction around a nourished planform, effects of nourishing with sands coarser and finer than the native, planform migration of nourishment in an area which is otherwise devoid of sediment, and the invariance of total evolution to the sequencing of storms. Moreover, the total plan area of a nourishment project on a long straight beach is shown to be dependent on the nourishment sand size relative to the native sand size and to evolve to larger and smaller total plan areas for nourishment sands coarser and finer than the native, respectively.

A short review of the general characteristics of detailed calculation methodology is presented. This methodology is carried out with numerical models, which are developing rapidly. All models of this type can be solved by an explicit, implicit, or a mixed explicit-implicit approach. Results are best simulated for systems which are substantially out of equilibrium, such as a large beach nourishment project, the imposition of a structure which interrupts a significant amount of longshore sediment transport, or a profile that is subjected to a storm surge. All models require one or more transport equations and a conservation equation. Two types of methodologies are available for profile evolution models involving cross-shore sediment transport. In this chapter, these models have been referred to as engineering models which specify a target profile to which the solution will converge if the forcing conditions are maintained constant and a 'research' methodology which is based on calculated wave and suspended sediment transport fields and bedload transport. This latter class is not considered to be appropriate yet for engineering applications. Profile response models represent the erosion mode of profile evolution much better than the recovery mode. The simplest planform evolution models are based on a total transport model and a continuity equation that assumes the profile to respond to gradients in sediment transport by being displaced landward or seaward without any change of form. More complex models are based on calculations of the hydrodynamics and suspended sediment fields and the bedload transport. Combined (three-dimensional) models compute the profile and planform evolution simultaneously and generally parallel the characteristics, limitations and features of the individual models for profile and planform evolution.

Acknowledgements

The author is indebted to the Florida Sea Grant Program and the Office of Beaches and Coastal Systems of the Florida Department of Environmental Protection, which have provided funding over the years for the work which he and his students have conducted on nearshore processes and related engineering applications.

REFERENCES

Ahrens, J.P. and Hands, E.B.: 'Velocity Parameters for Predicting Cross-Shore Sediment Movement', *Journal of Waterway, Port, Coastal and Ocean Engineering,* American Society of Civil Engineers, Vol. 124, No. 1, 1998, 6–20.

Andrassy, C.J.: 'Monitoring of a Disposal Mound at Silver Strand State Park', *Proc. Coastal Sediments '91*, American Society Civil Engineers, 1991, 1970–1984.

Bailard, J.A. and Inman, D.L.: 'An Energetics Bed Load Model for a Plane Sloping Beach: Local Transport', *Journal of Geophysical Research*, Vol. 86, No. C3, 1981, 2035–2043.

Bakker, W.T.: 'The Dynamics of a Coast With a Groyne System', *Proc. 11th International Conference on Coastal Engineering,* 1968, 492–517.

Birkemeier, W.A.: 'Field Data on Seaward Limit of Profile Change', *Journal of Waterway, Port, Coastal and Ocean Engineering*, American Society of Civil Engineers, Vol. 111, No. 3, 1985, 598–602.

Bodge, K.R.: 'A Literature Review of the Distribution of Longshore Sediment Transport Across the Surf Zone', *Journal of Coastal Research*, Vol. 5, No. 2, 1989, 307–328.

Bridges, M.H.: 'Analysis of the Processes Creating Erosional Hot Spots in Beach Nourishment Projects', MSc. thesis, Department of Coastal and Oceanographic Engineering, University of Florida, Gainesville, FL, 1995.

Browder, A.E. and Dean, R.G.: 'Perdido Key Beach Nourishment Project: Gulf Islands National Seashore', University of Florida Coastal and Oceanographic Engineering Department, Report UFL/COEL-97/014, 1997.

Bruun, P.: 'Coast Erosion and the Development of Beach Profiles', U.S. Army Corps of Engineers, Beach Erosion Board, Tech Memo No. 44, 1954.

Combe, A.J. and Soileau, C.W.: 'Behavior of Man-made Beach and Dune, Grand Isle, Louisiana', *Proc. Coastal Sediments '87*, ASCE, 1987, 1232–1242.

Dally, W.R. and Dean, R.G.: 'Suspended Sediment Transport and Beach Profile Evolution', *Journal of Waterway, Port, Coastal and Ocean Engineering,* American Society of Civil Engineers, Vol. 110, No. 1, 1984, 15–33.

Dalrymple, R.A.: 'A Finite Amplitude Wave or a Linear Shear Current', *J. Geophysical Research,* Vol. 79, No. 30, 1974, 4498–4504.

Dean, R.G.: 'Stream Function Representation of Nonlinear Ocean Waves', *J. Geophysical Research,* Vol. 70, No. 18, 1965, 4561–4572.

Dean, R.G.: 'Equilibrium Beach Profiles: U.S. Atlantic and Gulf Coasts', Department of Civil Engineering, Ocean Engineering Report No. 12, University of Delaware, January, 1977.

Dean, R.G.: 'Coastal Sediment Processes: Toward Engineering Solutions'. In: *Proc. Coastal Sediments '87,* Specialty Conference on Advances in Understanding of Coastal Sediment Processes, New Orleans, LA, ASCE, 1987, 1–24.

Dean, R.G.: 'Equilibrium Beach Profiles: Principles and Applications', *J. Coastal Res.,* 7, 1, 1991, 53–84.

Dean, R.G.: 'Beach Nourishment: Planform Considerations', *Proc., Coastal Dynamics '95,* American Society of Civil Engineers, 1995, 533–546.

Dean, R.G.: 'Interaction of Littoral Barriers and Adjacent Beaches - Effects on Profile Shape and Shoreline Change', *Journal of Coastal Research,* Special Issue, No. 23, 1996, 103–112.

Dean, R.G.: 'Models for Barrier Island Restoration', *Journal of Coastal Research,* Vol. 13, No. 3, 1997, 694–703.

Dean, R.G.: 'Beach Nourishment: a Limited Review and Some Recent Results', *Proc., 26th International Conference on Coastal Engineering,* Keynote Address, ASCE, Copenhagen, Denmark, 1998, 45–69.

Dean, R.G. and Yoo, C.-H.: 'Beach Nourishment Performance Predictions', *Journal of Waterway, Port, Coastal, and Ocean Engineering.,* American Society of Civil Engineers, Vol. 118, No. 6, 1992, 567–586.

Dean, R.G. and Yoo, C.-H.: 'Predictability of Beach Nourishment Performance', *Proc., Coastal Zone Management,* Special Volume on 'Beach Nourishment Engineering and Management Considerations', edited by D.K. Stauble and N.C. Kraus, American Society of Civil Engineers, 1993, 86–102.

Dean, R.G. and Yoo, C.-H.: 'Beach Nourishment in Presence of Seawall', *Journal of Waterway, Port, Coastal, and Ocean Engineering.,* American Society of Civil Engineers, Vol. 120, No. 3, 1994 302–316.

delValle, R., Medina, R. and Losada, M.A.: 'Dependence of Coefficient K on Grain Size', Technical Note No. 3062, *Journal of Waterway, Port, Coastal, and Ocean Engineering,* American Society of Civil Engineers, Vol. 119, No. 5, 1993, 568–574.

Dette, H.H., Führböter, A. and Raudkivi, A.J.: 'Interdependence of Beach Fill Volumes and Repetition Intervals,'' *Journal of Waterway, Port, Coastal, and Ocean Engineering,* American Society of Civil Engineers, Vol. 120, No. 6, 1994, 580–593.

Douglass, S.L.: 'Estimating Landward Migration of Nearshore, Constructed Sand Mounds', *Journal of Waterway, Port, Coastal and Ocean Engineering,* American Society of Civil Engineers, Vol. 121, No. 5, 1995, 247–250.

Gravens, M.B. and Rosati, J.D.: 'Numerical Model Study of Breakwaters at Grand Isle, Louisiana', Miscellaneous Paper CERC-94-16, U.S. Army Corps of Engineers, 1994, 75pp.

Grosskopf, W.G. and Behnke, D.H.: 'An Emergency Remedial Beach Fill Design for Ocean City, Maryland', *Shore and Beach,* Vol. 61, No. 1, 1992, 8–12.

Grosskopf, W.G. and Stauble, D.K.: 'Atlantic Coast of Maryland (Ocean City) Shoreline-Protection Plan', *Shore and Beach,* Vol. 61, No.l, 1992, 3–7.

Hallermeier, R.J.: 'Calculating a Yearly Limit Depth to the Active Beach Profile', Coastal Engineering Research Center, U.S. Army Corps of Engineers, Technical Paper, No. 77–9, 1977.

Hallermeier, R.J.: 'Uses for a Calculated Limit Depth to Beach Erosion', *Proc. 16th Intl. Conf. Coastal Engrg.,* American Society of Civil Engineers, Hamburg, 1978, 1493–1512.

Hallermeier, R.J.: 'Sand Motion Initiation by Water Waves: Two Asymptotes', *Journal of Waterway, Port, Coastal, and Ocean Engineering,* American Society of Civil Engineers, Vol. 106. No. 3, 1980, 299–318.

Hallermeier, R.J.: 'A Profile Zonation for Seasonal Sand Beaches from Wave Climate', *Coast. Engrg.,* 4(3), 1981, 253–277.

Hands, E.B.: 'Unprecedented Migration of a Submerged Mound Off Alabama Coast', *Proc. of the Twelfth Annual Conference of the Western Association and the Twenty Fourth Annual Texas A&M Dredging Seminar*, Las Vegas, NV, 1991.

Hands, E.B. and Allison, M.C.: 'Mound Migration in Deeper Waters and Methods of Categorizing Active and Stable Depths', *Proc. Coastal Sediments '91,* American Society of Civil Engineers, 1991, 1985–1999.

Hands, E.B. and DeLoach, S.R.: 'An Offshore Mound Constructed of Dredged Material', *Proc., Dredging and Dredged Material Disposal, Dredging '84,* New York, NY, American Society of Civil Engineers, Vol. 2, 1980, 1030–1038.

Hands, E.B., DeLoach, S.R. and Vann, R.: 'Post-construction Adjustment of an Offshore Mound'. *Abstracts, 21st International Conference on Coastal Engineering,* New York, NY, American Society of Civil Engineers, 1988, 262–263.

Hands, E.G. and Resio, D.T.: 'Empirical Guidance for Siting Berms to Promote Stability or Nourishment Benefits', *Proc., 2nd International Conference on Dredging and Dredged Material Placement, Dredging '94,* New York, NY, American Society of Civil Engineers, 1994, 220–228.

Hanson, H.: 'GENESIS – a Generalized Shoreline Change Numerical Model', *Journal of Coastal Research,* Vol. 5, No. 2, 1–28.

Hanson, H. and Kraus, N.C.: 'GENESIS: Generalized Model for Simulating Shoreline Change, Report 1, Technical Reference', Technical Report No. 89–19, Coastal Engineering Research Center, Waterways Experiment Station, Vicksburg, MS, 1998.

Hanson, H. and Kraus, N.C.: 'Optimization of Beach Fill Transitions', in *Beach Nourishment Engineering and Management Considerations*, D.K. Stauble and N.C. Kraus, eds., *Coastal Zone '93,* American Society of Civil Engineers, 1993, 103–117.

Horikawa, K., Sasaki, T. and Sakuramoto, H.: 'Mathematical and Laboratory Models of Shoreline Changes due to Dredged Holes', *Journal of the Faculty of Engineering,* University of Tokyo, Vol. 34, No. 5, 1977, 49–57.

Jensen, R.E. and Garcia, A.: 'Wind, Wave and Water Level Assessment for the January 4, 1992 Storm at Ocean City, Maryland', *Shore and Beach,* Vol. 61, No. 1, 1992, 13–22.

Komar, P.D.: 'Beach Sand Transport: Distribution and Total Drift', *Journal Waterway, Port, Coastal, and Ocean Division,* American Society of Civil Engineers, Vol. 103, WW 2, 1977, 225–239.

Komar, P.D.: 'Environmental Controls on Littoral Sand Transport', *Proceedings, 21st International Conference on Coastal Engineering,* American Society of Civil Engineers, 1988, 1238–1252.

Komar, P.D. and Inman, D.L.: 'Longshore Sand Transport on Beaches', *J. Geophys. Res.,* 1970, 5914–5927.

Kraus, N.C.: 'The January 4, 1992 Storm at Ocean City, Maryland', *Shore and Beach,* Vol. 61, No. 1, 1993, 2.

Kraus, N.C. and Wise, R.A.: 'Simulation of January 4, 1992 Storm Erosion City, Maryland', *Shore and Beach,* Vol. 61, No.l, 1992, 34–40.

Kriebel, D.L.: 'Beach and Dune Response to Hurricanes', M.Sc. thesis, University of Delaware, Newark, DE, 1982, 349pp.

Kriebel, D.L. and Dean, R.G.: 'Numerical Simulation of Time-dependent Beach and Dune Erosion', *Coastal Engrg.,* 9, 3, 1985, 221–245.

Larson, M., Hanson, H., and Kraus, N.C.: 'Analytical Solutions of One-Line Model for Shoreline Change Near Coastal Structures', *Journal of Waterway, Port, Coastal and Ocean Engineering,* American Society of Civil Engineers, Vol. 123, No. 4, 1997, 180–191.

Larson, M. and Kraus, N.C.: 'SBEACH: Numerical Model for Simulating Storm-induced Beach Change, Report 1: Empirical Foundation and Model Development'. U.S. Army Coastal Engineering Research Center, Waterways Experiment Station, *Technical Report CERC-89-9,* 1989.

Laustrup, C., Madsen, H.T., Sorensen, P., and Broker, I.: 'Comparison of Beach and Shoreface Nourishment, Torsminde, Tange, Denmark', *Proc., 25th International Conference on Coastal Engineering,* American Society of Civil Engineers, Orlando, FL, 1996, 2927–2940.

Leidersdorf, C.B., Hollar, R.C. and Woodell, G.: 'Beach Enhancement Through Nourishment and Compartmentalization: the Recent History of Santa Monica Bay', *Proc., Coastal Zone Management, Special Volume on Beach Nourishment Engineering and Management Considerations,* Edited by D.K. Stauble and N.C. Kraus, American Society of Civil Engineers, 1993, 71–85.

McDougal, W.G., Williams, A.N. and Furukawa, K.: 'Multiple Pit Breakwaters', *Journal of Waterway, Port, Coastal and Ocean Engineering,* American Society of Civil Engineers, Vol. 122, 1996, 27–33.

McLellan, T.N. and Kraus, N.C.: 'Design Guidance for Nearshore Berm Construction', *Proc., Coastal Sediment '91,* American Society of Civil Engineers, 1991, 2000–2011.

Moore, B.D.: 'Beach Profile Evolution in Response to Changes in Water Level and Wave Height', MCE thesis, Department of Civil Engineering, University of Delaware, Newark, DE, 1982, 164pp.

National Research Council: *Beach Nourishment and Protection*, National Academy Press, Washington, DC, 1995.

Nichols, R.J., Birkemeier, W.A. and Hallermeier, R.J.: 'Application of the Depth of Closure Concept', *Proc., 25th International Conference on Coastal Engineering,* American Society of Civil Engineers, 1996, 3874–3878.

Otay, E.N.: 'Long-term Evolution of Nearshore Disposal Berms', Ph.D. dissertation, Dept. of Coastal and Oceanographic Engineering, University of Florida, Gainesville, FL, 1994.

Pelnard-Considère, R.: 'Essai de Theorie de l'Evolution des Formes de Rivate en Plages de Sable et de Galets', *4th Journees de l'Hydraulique,* Les Energies de la Mar, Question III, Rapport No. 1, 1956 (in French).

Perlin, M. and Dean, R.G.: '3–D Model of Bathymetric Response to Structures', *Journal of Waterway, Port, Coastal, and Ocean Engineering,* American Society of Civil Engineers, Vol. 111, No. 2, March, 1985, 153–170.

Rijkswaterstaat: *Manual on Artificial Beach Nourishment*, Centre for Civil Engineering Research, Codes and Specifications, 1987.

Roelvink, J.A. and Hedegaard I.B.: 'Cross-Shore Profile Models', *Coastal Engineering,* Special Issue No. 21, Part 1/3, 1993, 163–192.

Southgate, H.N. and Nairn, R.B.: 'Deterministic Modelling of Nearshore Processes, Part 1: Waves and Currents', *Coastal Engineering,* Vol. 19, No. 1, 2, 1993, 27–56.

Southgate, H.N. and Nairn, R.B.: 'Deterministic Profile Modelling of Nearshore Processes. Part 2: Sediment Transport and Beach Profile Development', *Coastal Engineering,* Vol. 19, No. 1,2, 1993, 57–96.

Stauble, D.K. and Grosskopf, W.G.: 'Monitoring Project Response to Storms: Ocean City, Maryland Beachfill', *Shore and Beach,* Vol. 61, No.1, 1992, 23–33.

Trindell, R., Arnold, D., Moody, K., and Morford, B.: 'Post-Construction Marine Turtle Nesting Monitoring Results on Nourished Beaches', *Proc. Natl. Conf. Beach Pres. Tech.,* Florida Shore and Beach Pres. Assn., Tallahassee, FL, 1998, 77–92.

U.S. Army Corps of Engineers: 'Beach Fill Transitions', Coastal Engineering Technical Note CETN-II-6, U.S. Army Waterways Experiment Station, Coastal Engineering Research Center, 1982.

Walton, T.L.: 'Shoreline Solution for Tapered Beach Fill', *Journal of Waterway, Port, Coastal, and Ocean Engineering,* Technical Note, American Society of Civil Engineers, Vol. 120, No. 6, 1994, 651–655.

Watanabe, A.: 'Modeling of Sediment Transport and Beach Evolution', In: *Nearshore Dynamics and Coastal Processes; Theory, Measurement and Predictive Models,* K. Horikawa, (ed.). University of Tokyo Press, 1988, 292–302.

Wiegel, R.L.: 'Dade County, Florida, Beach Nourishment and Hurricane Surge Protection', *Shore and Beach,* Vol. 60, No. 4, 1992, 2–28.

Yoo, C.H.: 'Realistic Performance of Beach Nourishment', Ph.D. dissertation, Department of Coastal and Oceanographic Engineering, University of Florida, Gainesville, FL, 1993, 150pp.

INDEX